# Viruses of Invertebrates

# Viruses of Invertebrates

*Edited by*

EDOUARD KURSTAK
*Faculty of Medicine*
*University of Montreal*
*Montreal, Quebec, Canada*

MARCEL DEKKER, INC. New York • Basel • Hong Kong

**Library of Congress Cataloging-in-Publication Data**

Viruses of invertebrates / edited by Edouard Kurstak.
p. cm
Includes bibliographical references and index.
ISBN 0-8247-8469-3
1. Invertebrates--Viruses. Kurstak, Edouard
QR325.V57 91-8669
592'.0234--dc20 CIP

*This book is printed on acid-free paper*



**MARCEL DEKKER, INC.**
270 Madison Avenue, New York, New York 10016

Current printing (last digit):
10 9 8 7 6 5 4 3 2 1

**PRINTED IN THE UNITED STATES OF AMERICA**

# Preface

Although more than 300 species of invertebrate viruses have been described, mainly infecting insects, in comparison with what we know of vertebrate viruses, our knowledge of these viruses remains relatively obscure. However, this situation changed with the advent of invertebrate baculovirus expression systems for the development of recombinant subunit vaccines against several viruses and parasitic diseases. Of special interest in the field of biotechnology of vaccine production is the baculovirus *Autographa californica* nucleopolyhedrosis virus, a helper-independent viral expression vector that has proved to be an excellent expression system for the efficient production of immunogenic proteins in cultured insect cells. This baculovirus expression system was also used to secrete human interleukin 2; to synthesize, glycosylate, and secrete human alpha-interferon; and to produce human immunodeficiency virus, rabies, hepatitis B, influenza viruses, and parasitic coccidiosis recombinant subunit vaccines, among others.

Before the biotechnology era, baculoviruses were experimented with as more selective and more environmentally acceptable agents for the control of insect pests. In fact, between 1970 and 1980, baculoviruses gained high appreciation as potential viral pesticides to be used in biocontrol, an alternative to the environmental problems caused by chemical pesticides. As a result, in 1982, I edited *Microbial and Viral Pesticides,* published by Marcel Dekker, Inc.

Since that time increasing interest has been focused on several groups and families of invertebrate viruses. This volume contains an accumulation of knowledge on viruses of invertebrates. Important information on the structure, molecular biology, genetics, pathogenesis, ecology, and diagnosis of these viruses is contained within these pages. Some chapters stress new findings on the use of viruses in biocontrol and biotechnology. For example, there is a chapter on the possibilities offered by the baculovirus expression system to produce a vaccine against a coccidiosis intestinal disease caused by protozoan parasites of the genus *Eimeria,* affecting many mammalian and avian species of economic importance.

This volume contains 12 timely chapters on major invertebrate virus groups. Recent research data and unpublished material by contributors who are well known in their field are included. Special attention was paid to baculoviridae genome organization and expression. Other families of viruses, in which recent advances were made, are not neglected; cases in point are the Polydnaviridae a most unusual and unorthodox group of viruses, and Nodaviridae, the simplest class of multipartite viruses.

The volume particularly addresses virologists, microbiologists, invertebrate pathologists, geneticists, and professional workers employed in biotechnology and viral biocontrol of pests. It is hoped that postgraduate students in applied microbiology and virology will also find it a valuable source of information.

I extend my sincere gratitude and thanks to the contributors for the effort and care they have taken in the preparation and presentation of their manuscripts and to the staff of Marcel Dekker, Inc., for their part in the production of this volume.

**Edouard Kurstak**

# Contents

# Contributors

**Basil M. Arif** Forest Pest Management Institute, Forestry Canada, Sault Ste-Marie, Ontario, Canada

**Shänti L. Bilimoria** Department of Biological Sciences, Texas Tech University, Lubbock, Texas

**Susan K. Braun** Insect Pathology Laboratory, Agricultural Research Service, U.S. Department of Agriculture, Beltsville, Maryland

**John P. Burand** Department of Entomology, University of Massachusetts, Amherst, Massachusetts

**Norman E. Crook** Horticulture Research International, Littlehampton, West Sussex, England

**Peter Dobos** Department of Microbiology, College of Biological Science, University of Guelph, Guelph, Ontario, Canada

**Edward M. Dougherty** Insect Pathology Laboratory, Agricultural Research Service, U.S. Department of Agriculture, Beltsville, Maryland

**Roy Duncan** Department of Microbiology and Infectious Diseases, University of Calgary Health Sciences Centre, Calgary, Alberta, Canada

**Donald A. Hendry** Department of Microbiology, Rhodes University, Grahamstown, South Africa

**Mark C. Jenkins** Protozoan Diseases Laboratory, Agricultural Research Service, U.S. Department of Agriculture, Beltsville, Maryland

**James Kalmakoff** Department of Microbiology, University of Otago, Dunedin, New Zealand

**Shigemi Kawase** Faculty of Agriculture, Nagoya University, Chikusa, Nagoya, Japan

**Peter John Krell** Department of Microbiology, College of Biological Science, University of Guelph, Guelph, Ontario, Canada

**Edouard Kurstak** Department of Microbiology and Immunology, Faculty of Medicine, University of Montreal, Montreal, Quebec, Canada

**Norman F. Moore** Ministry of Defence, London, England

**Éva Nagy** Department of Veterinary Microbiology and Immunology, Ontario Veterinary College, University of Guelph, Guelph, Ontario, Canada

**V. K. Ward** Department of Microbiology, University of Otago, Dunedin, New Zealand

# Viruses of Invertebrates

# 1

# The Biology of Nuclear Polyhedrosis Viruses

SHÄNTI L. BILIMORIA
*Texas Tech University, Lubbock, Texas*

## INTRODUCTION

The description of jaundice, or polyhedrosis, of the silkworm provided by Nysten in 1808 was probably the first scientific account of a disease caused by a nuclear polyhedrosis virus (NPV). Nearly half a century later, Maestri (1) and Cornalia (2) carried out microscopic studies and observed highly refractive, crystallike bodies in the nuclei of cells of jaundiced silkworms. Bolle considered these crystals to be proteinaceous and named them "polyedrische Könchen" (polyhedral granules). In 1913 Glaser and Chapman suggested that the cause of wilt disease of the gypsy moth was viral, and in 1918 Acqua provided convincing proof that the jaundice of silkworms had a viral etiology (3).

The majority of the more than 300 baculoviruses (family Baculoviridae) isolated from several orders of insects and other arthropods are NPVs. The Baculoviridae are rod-shaped, enveloped viruses containing a circular double-stranded DNA genome (4,5). The term "polyhedrosis" is derived from the observation (1,2) that the nuclei of infected cells are packed with polyhedral inclusion bodies, or PIBs, and this characteristic is utilized to place the NPVs in subgroup A of the Baculoviridae.

Infection in vivo occurs per os. PIBs disintegrate in the insect gut, and the released virions initiate infection of the gut epithelial cells. Virus is transferred, in most cases, to the fat body tissue via the hemolymph (6). Toward the end of the infection cycle, millions of PIBs are formed in the fat body tissue; the flaccid larvae disintegrate, releasing the PIBs, which persist in the environment. The fact that most host larvae stop feeding within two days of ingesting virus has allowed the use of NPVs as biological control agents, and at least four NPVs are registered as biological control agents in the United States (7).

Since the mid-1970s an impressive amount of information has accumulated on the replication of the NPVs. Two phases have been defined in the infectious cycle: budded virus (BV) is produced as a result of nucleocapsids, which are assembled in the nucleus, acquiring envelopes at the nuclear or plasma membrane; occluded virus (OV) is produced when one to several nulceocapsids acquire envelopes through a de novo nuclear process and are then occluded into PIBs. While PIBs themselves are not infectious under cell culture conditions, polyhedra-derived virus (PDV) are infectious, albeit less efficiently than BV. Thus BV is the inoculum of choice for cell culture infections. These advances together with the development of a plaque assay by Hink (8) have allowed researchers to show that NPV replication cycle is regulated by a coordinately controlled, temporal cascade similar to that delineated for the herpesviruses. It has also been shown that the gene for the major protein comprising the PIB, polyhedrin, is expressed so strongly that 25% of the protein produced in cell culture infections is polyhedrin. This property of the NPVs has been exploited to develop powerful expression vectors for the efficient production of commercially important polypeptides in a eukaryotic system (9, 10).

Given the uniqueness of nuclear polyhedrosis viruses as a basic system for the elucidation of gene expression in eukaryotes, the prospects for development of expression vectors, and the potential impact of genetically engineered and natural viruses as pest control agents, it is clear that these agents are going to receive a great deal of attention from virologists, molecular biologists, and entomologists in the years to come. This review summarizes the salient aspects of the literature on NPVs with particular emphasis on developments during the 1980s. For more detailed recent reviews on specific aspects of the NPV, the reader is referred to an excellent work, *The Biology of Baculoviruses* (11).

## CLASSIFICATION

The nuclear polyhedrosis viruses belong to subgroup A of the genus *Baculovirus*, which, thus far, is the only genus described in the Baculoviridae. All baculoviruses manifesting a polyhedral, proteinaceous inclusion body phase in the nuclei of infected cells are placed in subgroup A. There are two morphological types of viruses (Figure 1) in this subgroup: single nucleocapsid NPV (SNPV; type species, *Bombyx mori* (SNPV) in which only one nucleocapsid is found per occluded virion and multiple nucleocapsid NPV (MNPV; type species, *Autographa californica* MNPV) in which occluded virions may contain a variable number of nucleocapsids per virion. It should be noted that budded virus as opposed to occluded virus is always found to possess one nucleocapsid per virion, irrespective of whether it is an SNPV or an MNPV (5).

Beyond categorization into SNPV and MNPV, morphology is of little im-

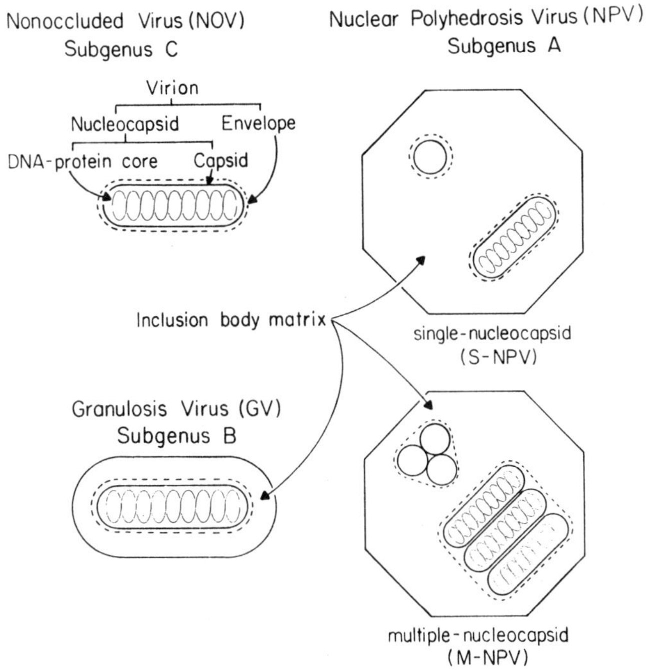


**Figure 1** Schematic representation of the Baculoviridae. The SNPVs and MNPVs contain numerous virions per inclusion body; MNPV virions may contain as many as 17 nucleocapsids per envelope. Not drawn to scale. (From Ref. 5, used with permission.)

portance in the further classification or identification of the NPVs, and modern methods such as polyacrylamide gel electrophoresis, restriction endonuclease fragment profiles of viral DNA, and Western blots utilizing either polyclonal or monoclonal antibody are very useful for the identification of specific viruses within subgroup A. Clusters within the subgroup are beginning to emerge as sophisticated molecular and immunological techniques are applied to the problem. A detailed account of this topic appears in the recent review of baculovirus taxonomy by Bilimoria (5).

## Features for Identification

### Gross Pathology

Gross pathological changes are observed late in NPV infection and are not of significant diagnostic value. The infection is frequently generalized in lepidoptera but is limited in other hosts such as mosquitoes (6).

### Microscopy

The highly refractile PIBs of the NPV shine characteristically when observed by phase-contrast or dark-field microscopy. As many as 100 PIBs may be found in the nucleus, which is often hypertrophied. Staining methods or the detection and positive identification of crystals as PIBs (12) are useful for distinguishing NPVs from cytoplasmic polyhedrosis viruses (family Reoviridae) but are unreliable for differentiation within the NPV. Several workers have reported characteristic shapes for the PIBs of particular NPVs; however, shape cannot be recommended as a particularly reliable criterion for identification.

Electron microscopy is essential for distinguishing SNPVs from MNPVs, but on its own, this technique has little value for further identification or classification of the NPVs.

### Biochemical Approaches

By far the greatest promise for the classification and identification of the NPVs is in the use of biochemical approaches. In this respect, the molecular taxonomy of the Baculoviridae is off to a good start. Initial approaches targeted viral structural proteins and utilized polyacrylamide gel electrophoresis (PAGE) or serology. Later, restriction endonuclease digestion, nucleic acid hybridization, Western blotting, and monoclonal antibody methods were applied to the problem. The general picture emerging from these types of studies is that viruses from related hosts are more similar than those from distant hosts.

The MNPVs related to the type species *Autographa californica* MNPV (AcMNPV) have been studied intensively. Cibulsky et al. (13) and Summers and Smith (14) showed that the MNPVs of *A. californica*, *Galleria mellonella*, *Rachiplusia ou*, and *Trichoplusia ni* had very similar virion polypeptide profiles. Miller and Dawes (15) compared the restriction endonuclease (REN) fragment profiles of *A. californica* and *T. ni* MNPVs and showed that these viruses were closely related but differed sufficiently to be called variants. Summers and Smith showed that *A. californica*, *T. ni*, and *G. mellonella* MNPVs were related but distinct strains, and Knudson and his colleagues (16) recently demonstrated that *Spodoptera exempta* MNPV (SeMNPV) is also a member of this group of viruses. It was clear from physical maps of the viral genomes that these NPVs had undergone deletions and insertions relative to one another. The data also suggested that *R. ou* MNPV is less closely related to any one of the remaining strains than they are to one another. Sequence homology studies have shown that *A. californica* and *R. ou* MNPVs share extensive homologies. These agree with PAGE studies, which show only minor differences between the structural polypeptides of these variants.

It is clear from REN analysis of viral clones separated by plaque purification of field isolates that each field isolate consisted of several strains having different REN profiles but apparently identical structural polypeptides. Variation among "predominant" strains obtained by different laboratories may be a function of the geographical location of the isolate, but selection due to differences in culture conditions from one laboratory to another cannot be ruled out (17). These phenomena and the production of recombinant virus by mixed infection and recombinant DNA methods pose significant problems for the classification and nomenclature of the NPVs, particularly with respect to the designation of types or variants.

Studies of virus isolates from related hosts are beginning to reveal interesting clusters of NPVs. One such group consists of MNPVs of the genus *Spodoptera*. Reassociation kinetic studies showed that the genomes of *S. exempta* and *S. exigua* MNPVs were more closely related to each other than either was to the MNPVs of *S. frugiperda* and *S. litoralis*, respectively (18). Other studies utilizing the less quantitative, filter-bound assays are in general agreement with these data (19,20). The serological results of Harrap et al. (21) are thus supported by the DNA studies. Recently, Kislev (22) showed that *S. littoralis* MNPV (SlMNPV) has extensive DNA homology with *S. frugiperda* MNPV (SfMNPV) and *S. exigua* MNPV.

The SNPVs from plusiine hosts constitute another interesting group of viruses from related hosts. Cibulsky et al. (13) and Smith and Summers (23) demonstrated similarities between the SNPVs of *T. ni* and *Pseudoplusia includens*. Bilimoria (24) showed that the REN profiles of these viruses were similar, and the apparent homology as determined by Upholt's method was 92%. Interestingly, the pattern for the *Syngrapha selecta* SNPV genome was very similar to three geographical isolates of *T. ni* SNPV, and these differences paralleled the in vivo host range of these viruses reported by Harper (25). Evidence for the existence of a similar cluster was also obtained by Smith and Summers (19,23).

Vlak and Gröner (26) and Brown et al. (27) showed that five European isolates of *Mamestra brassicae* MNPVs fell into two groups based on REN patterns, PAGE, and serological analysis. However, these groups were identical in their virulence, host range, and tissue tropism. In an extensive study utilizing analysis of restriction fragments, blot hybridization, and semiquantitative analysis of DNA dot hybridizations, Smith and Summers (19) identified clusters of *M. brassicae* and *Heliothis armigera* MNPVs with 90% homology. Less striking clusters (6–12% homology) consisted of (1) MNPVs from *H. armigera, S. exigua, S. frugiperda,* and *M. brassieae* and (2) *Anticarsa gemitalis* MNPV and *A. californica* MNPV and its variants. An important general conclusion to be derived from the work of Summers and Smith (19) is that, with few exceptions, there is more homology within the MNPVs and SNPVs than between them, suggesting that morphology is a valid criterion for separating these two types of

NPVs. It must be emphasized that homology values based on hybridization data are underestimates of the true relationship among the viral DNAs since no homology under the least stringent conditions does not rule out the possibility that the DNAs are 50% related in nucleotide sequence. The reason for this is that even such low conditions of stringency do not allow hybridization when the base mismatch is greater than 33%.

Recent research has confirmed earlier conclusions, based on serological and peptide analysis, that the major protein component of NPV inclusion bodies (i.e., the polyhedrin polypeptide) is highly conserved (28–30). It is fairly clear today that elements of polyhedrin are the "group antigens" referred to in much of the early literature on NPV (5,31,32). Two-dimensional peptide mapping studies on polyhedrin from various NPVs have shown that these polypeptides are of a family or class with only minor differences (33); certain basic, acidic, and neutral polypeptides were common to all polyhedrins tested and accounted for the serological cross-reactions observed among them (34). The amino acid sequence of several polyhedrin molecules as well as the base sequence of the polyhedrin gene are now available (28).

Rohrmann (28) utilized the conserved nature of polyhedrin in an attempt to construct a baculovirus phylogeny. Pointing to lack of serological cross-reactivity between the SNPV from the dipteran *Tipula paludosa* (TpSNPV) and the MNPV from the lepidopteran *Orgyia pseudotsugata* (OpMNPV) as well as the substantial differences in the *N*-terminal sequence of TpSNPV and several other MNPVs from lepidopteran hosts, Rohrmann suggests that the dipteran SNPVs diverged from other baculoviruses very early in the evolution of the Baculoviridae. The granulosis viruses (GVs; subgroup B) appeared after the divergence of the lepidopteran progenitor virus because GVs are found only in the lepidoptera and because GV polyhedrins demonstrate a much closer relatedness to the lepidopteran NPV polyhedrins than to the dipteran NPV. The hymenopteran NPVs are most closely related to the lepidopteran NPVs and are speculated to have arisen by cross-infection into the hymenopteran suborder Symphyta (sawflies). From the close relationship between the polyhedrins of the SNPVs and MNPVs of the lepidoptera, Rohrmann suggests that the evolution of MNPVs is a relatively recent event. Possible evolutionary advantages of multicapsid NPVs include increase in infectious potential to offset damage by UV radiation as well as selection for recombination leading to persistence of genetic variation within the viral population (28). A mechanism of this type might also explain the much broader host range of the MNPVs compared to the SNPVs and GVs.

It is also becoming clear that there are other conserved proteins in the NPVs in addition to polyhedrin. Kelly et al. (35) reported a major nucleocapsid protein in *H. zea* SNPV. This is a DNA binding protein present in the core of the capsid and is rich in arginine and helix-destabilizing residues. Similar proteins have been shown in *T. ni* NPV, but other NPVs appear to lack them (36).

### Serology

Recent studies have confirmed, earlier findings [see review by McCarthy and Gettig (31)] that there is a high degree of serological cross-reactivity among the polyhedrins; as mentioned earlier, the classical concept of a "group antigen" is quite valid. Nevertheless, the new sophistication in immunochemical methodology has shown that discrimination between the polyhedrins of SNPVs, MNPVs, and GVs can be made. Even polyhedrins of closely related virsuses such as AcMNPV and *T. ni* MNPV can be distinguished with the use of monoclonal antibody (37). Huang et al. (32) have documented evidence for the presence of group and subgroup epitopes and the existence of multiple copies of a single epitope or a series of closely related epitopes on the polyhedrin molecule.

Studies using conventional sera in Western blot assays have confirmed biochemical evidence that differences between GVs and NPVs are greater than those between SNPVs and MNPVs. As for the presence of group antigens in the nucleocapsid, Hohmann and Faulkner (37) showed that monoclonal antibody can be used to identify common determinants in this element of the virus.

PAGE, REN analysis, DNA hybridization, and immunochemistry are useful, to varying extents, in the identification of NPVs and clearly have been helpful in establishing tentative clusters of related NPVs. However, Rohrmann (28) points out that "serology and DNA hybridization have proven inadequate for quantifying baculovirus relatedness. The only reliable method for quantifiable data is through DNA sequence analysis of genes coding for orthologous proteins." The latter approach has much merit, but sequencing the polyhedrin gene alone may not be sufficient. Given the number of NPV isolates that need to be processed, the lack of knowledge about the genes of these viruses, and the time-consuming nature of DNA sequencing, it is very likely that we shall be utilizing serology (particularly Western blotting) and DNA hybridization until sufficient progress has been made in the characterization of NPVs other than those from *A. californica* and *O. pseudotsugata*.

## STRUCTURE

The major structural components of the nuclear polyhedrosis virus are the virions and the occlusion body in which the virion is embedded.

### The Occlusion Body

The OBs are referred to as polyhedra because in most NPV, they manifest a polyhedral shape when viewed by microscopy (Figure 2). The size of NPV OBs ranges from 0.5 to 15 $\mu$m in diameter. The actual mean size of mature polyhedra for most NPVs is 0.8–2 $\mu$m. Aside from the virions, the two major components of the OB are the inner matrix and the envelope. The inner matrix of the OBs is

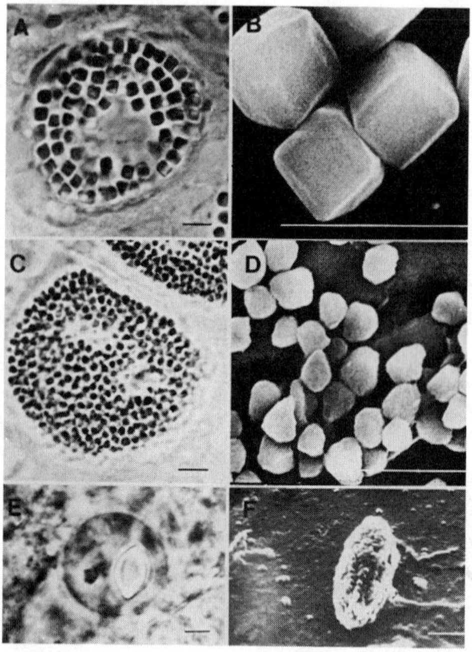


**Figure 2** Light and scanning electron micrographs illustrating shapes characteristic of representative nuclear polyhedrosis virus occlusion bodies. (A and B) *Autographa californica* MNPV; (C and D) *Trichoplusia ni* SNPV; (E and F) *Aedes sollicitans* SNPV. Note the beveled edges on the polyhedra of the AcMNPV (B). The rough surface on spindle-shaped OBs of the *Aedes sollicitans* SNPV (F) indicates it is immature. Bar = 5 $\mu$m. (From Ref. 45, used with permission.)

composed of a single protein (38) that appears to have the shape of a six-armed nodal unit, and the molecules are oriented in a face-centered cubic lattice (39,40). For a long time, the occlusion body envelope was thought to be an artifact (38), although the consensus now is that it is a legitimate viral component (40). Minion et al. (41) showed the occlusion body envelope of *H. virescens* NPV is composed primarily of carbohydrates, particularly hexose and pentose with smaller amounts of uronic acid and hexosamine being present. These workers also showed that there was no significant amount of protein in the envelope and that nonpolar solvents did not affect the integrity of the envelope structure. These data appear to rule out lipids as a component, and thus a unit membrane structure, of the occlusion body envelope. However, since enzymatic degradation of lipids and proteins could have occurred during the purification of the envelope material, the existence of small amounts of lipids and proteins could not be ruled out completely. Indeed, very recently, Whitt and Manning (42) used immunofluorescence microscopy and monoclonal antibodies to show that the envelope or, as they term it, the "calyx" of AcMNPV OBs contains a 34 kD phosphorylated protein (pp34) together with a subpopulation of polyhedrin protein. The expression of pp34 late in the replication cycle and the observation that it continues to be phosphorylated until 60–70 h postinfection further support its role in the occlusion process. The effect of reducing agents on pp34 and polyhedrin suggests that these proteins are associated with the calyx via thiol linkages. Several workers have suggested that the fibrous material variously described as electron-dense fragments, condensed fibrous sheets, membranelike profiles, and so on (Figure 3B), and observed very late in the infection cycle as polyhedra matured, is the same as the membranelike material constituting the occlusion body envelope (43,44). However, there is no direct evidence linking this fibrous material with the OB envelope.

## Virion

The basic structure of the virion is the same in all baculoviruses. The virion is an enveloped, rod-shaped nucleocapsid, and there is an amorphous but definite layer between the nucleocapsid and the envelope (Figure 3). In the SNPVs, each virion typically contains only one nucleocapsid, whereas in the MNPVs, each virion usually contains several nucleocapsids per envelope.

A currently accepted model for the structure of the virion is seen in Figure 4 (45). The nucleocapsid is a cylindrical core of DNA and protein surrounded by a tubular capsid capped at both ends (38,39,46). The diameter of the capsid in ultrathin sections averages 40–50 nm and that of the core averages 38–46 nm (38,46). The capsid is composed of rings of subunits stacked perpendicular to the longitudinal axis, and thus capsid symmetry is not helical. The periodicity of the rings is approximately 4.5 (47,48). The caps of the nucleocapsid are composed of short series of rings of diminishing diameter. Several workers have suggested that the nucleocapsid has polarity in that there are "claw and nipple" structures at opposite ends of the rods (49–51). At this time, there is insufficient evidence in support of the hypothesis that polarity has a role in nucleocapsid alignment at the nucleopore. The nucleocapsid core contains the DNA and a highly basic protein which is bound to the DNA. This basic protein is thought to maintain the DNA in a condensed and packaged state by neutralizing the phosphate moieties (35,52).

The NPV envelope is a lipoprotein unit membrane possessing the characteristic trilaminar structure, i.e., a central lipid layer sandwiched between two protein layers. The three layers of this membrane can be clearly resolved in ultrathin sections (Figure 3) (45). Stoltz et al. (53) provided strong support for Hughes's hypothesis (54) that the envelope is formed de novo in the nuclei of infected cells. The extracellular (i.e., budded) virions that spread the infection

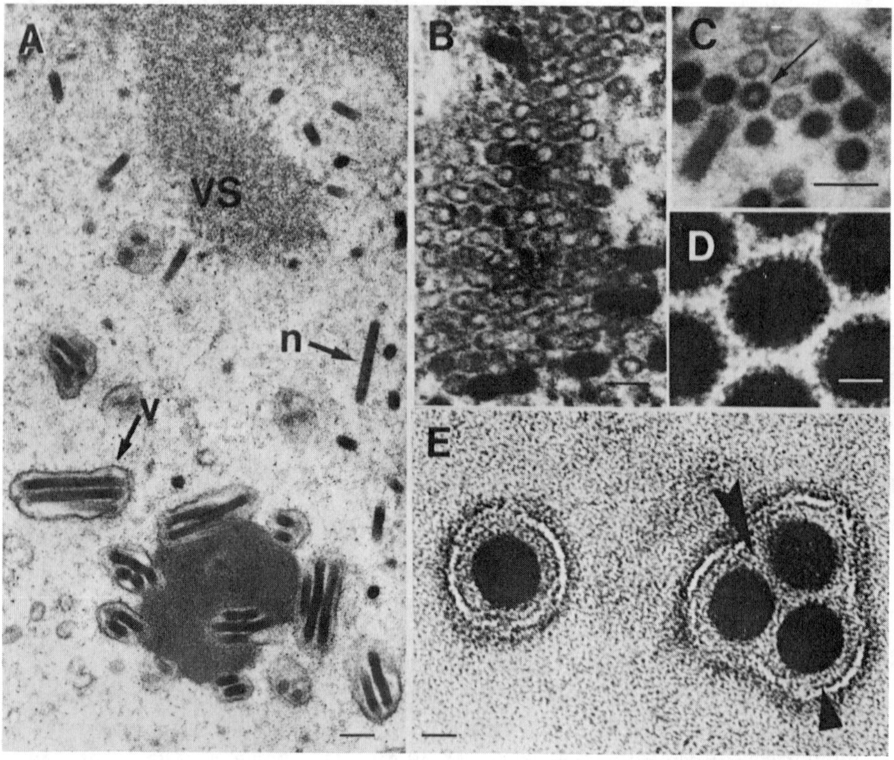


**Figure 3** Electron micrographs illustrating major components of the NPV virion. (A) Section through an infected fat body cell nucleus adjacent to the virogenic stroma (VS) illustrating rod-shaped nucleocapsids (n), multinucleocapsid virions (v), and a developing polyhedron of an NPV in the noctuid *Scotogramma trifolii*. (B–D) Cross-sections through empty capsids (B) and developing (C) and complete nucleocapsids (D) of the SNPV isolated from *Aedes sollicitans*. Note the electron-translucent central area (arrow) in the developing nucleocapsid (C) and the portion of the capsid extending beyond the dense core in the nucleocapsids (D). (E) Cross section through occluded virions of the *Autographa californica* MNPV. In the occluded virions the capsid and intermediate layer cannot be distinguished from one another. They appear as a relatively thick layer surrounding the dense DNA-protein core of the nucleocapsid (large arrowhead). This layer is in turn surrounded by the trilaminar envelope. The inner layer of the envelope is visible as a thin dense layer appressed to the intermediate layer of the virion (small arrowhead). Bar = 100 nm (A) and 20 nm (B–E). (From Ref. 45, used with permission.)

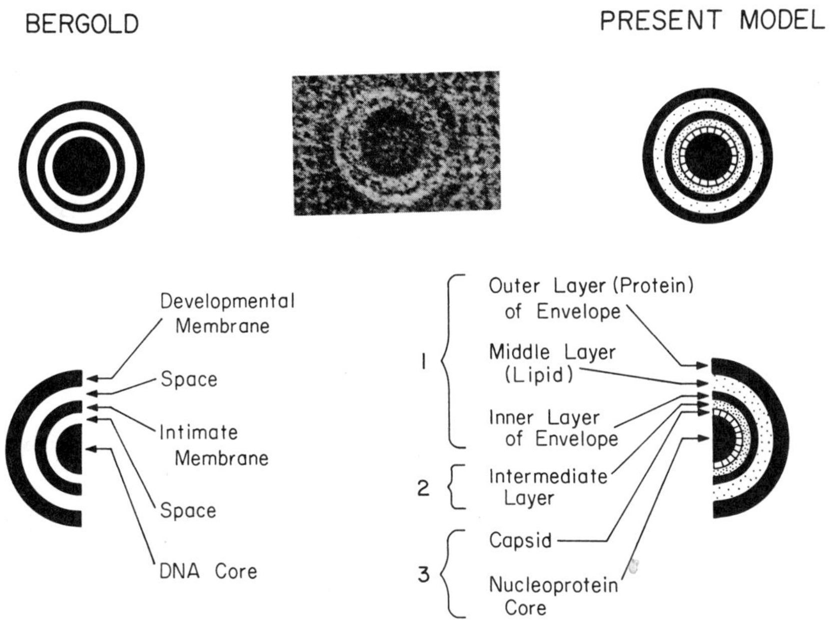


**Figure 4** Comparison of Bergold's model for the baculovirus virion with a cross section through a virion and a current model of the virion. In the present model, 1, 2, and 3, are respectively, the envelope, intermediate layer, and nucleocapsid. In Bergold's model the intimate membrane corresponds with the inner layer of the envelope. The space that Bergold illustrated concentrically surrounding the DNA core most likely corresponds to a portion of the capsid and intermediate layer. (From Ref. 45, used with permission.)

within the host and that are used as inoculum in infecting cell cultures are enveloped in membranes derived from the host plasma membrane. Adams et al. (55) have shown that BV envelopes have a distinctive hemispherical cap of peplomers, readily resolved in ultrathin sections, on one end of the virion.

Federici (45) defines the intermediate layer as "the viscous layer of low to moderate electron density that exists between the nucleocapsid and envelope in the mature virion." This layer ranges in thickness from 6 to 12 nm and also occurs in MNPVs. The composition of the amorphous or viscous layer is not clear. Kawamoto et al. (56,57) have shown that virus-specified structures are present in this layer, and it has been suggested that numerous proteins that are detected in intact virions but not in nucleocapsids may well be from the intermediate layer (14,58). The intermediate layer may have a parallel in the amorphous "lateral bodies" of poxviruses (59,60).

Budded virions from both larvae and cell culture have a pleomorphic appearance in contrast to the tight-fitting envelope of occluded virions observed in thin sections. The morphology of OVs in the MNPV will vary with the number of nucleocapsids in the virion.

## PHYSICOCHEMICAL PROPERTIES

The complex organization of NPVs into OV and BV was described in the last section. In this section, the biophysical and biochemical properties of polyhedral inclusion bodies (PIBs), polyhedra-derived virions (PDVs), and budded virus are discussed.

### Polyhedral Inclusion Bodies

The large size of the polyhedral inclusion body of the NPVs presents some problems in purification. Nevertheless, acceptable methods for purifying PIBs have been developed. Larva- or tissue culture-derived PIBs are released by homogenization or sonication, respectively, and are recovered by low-speed centrifugation. The crude preparations are routinely purified on 40–63% w/w sucrose gradients at 100,000 g for 30 min. The PIBs, which band at 54–56% sucrose, are harvested and washed with deionized water.

In the laboratory, PIBs are quite stable and retain infectivity for larvae for years even when stored at room temperature. Infectivity in the field is lost rapidly due to the ultraviolet component of sunlight. For example, the half-lives of pure and impure *Gilpinia hercyniae* NPV on spruce foliage were 38 and 55 days, respectively (61).

Early experiments designed to determine the subunit structure of the PIB matrix showed that protein multimers are 276–378 kD in size (62, 63). Harrap and other workers have reported sedimentation coefficients of 11.1–11.3 S for these multimers; these values are in reasonable agreement with Bergold's value of 12.5 S. Kozlov et al. (64, 65) obtained an $M_r$ of 28,000 for a matrix subunit extracted under reducing conditions. These workers also detected low molecular weight polypeptides, presumably degradation products. This was later confirmed by a number of laboratories, and the 28 kD polypeptide, subsequently named polyhedrin, was shown to be viral coded (16, 66). In view of these results and electron microscopic studies of the PIB matrix [see review by Federici (45)], it appears that the matrix of the NPV inclusion body consists of multimers forming a six-armed nodal unit of perhaps seven polyhedrin molecules.

### Virion

In discussing the physicochemical properties of the NPVs one must keep in mind that two types of virions, polyhedra-derived virus and budded virus, are produced. While there are real differences between the properties of PDVs and BVs, many are artifacts of the techniques required to extract and purify these two types of virions.

### Polyhedra-Derived Virus

PDVs are purified by solubilization of polyhedra (occlusion bodies) under alkaline conditions. Polyhedra are solubilized at 5 mg/ml in dilute alkali saline, pH 10.9, (0.1 *M* sodium carbonate, 0.17 *M* sodium chloride, and 0.01 *M* EDTA), at room temperature or 4°C for 30 min [Whitt and Manning (67) showed recently that in addition to previously documented hydrophobic considerations, electrostatic interactions are important for occlusion body stability.] The preparation is then layered on 1.17–1.16 g/ml sucrose gradients and centrifuged at 100,000 g for 1 h. Soluble polyhedrin remains at the top of the gradient and can be purified further if needed. The virions form multiple bands depending on the number of nucleocapsids per envelope in the case of the MNPVs. SNPVs, of course, would yield only two bands: virion and nucleocapsid. The densities at which the virions band in the gradient range from 1.20 to 1.25 g/ml. The bands are removed, diluted with tris buffer, and centrifuged at 100,000 g for 30 min. Kobayashi and Nakagaki (68) have shown that, for *Bombyx mori* NPV, the use of low temperature during dissolution and distilled water (rather than a buffer) for the preparation of sucrose gradients maximizes virus yield. The physical properties of NPV virions may be summarized as follows: dimensions, 40–140 × 250–400 nm; sedimentation coefficient ($S^{w}20$), 1228–1530; buoyant density, 1.54 g/ml; particle weight 440 × $10^6$ (36).

Yamafugi et al. (69) and subsequently other workers (70–73) documented the association of an alkaline protease with larva-derived PIBs. This enzyme is a serine protease with an optimal activity at about pH 10.6. Although a clear role for this protease (or these proteases) is not established, it is thought that the enzyme functions in the dissolution of the PIBs in the insect gut, which in the case of the lepidoptera has a pH of around 10.5. Several workers have alluded to the detrimental effects of the alkaline protease on polyhedrin and PDV. Great care must be exercised when purifying PDV from larva-derived polyhedra, particularly if the virus is to be used for the characterization of structural polypeptides. The alkaline protease is inactivated by heating at 70–80°C for 20–60 min. Protease inhibitors such as mercuric chloride, phenylmethylsulfonyl fluoride, and diisopropyl fluorophosphate are also effective but are toxic and can be avoided (74).

### Budded Virus

BV is purified from media of infected cell cultures 40 h after infection and before the cells have lysed (75). Cells and PIBs are removed by low-speed centrifugation and the supernatant is passed through a 1.08 g/ml w/w sucrose cushion at 100,000 g for 30 min. The resulting pellet is fractionated on 1.08–1.24 g/ml w/w sucrose gradients at 100,000 g for 1 h. The virus band at 1.17–1.18 g/ml is removed from the gradient and pelleted at 100,000 g for 30 min.

### Nucleocapsid

Nucleocapsids may be purified from either PDV or BV by removing the envelope with Nonidet P40 (NP40) detergent. Depending on the specific virus, treatment with 1–3% NP40 for 30 min at 30°C should be sufficient (14, 21). Nucleocapsids have been purified on 10–50% sucrose gradients at 45,000 g for 45 min or banded on 40% (w/w) cesium chloride at 62,000 g for 64 h (21). Nucleocapsids range in size from 40 × 200 nm to 50 × 300 nm, and their buoyant density in cesium chloride is 1.47 g/ml (76).

### Virion Polypeptides

Eleven to twenty-six polypeptides, ranging in $M_r$ from 160,000 to 7,500, have been detected in purified NPV preparations (74). Given the presence of alkaline protease activity in larva-derived PIBs and the expected variation in molecular weight determinations in different laboratories, caution should be exercised in the interpretation of the data on NPV structural polypeptides. Not all investigators report the same set of polypeptides as the structural proteins of any particular virus. For example, there is disagreement on the number of structural proteins in the MNPV type strain AcMNPV (74). Significant differences also have been detected between the PDV and BV of AcMNPV, VP46, VP22, VP19, and VP18 are more prominent in PDV than in BV. On the other hand, VP85, VP64, VP26, and VP17 are more prominent in BV than in PDV (74). VP64 is a major component of the viral envelope (77).

A number of investigators have studied the glycosylation and phosphorylation of NPV structural polypeptides. Stiles and Wood (77) conducted elegant studies on the glycoproteins of AcMNPV. Using tunicamycin to demonstrate specific incorporation of particular precursors, these workers showed that there are 11 structural glycoproteins in BV but only 1 glycoprotein in PDV. Maruniak and Summers (78) and Dobos and Cochran (79) showed that there are 10 to 14 phosphoproteins in PDV and 9 phosphoproteins in BV of AcMNPV.

The polypeptide profiles of SNPVs and MNPVs are generally similar, and there is no correlation between polypeptide size profile and morphological type among the NPVs.

## DNA

The DNA of baculoviruses is double-stranded. The base composition of the NPVs ranges from 37 to 59% G:C. Values ranging from 42 to 47% G:C have been obtained for AcMNPV and G:C contents of 42.7 and 45% have been reported for BmNPV.

Kok et al (80) first showed circular molecules in preparations of BmNPV DNA examined by electron microscopy. Later studies revealed the presence of a large proportion of open circular (form II) DNA as well as smaller amounts of supercoiled (covalently closed; form I) and linear (form III) molecules (81, 82). Summers and Anderson (83) and other workers have suggested a similar structure based on (1) hydrodynamic properties in neutral and alkaline sucrose density gradients and (2) separation of supercoiled molecules from linear and open circular forms on cesium chloride gradients containing the intercalating dye ethidium bromide or propidium iodide (36). Physical maps (discussed later) generated with a large number of restriction endonucleases confirm the circular nature of NPV DNA.

Studies on the size of NPV DNA have yielded a tremendous range of values due to the susceptibility of the molecule to shearing. Direct measurements based on electron microscopy have yielded values of $58 \times 10^6$ to $80 \times 10^6$ D. Molecular sizes derived from sedimentation coefficient values range from $63 \times 10^6$ to $119 \times 10^6$. Kelly used reassociation kinetics to show that the molecular weights of DNA from NPV isolated from *Spodoptera* sp. ranged from $62 \times 10^6$ to $83 \times 10^6$. Restriction endonucleases (RENs) have provided yet another method for determination of the molecular weight of viral DNA. Since the mobility, in agarose or acrylamide gels, of DNA fragments resulting from REN treatment is a function of the molecular weight of the fragment, fragment size can be determined relative to known standards. The sum of the sizes of all the fragments generated by a particular REN then yields the molecular weight of the entire DNA molecule. DNA size determined by this approach ranges from $68 \times 10^6$ to $109 \times 10^6$ D(24,36). The molecular weight values obtained for AcMNPV DNA range from $72 \times 10^6$ to $104 \times 10^6$D, depending on method used, suggesting a considerable amount of variation based on methodology. It is clear from the data compiled by Harrap and Payne (36) that approaches utilizing REN or electron microscopy are much more reliable than analysis in sucrose gradients. There is no correlation between DNA size and taxonomic grouping of NPVs.

## REPLICATION IN CELL CULTURE

The vast majority of the literature on NPV replication and molecular biology is on the AcMNPV because of the relative ease with which this virus can be grown in cell culture. Some information on SfMNPV and TnMNPV is available. Propagation of SNPVs in cell culture has only recently been established.

### MNPVs

#### Effect of the Culture Environment on Virus Growth

The growth of NPVs in cell culture is affected by a number of factors. Included among these are temperature (84–86), ionic and nutritional composition of the culture medium (86–88), multiplicity of infection (MOI), cell type (89–91), cell density (85,92,93), and rate of cell division (94). NPVs grow at an optimum temperature of 28°C, which is the same as the optimum temperature of most insect cell lines (84). The pH optimum of 7, however, is different from the optima of 6.1–6.2 for lepidopteran cell cultures (86,95,96). Goodwin and Adams (88) have reported that efficient replication of *Lymantria dispar* MNPV in the homologous cell line IPLB-LD-652 requires higher levels of glutamine than those required for normal growth of uninfected cultures. These workers also suggested that sterols and lipids were important for viral assembly. MOI affects both the percentage of cells infected and the timing of replicative events (86,97). It has been shown by several workers that the efficiency of infection and replication of NPVs vary with cell lines. For example, the SF-IPLB-21 cell line is three times less susceptible to AcMNPV than is the TN-368 cell line (98). Cloned cell lines derived from the parent TN-368 cell line have manifested similar differences (94). However, the differences in susceptibility of cell lines from different hosts to any particular virus are even greater (89,91,99,100). Cell density has a marked effect on MNPV replication. Wood et al. (93) showed that in high-cell-density TN-368 cultures infected with AcMNPV, there was a decrease in virus replication, an 18 h delay in the synthesis of one of the four early proteins, and inhibition of both cellular and viral DNA synthesis. Volkman and Knudson (101) showed that the size of AcMNPV plaques was dramatically reduced with increasing density of IPLB-SF-21 cells. Brown and Faulkner (102) suggested that host susceptibility and virus growth may be influenced by phases of the cell cycle. Lynn and Hink (103) showed that cells in S phase were more susceptible to infection than those in $G_2$ phase, although the virus yields were not affected.

### Cytopathic Effect

The cytopathic effect (CPE) of the NPV in cell culture varies slightly with the cell line. The CPE in TN-368 cells infected with AcMNPV was studied by Volkman and Summers (94) (Figure 5). When infected at an MOI of 10–20, these cells manifest the initial stages of CPE at 8–10 h postinfection (p.i.). These cells are normally spindle-shaped but become rounded upon infection. There is nuclear hypertrophy, disappearance of condensed chromatin, and the formation of a virogenic stroma which appears as a stippled area that becomes more condensed as the infection progresses. Polyhedra are observed in 5% of the infected cell nuclei at 12–14 h p.i. and in about 40% of the nuclei by 24 h p.i. BV is produced mainly during the prepolyhedral CPE stage; BV production does not occur after the appearance of polyhedra (75).

### Viral Morphogenesis

Viral attachment can occur within 5–10 min after inoculation (86,104). Penetration and uncoating are accomplished within 3 h (105,106). Electron microscopic

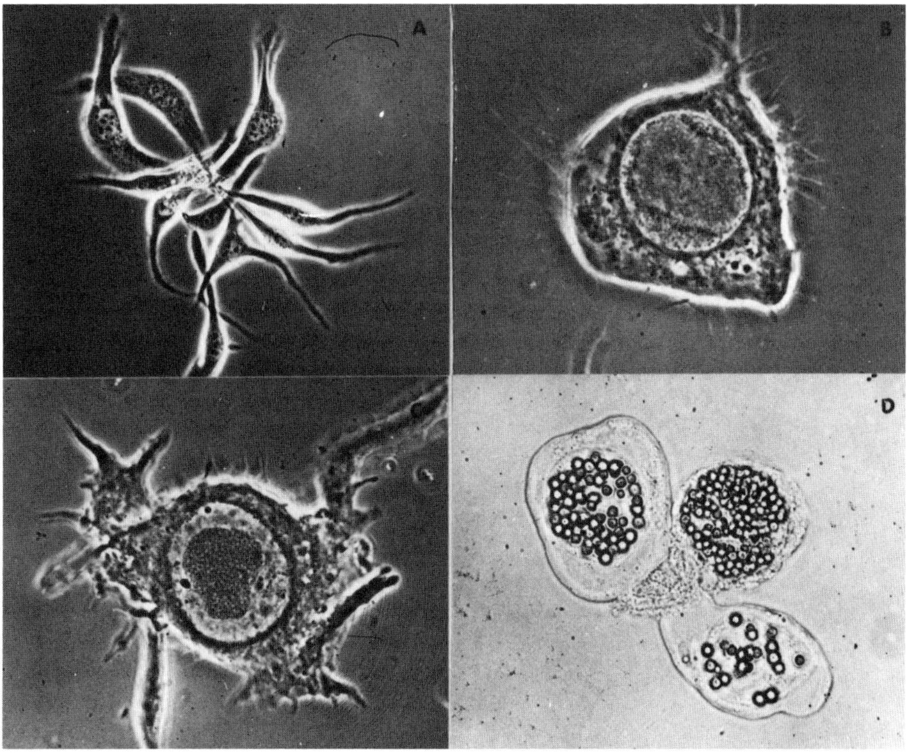


**Figure 5** Progression of CPE in TN-368 cells infected with AcMNPV. Uninfected cells are represented in (A). The first stage of CPE (B) is represented by rounding of cells, enlargement of nucleus, disappearance of nucleoli, and formation of virogenic stroma. Later, the virogenic stroma appears to become more compact (C) with Brownian motion around the internal periphery of the nuclear membrane often seen. The final stage (D) is characterized by lysed cells, with polyhedra found in the medium and by cells containing polyhedra. (From Ref. 94, used with permission.)

studies of MNPV infections have suggested that BV enters the cell cultures by viropexis and/or fusion and PDV enters by viropexis (55,104,105,107,108). Recently, Volkman's group (109,110) showed that AcMNPV BV entered IPLB-SF-21 cells by adsorptive endocytosis. Neutralizing monoclonal antibody, AcV1, prevented AcMNPV from using this pathway, but not all infectivity could be neutralized. The residual infectivity was shown to be due to AcV1-linked AcMNPV and not aberrant particles lacking the AcV1-reactive BV epitope. Fusion at the plasma membrane, but not phagocytosis, seemed to be an additional, albeit less efficient, pathway for entry. The major protein immunoprecipitated from solubilized, radiolabeled BV had a molecular weight of 64,000, suggesting that this glycoprotein is important in the ability of the virus to enter host cells. Inhibition of glycosylation with tunicamycin showed that the sugar moiety of the 64 kD protein was not essential for normal entry of virus into cells (111).

The immediate fate of the virus particle upon entry is not clear; Granados (112) has suggested that the nucleocapsid is released from vacuoles into the cytoplasm. Two mechanisms have been described for uncoating: attachment of naked nucleocapsids to the nuclear pores (107) and entry of the entire nucleocapsid into the nucleus (104). The second mechanism is favored for MNPV infections, although direct evidence that it results in a viable infection is not available.

Once uncoating occurs, virus particles are not found in infected cells until after the development of CPE (107). Virogenic stroma is observed by 8–9 h p.i., and by 10 h p.i. virus particles are seen in the stromal network (Figures 6A and 6B) (105).

The precise mechanism of nucleocapsid assembly is not known. Bassemir et al. (108) have suggested that the capsid is constructed first and subsequently filled with the nucleoprotein core. Fraser (113) studied the assembly of AcMNPV nucleocapsids in detail. Nucleocapsid maturation appeared to begin with the assembly of capsid sheaths on base structures in the pockets of the virogenic stroma. The partially assembled precapsid then appeared to associate end on with the matrix of the virogenic stroma during acquisition of nucleocapsid core complex. Packaging of the nucleoprotein complex appeared to be associated with the cap structure. Volkman (114) showed that cytochalasin D interferes with proper nucleocapsid assembly and suggested that microfilaments might be involved in this process. Viral DNA replication and synthesis of the major capsid antigen and its transport to the nucleus were not inhibited by cytochalasin D.

Following nucleocapsid assembly, and by 12 h p.i., there is de novo synthesis (53) and assembly of envelope material in the nuclei (105) (Figure 6B). From this point in the infection process, the virus matures along two distinct pathways. Those virions destined to become PDVs acquire an envelope in the nucleus. On the other hand, the BVs migrate to the plasma membrane and bud with single nucleocapsids per envelope (104,107). Each of these maturation processes appears to be initiated by attachment of the cap structure of the naked nucleocapsids to the enveloping membranes. The envelope appears to be attached to the pointed or "nipple" end of the nucleocapsid (55); this attached portion of the membrane contains peplomers and seems to have been derived from regions of the plasma membrane that have an electron-dense "spiny coat" appearance (53,55). Volkman et al. (115) have shown by immunoelectron microscopy that antigens reactive to neutralizing monoclonal antibody were more highly concentrated at the end with peplomers. The BV envelope is very "loose fitting" compared to the envelope of virions that are occluded (Figure 3A) (53,116).

PIBs begin to be formed (Figure 6C) by 14–16 h p.i. (105). Only enveloped virions are occluded within the PIBs, and the virions do not appear to disrupt the crystalline array of the polyhedrin lattice (45). The polyhedra are ultimately surrounded by an electron-dense polyhedral envelope (discussed earlier) which appears to be formed from condensation of a fibrous, electron-dense substance associated with immature PIBs (105,117).

### Virus Assay

Both qualitative and quantitative procedures have been developed for the assay of NPV in cell culture [see Hughes and Wood (118) for a recent review]. Plaque assays are available for several NPVs using the TN-368, SF-IPLB-21, and IPLB-HZ-1075 cell lines as indicator cells (8,119–123). Most procedures rely on the presence of PIBs within infected cells to detect plaques, but systems producing cell lysis have also been utilized, particularly for the assay of viral mutants unable to produce PIBs (124). A quantal ($TCID_{50}$) assay for NPV titration developed by Brown and Faulkner (102) is widely used. It is important to note that both plaque assay results and $TCID_{50}$ titrations of NPV will vary with the density of cells seeded per well and the incubation period of the virus required to identify infection (119). Volkman and Goldsmith (99) have developed useful immunoperoxidase assays (125) for detecting NPV infections in the absence of polyhedra production.

### Persistent Infections

Persistent viral infections are long-term infections in which the virus, sometimes highly cytocidal, continues to multiply while the host organism or cell cultures continue to grow or divide. Mims (126) classified three forms of viral persistence: (1) virus persists without shedding infectious virus, (2) infectious virus is shed but only periodically, and (3) virus replication and release occur continuously. The recent review by Burand et al. (127) gives a detailed account of baculovirus persistence.

Several investigators have provided tantalizing evidence suggesting that NPV might persist in an occult or "latent" state in its host (128,129). It has been suggested that acute infection can be induced in such hosts with stimuli such as heat, crowding, a change in diet, or ingestion of a heterologous virus (130–133). The validity of baculovirus persistence in vivo is not clearly proven. However, persistent NPV infections have been established in cell culture, and these are outlined next.

McIntosh and Ignoffo (133) established a persistent infection of SfMNPV in its homologous cell line, IPLB-SF-21, by infecting cell cultures with low multiplicity of infection. These cells manifested typical NPV cytopathology, including production of PIBs, at about one week after infection. Cells that appeared to be uninfected (SFP) were subsequently cultured. The SFP did not remain attached to the culture vessel and continued to shed virus. Although 70% of the cells tested positive for SfMNPV antigens, only 0.8% of the population

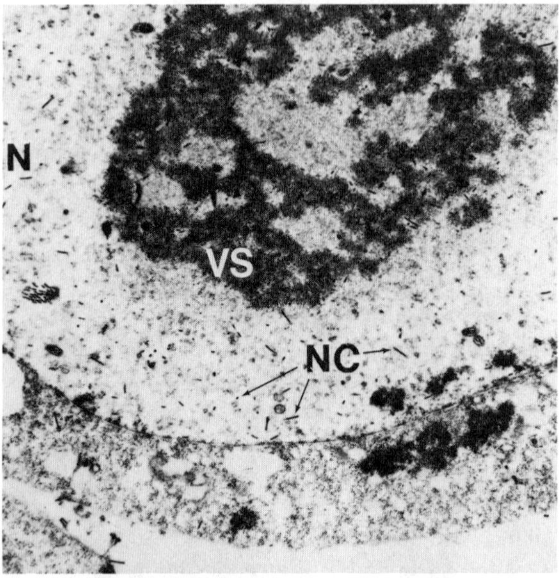


(A)

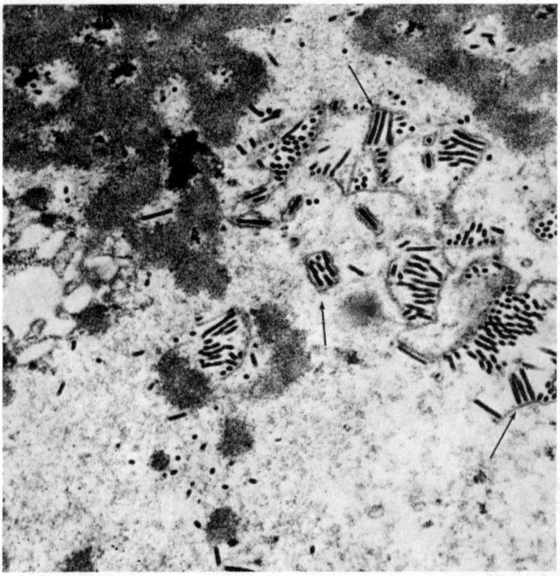

(B)

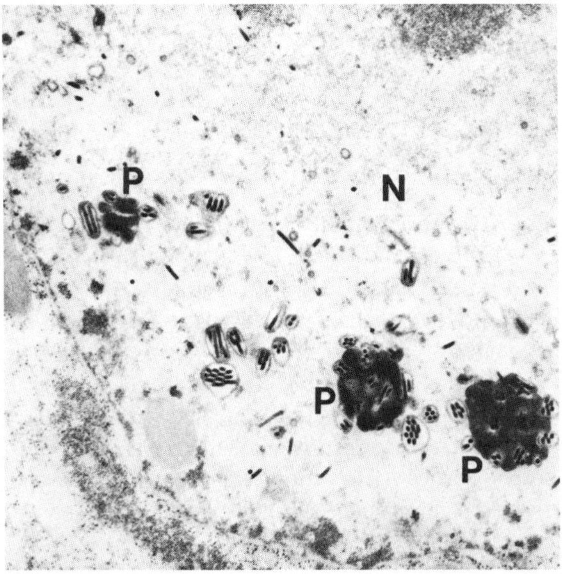

(C)

**Figure 6** (A) Formation of virogenic stroma (VS) and nucleocapsids (NC) in an AcMNPV-infected *S. frugiperda* (SF-IPLB-21AE) cell. The nucleus (N) is hypertrophied. A few enveloped virions are found at the periphery of the nucleus. Magnification 11,000×. (B) Envelopment of nucleocapsids in the nucleus of an AcMNPV-infected *S. frugiperda* (SF-IPLB-21AE) cell. Several large groups of nucleocapsids have acquired or are acquiring envelopes (arrows). Magnification 28,000×. (C) Formation of polyhedra (P) in the nucleus (N) of an AcMNPV-infected *S. frugiperda* (SF-IPLB-21AE) cell. Enveloped nucleocapsids are in the process of being occluded within polyhedra. Magnification 25,000×. (From W. M. Carpenter and S. L. Bilimoria, unpublished.)

was producing virus and 3% contained PIBs, which were three times less infectious for larvae than those produced in acute infections. The persistently infected cells were smaller, grew more slowly, and, compared to the normal cells, reached a higher cell density. These cells could not be superinfected with SfMNPV and were also somewhat resistant to AcMNPV. On the basis of the fact that homology between SfMNPV and AcMNPV is less than 1% (19), it has been suggested that the resistance to infection by AcMNPV is mediated by an "interferonlike" antiviral factor rather than by a virus-specific factor. This contrasts with the lack of resistance of TN-368 cells persistently infected with the nonoccluded virus (NOV, subgroup C), Hz-IV to AcMNPV. Since the Hz-IV and AcMNPV also share less than 1% DNA homology, Burand et al. (127) postulated a virus-specific mechanism for resistance to superinfection of their persistent system with homologous NOV.

An understanding of the molecular biology of persistence for the NPVs will be important for elucidation of the mechanisms of virus specificity and latency. Currently, there is clearly a dearth of information on the biochemistry of persistent NPV infections.

## SNPVs

Thus far, only three SNPVs, *Heliothis zea* SNPV (HzSNPV) (134), *Orgyia leucostigma* SNPV (OISNPV) (135), and *Trichoplusia ni* SNPV (TnSNPV) (6) have been successfully propagated in cell culture. Ignoffo et al. (134) infected *H. zea* cells with HzSPNV and showed that whereas free virions or PIBs could not be detected through seven passages, an increase in infectivity was detected by bioassay in neonate larvae. Goodwin et al. (87) utilized electron microscopy to show that the *H. zea* cell line IPLB-HZ-1079 could be infected with HzSNPV. Recent studies on HzSNPV have been performed using the IPLB-HZ-1075 cell line (133,136–139). The morphogenesis of infection was very similar to that observed with MNPVs (discussed previously). Both extracellular and intracellular infectivity reached maximum levels at 3–4 days p.i. (137,140). The infectivity of BV for larvae remained stable with continued passage; however, the infectivity of polyhedra decreased with passage (138). Yamada and Maramorosch established a plaque assay for HzSNPV. McIntosh et al. (142) recently showed that HzSNPV replicated in *H. virescens* cells, but not in *H. armigera*, *T. ni*, or *S. frugiperda* cells.

Sohi et al. (135) reported the replication of OISNPV in four established cell lines developed from minced larvae of *O. leucostigma*. The level of infection was 65–90% in two of these cell lines (IPRI-OL-12 and IPRI-OL-13) and 3% in the other two lines (IPRI-OL-4 and IPRI-OL-9). Infected cells developed PIBs within 24 h p.i. Viral cytopathic effect and morphogenesis, as detected by electron microscopy, were similar to observations made with HzSNPV by other workers. Rate zonal centrifugation of alkali-released virions (PDVs) revealed only one virion band, confirming the single-nucleocapsid nature of the progeny virion. Sixty serial passages of the O1SNPV through the cell cultures showed that the percentage of infected cells and PIB production stayed generally high in IPRI-OL-13 cells but declined sharply after the fourth passage in IPRI-OL-12 cells.

Recently, Granados et al. (143) infected newly established *T. ni* embryonic cell lines with TnSNPV. Phase-contrast and electron microscopy showed that 29 of 36 cell lines tested were infected with SNPV; the percentage of infected cells ranged from 1 to 60%.

The development of cell culture infections of SNPVs is a significant advance in the biology of the NPVs since a number of very important viruses are of this morphological type; for example, HzSNPV was the first baculovirus to be awarded a license for commercial use. Comparative studies of SNPV and MNPV gene expression should help establish general principles of NPV replication. Since individual isolates of SNPVs appear to have a much narrower host range than the MNPVs, comparative studies will be important for a clear understanding of the biochemical basis of NPV host specificity.

## GENETICS

The majority of the NPV mutants studied are of AcMNPV and are defective in occlusion. These OV mutants include few-polyhedra (FP) mutants, OV morphology mutants, polyhedrin deletion mutants, and an overproducer of OV, the high-occlusion body (HOB) mutant (144).

### Serial Passage Mutants

Few-polyhedra virus was first reported by Hink and Vail when they described the plaque assay for AcMNPV (8). Visualization of plaques by oblique lighting and a dissecting microscope revealed two phenotypes described as "many polyhedra" (MP) and "few polyhedra" (FP). Ramoska and Hink (145) showed that MPs had 15 to 35 PIBs per nuclear section while most FPs had none and seldom more than seven. MacKinnon et al. (44) reported that serial passage of MNPVs in cell culture led to the appearance of aberrant viruses in late-passage virus. Potter et al. (146) showed that FPs arise upon undiluted (therefore high-multiplicity) serial passage of BVs in cell culture. These workers noted that when FPs were plaque-purified rigorously, they remained homogeneously FP upon serial passage. FP and MP BVs were equally pathogenic upon injection into larval hemolymph. FPs arose from MPs both in larvae and in cell culture. Miller (144) views FPs "as mutants which arise spontaneously upon serial passage and are selected by passage in cell culture owing to their increased production of BVs, possibly at the expense of polyhedra formation." Most FPs appear to be relatively stable mutants of MP viruses (144).

Much progress has been made recently in elucidating the nature of FP mutants. Potter and Miller (147) showed that anomalies in the REN fragment patterns of their FP-D mutant were due to insertion of DNA, possibly of host origin, into the viral DNA. Miller and Miller (148) showed that this insertion was in fact a *copia*-like transposable element derived from the *T. ni* cells in which the virus had been serially passaged. They also showed that this transposable element, designated TE-D, was present in multiple copies (approximately 50) in *T. ni;* TE-D was dispersed throughout the genome and was mobile in it. Like the retroviruses, *copia*-like elements have long directly repeated sequences at their termini. These are referred to as long-terminal repeats (LTRs). The LTRs of TE-D, called ID sequences, are 0.27 kb in length. The length of the entire element is 7.3 kb. FP-D preparations contain two different viral DNA genomes: the FP-DL (large) DNA contains the entire TE-D insert and the FP-DS (small) contains only the ID sequence. In both cases, the insertions are in the identical location in the HindIII-K fragment I of the AcMNPV genome (see Figure 7 for restriction maps). Attempts to plaque purify FP-D result in the isolation of either FP-DS or a combination of both FP-DL and FP-DS. Miller (144) believes that this is due to general genetic recombination between the two terminal ID sequences or to a transposition-specific event which leaves a single ID sequence following transposition. Why or how the internal segment of TE-D is lost or retained is not known.

Fraser et al. (149) isolated FP mutants from AcMNPV and *Galleria mellonella* MNPV (GmMNPV) serially passaged three times through *T. ni* cells. These workers made the interesting observation that eight of the nine FP mutants isolated had insertions in the HindIII-I region of the viral genome. All of the FPs isolated were defective in the synthesis of a 25 kD protein. Marker rescue experiments correlated the synthesis of this protein with the HindIII-I fragment of AcMNPV and established the role of the 25 kD protein in the FP phenotype. These results suggest that all FP mutants arise by some type of mutation in the AcMNPV HindIII-I region which affects the synthesis of the 25 kD protein, and that the majority of spontaneous mutants arising upon serial passage are due to host DNA insertions (144). In a subsequent report, Fraser et al. (150) showed that the insertion (located at 35.0 to 37.7 map units) has sequence properties of a transposon of *T. ni* host cells. Recently, Beames and Summers (151) described new FP mutants with alterations at 33.8–37.7 map units which were either deletions or insertions of *S. frugiperda* host DNA sequences. Northern and S1 nuclease analysis showed that both deletions of viral DNA sequences and insertion of host DNA sequences produce altered transcripts. These workers also detected an open reading frame for a 25 kD protein at the site in the wild-type genome where the DNA insertions had been mapped and point out that this protein is the same size as the infected cell protein missing from the FP mutants of Fraser et al. Since Potter and Miller (152) did not observe HindIII-I insertions in the five FP mutants they analyzed, and Wood found no change in the 25 kD protein of the FP mutant he analyzed, Miller (144) suggests that Fraser et al. (149) may somehow have been selecting a special class of FP mutants; i.e., the findings are not "universal."

Further research clearly is required to fully understand the nature of FP mutants. The implications of the "natural" insertion of host elements into viral genomes could be very important in the use of NPVs as pesticides and also for the evolution of these viruses and their hosts (144).

## Temperature-Sensitive Mutants

Using NTG (*N*-methyl-*N*'-nitro-*N*-nitrosoguanidine) and BUdR (5-bromodeoxyuridine) as mutagens, Brown et al. (153) and Lee and Miller (154) isolated temperature-sensitive (*ts*) mutants of AcMNPV at frequencies ranging from 0.5% to 4.8%. While both groups utilized a restrictive temperature of 33°C, the permissive temperatures were 27° and 23.5°C, respectively. Miller (144), however, cautions against the use of 33°C as a restrictive temperature since the wild-type virus produces defective polyhedra at this elevated temperature. Preadaptation of cell lines to the restrictive temperature is also recommended.

Three broad categories of *ts* mutants are defined based on their ability to produce BV or OV at the restrictive temperature (144). The first class is defective in both BV and OV synthesis ($BV^- OV^-$). The second class is defective in OV synthesis only ($BV^+ OV^-$). The third class is $BV^- OV^+$, but the polyhedra are defective (noninfectious per os). Brown et al. (153) found reversion frequencies for 10 *ts* mutants to range from $2.1 \times 10^{-5}$ to $7.8 \times 10^{-4}$, indicating stable mutants. Leak frequencies were generally 10- to 30-fold higher than reversion frequencies. Leak frequencies higher than $10^{-3}$ can interfere with mapping of mutations by marker rescue (144).

Miller (144) recommends ability to form plaques at the restrictive temperature as the simplest and most direct means of assessing the production of BV. Preadaptation of indicator cells to growth at the high temperature is critical. A tenfold or greater reduction in plaque-forming ability against high-temperature *wt* infections is suggested as a "reasonably clear indication of an impairment in infectious BV production."

Miller and Dawes (155) utilized pulse-labeling (with $^{35}S$-methionine) of infected cells at 24 h p.i. to screen for defects in protein synthesis. At least 4 of the 15 mutants were defective in polyhedrin production at this time. At least five infected cell-specific proteins showed varying levels of label uptake in the various mutant-infected cells. Three mutants had alterations in specific polypeptides. In light of the findings of Fraser et al. (149) it is interesting that one mutant (*ts*B305) which has genotype alterations in the HindIII-I region also had an altered 25 kD protein.

At least two DNA replication-defective mutants have been isolated. The *ts*8 mutant isolated by Brown et al. (153) is defective in DNA replication. Gordon and Carstens (156) showed that *ts*8-infected cells synthesized only early virus proteins at the nonpermissive temperature, and in contrast to *wt*-infected cells, did not inhibit host cell protein synthesis. Failure of the rate of accumulation of viral DNA to increase after a shift from permissive to nonpermissive temperature indicated that *ts*8 was involved in the synthesis of proteins required for viral DNA synthesis. The mutation was mapped by marker rescue to the region between 60.1 and 62.0 of the AcMNPV physical map. A similar mutant (*ts*B821) has also been described by Miller et al. (157). Pulse-labeling with $^{35}S$-methionine showed that *ts*B821 is defective in a very early virus function apparently related to the induction of delayed-early genes.

Complementation analysis of *ts* mutants has been carried out by two laboratories. Using a complementation test based on measuring polyhedrin synthesis by radioimmunoassay, Brown et al. (153) were able to distinguish five complementation groups for AcMNPV. Lee and Miller (154) used a complementation test based on the percentage of cells containing polyhedra and distinguished 14 groups of 15 mutants. Duncan and Faulkner (158) used a test similar to that of Lee and Miller to analyze the 10 mutants isolated by Brown et al. as well as 8 new mutants and distinguished a total of 10 complementation groups. The potential utility of complementation analysis has been superseded by recent advances in molecular genetics, which allow a more direct and precise analysis of NPV genes.

Three different groups have independently established the ability of NPVs to recombine: Brown and Faulkner (159) demonstrated recombination between *ts* mutants; Summers et al. (160) and Crozier and colleagues (161,162) showed that recombination between closely related but genotypically distinct AcMNPV variants could occur. Crozier's group utilized larvae rather than cell culture systems in their studies.

NPV mutants have been mapped using both classical recombination approaches and marker rescue. Brown and Faulkner (159) performed two-factor crosses and ordered six $BV^{-}OV^{-}$ mutants on a linear recombinational map. Their results implied a circular map but this was not clearly established. Miller's laboratory developed the marker rescue technique for NPV mutations and mapped seven *ts* mutants (147,163). The mutant DNA is cotransfected with a series of wild-type restriction fragments. One fragment should rescue the mutant and convert it to wild type such that plaques are produced at the restrictive condition.

## Occluded Virus Mutants

A number of OV morphology mutants have been described. Brown et al. (164) isolated occlusion body morphology (om) mutants which produce one to two large (10–15 $\mu$m) cuboidal inclusions per infected cell nucleus. Electron microscopy showed that the paracrystalline lattice of the inclusions was similar to that of wild type, but more than 99% of the inclusion bodies did not contain virions. Furthermore, the $LD_{50}$'s of these cuboidal inclusions are 100-fold higher than those of wild-type polyhedra. The two mutants that were studied (M5 and M6) had polyhedrin mobility in SDS-PAGE and *Staphylococcus aureus* V8 digestion patterns for polyhedrin that were distinct from wild-type polyhedrin, clearly establishing that the mutations are in the polyhedrin gene. Carstens et al. (168)

further showed that there were at least three mutations in M5 DNA. They mapped two 400 bp inserts at 0.0 to 3.0 and 46.5 to 48.6 on the AcMNPV genome. The mutant DNA consisted of a combination of two DNAs: one had the entire viral genome and the other had the regions between 0.0 to 3.0 and 46.5 to 48.6 deleted. Although this situation may be somewhat similar to the FP-D and TE-D elements described earlier, the origin of the M5 400 bp inserts has not been established. Carstens et al. (166) showed that one of the three mutations in M5 is due to a single base-pair change in the BamHI site within the polyhedrin gene. As a result of this, a proline codon for amino acid 58 (CCT) is converted to a leucine codon (CTT). Computer-predicted secondary-structure analysis indicated that the amino acid at position 58 could be critical to the proper folding of polyhedrin. Duncan et al. (167) have studied another om mutant, M29, which produces many small occlusion-body-like particles (95–180 nm) but does not manifest virions, paracrystalline lattice, or polyhedral membrane. Polyhedrin and all other infected cell-specific proteins in M29 infections appeared to be normal, but some proteins were delayed by about 3 h. Carstens et al. (168) have recently shown by marker rescue experiments that this mutation is located in the 0.0–5.9% region of the AcMNPV genome. Direct sequencing revealed a single-point mutation at position 253 of the polyhedrin gene. Thus the mutation causes a substitution of phenylalanine for leucine at amino acid 84 of the M29 polyhedrin protein.

Polyhedrin deletion mutants have been produced by site-directed deletion (124). Deletions were introduced into cloned EcoRI-I fragment (which contains the polyhedrin gene) of AcMNPV and cotransfecting this clone with wild-type, circular AcMNPV DNA. The cell replaces the wild-type with the homologous plasmid region, thus introducing the deletions into the viral genome. Smith et al. (124) were able to select the mutant by using the technique of "oblique lighting" to distinguish $OV^-$ from $OV^+$ plaques. SDS-PAGE analysis of infected cell-specific proteins showed that $OV^-$ mutants failed to synthesize only the 33 kD polyhedrin protein. Electron microscopic analysis of mutants that failed to produce polyhedrin showed normal development of virus fibrous material and fibrous sheets, indicating that polyhedrin is not a constituent of these structures. No significant differences were found in the rate of BV release compared to wild-type infections (although $OV^-$ infections did undergo premature lysis) at one to three days versus the four to five days required for wild-type infections. These results clearly show that BV production is independent of polyhedrin synthesis and that polyhedrin is not required for the envelopment of nucleocapsids or for the formation of fibrous material (144).

Wood et al. (169) have isolated a high-occlusion body (HOB) mutant that is an overproducer of occluded virus. The HOB mutant had significantly shorter $LT_{50}$ than the parental 1A strain, but the $LC_{50}$ did not differ significantly.

# MOLECULAR BIOLOGY

## Genome Organization

### Physical Maps

In addition to their usefulness for the identification of NPVs and other DNA viruses, RENs are indispensable for the construction of physical maps (170). A physical map is a scale diagram of the relative locations of cleavage sites in the viral genome. Individual REN fragments are fractionated by gel electrophoresis and their sizes are determined from relative mobility during electrophoresis (171) or by electron microscopy (172). The fragments are numbered (largest to smallest) in alphabetical order. Several approaches may be used to determine the order of the fragments in the intact genome (173–176). The method used depends on the length of the DNA and the degree of detail required. Physical maps are essential for the study of genetic organization, gene transcription and regulation, and gene replication. Maps using several RENs have been established for AcMNPV (26,155,177–180) (see Figure 7) and a number of other NPVs (181–186).

Given the circular structure of the NPV genome, site 1 in the physical maps of AcMNPV is, by convention, at the EcoRIF/I junction (Figure 7). Genetic variability, resulting from the loss or addition of restriction sites, among the different strains of AcMNPV used in various laboratories alters the REN fragment designation. Cochran et al. (16) have pointed to the need for a composite physical map of all the AcMNPV strains to overcome problems resulting from variability.

Several laboratories have established restriction fragment libraries in a variety of cloning vectors (19,177,180,187). These reagents have already proved extremely useful as probes in transcriptional mapping and in functional mapping of NPV genes by techniques such as marker rescue.

### Sequence Homology

As with the DNA viruses of vertebrates (188–193), the NPV also have a highly interesting example of intragenomic sequence homology. Cochran and Faulkner (194) showed that their HR3 strain of AcMNPV contained five homologous regions (*hr*'s) interspersed throughout the viral genome. These regions (designated *hr*1 through *hr*5) occurred at or near EcoRI junctions (195) and at map positions shown in Figure 7. Subsequently, homologous regions have been detected in other AcMNPV strains or variants such as AcMNPV-E, *S.exempta* MNPV(SeMNPV), and *Choristoneura fumiferana* MNPV (CfMNPV) (only four homologous regions were found in the last two variants), suggesting that *hr*'s play an important role in the life cycle of the NPV. The *hr* sequences range in size from 500 to 800 bp.

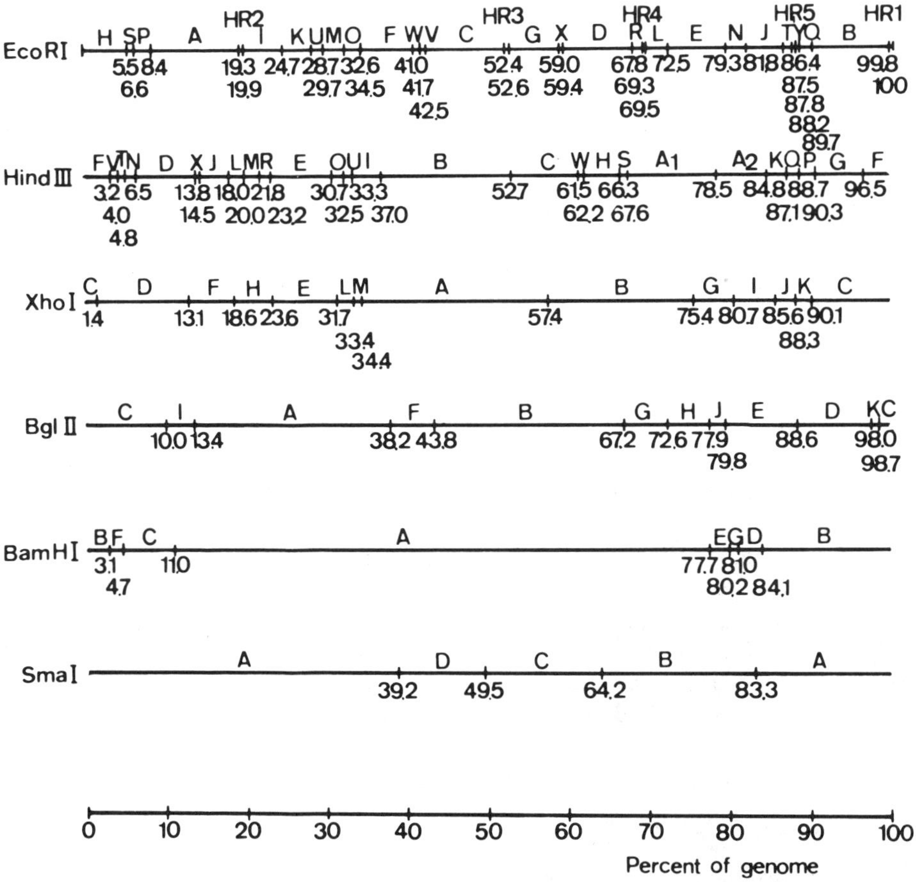


**Figure 7** Physical map of the E variant of AcMNPV DNA. The circular map is linearized for presentation. HR1, 2, 3, 4, and 5 refer to the homologous repetitive regions. (From Ref. 220, used with permission.)

Guarino and Summers (196) have shown that *hr*5 of AcMNPV-E is a strong enhancer of delayed-early gene expression. The nucleotide sequence of *hr*5 revealed two conserved repeated sequences: a 30 bp sequence is repeated three times, and a 34 bp sequence is imperfectly repeated six times. The EcoRI sites are part of a 30 bp imperfect palindrome contained in the 34 bp imperfect direct repeat. These two conserved repeats are separated by nonhomologous DNA. Similar results have been obtained for *hr*5 by Liu et al. (197). Guarino et al. (195) have completely sequenced the remaining four *hr*'s in AcMNPV and have shown that *hr*'s functioned as enhancers in a transient expression system. The homology within an *hr* was 65–87%, and the consensus sequences for each region were 88% homologous to each other. Sequences proximal to the EcoRI sites were most conserved, while the distal sequences were least conserved. The EcoRI sites formed the core of a 28 bp imperfect inverted repeat. Deletion analysis of *hr*5 showed that the minimal sequence requirement for the enhancer function is one copy of the inverted repeat. Both the conserved and nonhomologous sequences of the *hr*'s are rich in adenine and thymine.

There is some evidence for the presence of *hr*'s in other NPVs including OpMNPV, AgMNPV, and TnSNPV (16). The *hr* regions of CfMNPV are not homologous to those of the AcMNPV variants (182). It is tempting to speculate that the apparent universality of distinct *hr*'s in NPVs and their role (together with the requirement of a host factor) in the enhancement of the temporal cascade of NPV gene expression and possibly also in the initiation of viral DNA replication (194) might constitute a common mechanism by which the NPVs govern their host specificity. If this is the case, then NPV infections of a substantial number of nonpermissive hosts should be restricted during the transition from immediate-early to delayed-early expression.

Intergenomic sequence homology among the NPVs is discussed under "Classification." However, an aspect of intergenomic sequence homology relevant to NPV gene organization is evidence suggesting that the genomes of a number of NPVs might be arranged collinearly (16). The genomes of AcMNPV and OpMNPV appear to be collinear (186), and the same appears to be true for AcMNPV and CfMNPV (198) and also for AcMNPV and HzSNPV.

## Genome Expression

### Viral Protein Synthesis

Several laboratories (28,58,78,79,199,200) have documented the kinetics of AcMNPV- and TnMNPV-specific protein synthesis and demonstrated the sequential and usually transient synthesis of 30–40 polypeptides (74) (Figure 8). On the basis of their time of appearance and the effect of reversible inhibitors of protein synthesis such as cycloheximide and inhibitors of DNA replication (e.g., cytosine arabinoside), viral protein synthesis has been divided into three to four temporal classes. These are designated variously in different laboratories. The following designations are in vogue (the times in parentheses refer to the first observation postinfection of polypeptides): six immediate-early or alpha polypeptides (2 h); 23–25 delayed-early or beta polypeptides (6 h; this phase defines the onset of DNA synthesis); three late or gamma polypeptides (10–12 h); and two delayed-late or delta polypeptides (15 h). During the delta phase, polyhedrin is synthesized at a high rate, and the production of other polypeptides is reduced (58,78,199,200). Except for the immediate-early polypeptides, each subsequent class appears to require the proper expression of proteins of the

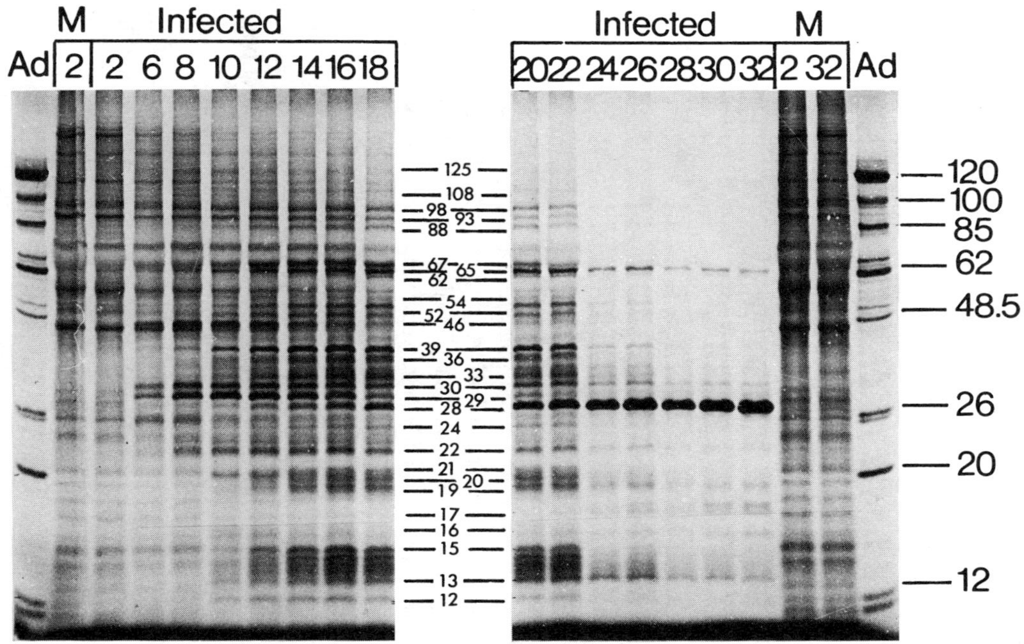


**Figure 8** SDS-polyacrylamide gel fluorogram of *S. frugiperda* cells pulse-labeled with $^{35}S$-methionine for 1 h at the indicated times after infection with AcMNPV. Adenovirus type 2-infected cell extracts (Ad) were included as molecular weight markers. Mock-infected cells pulse-labeled at 2 and 32 h p.i. were also included (M). The numbers along the top of the electropherogram refer to the hour postinfection when the cells were pulse-labeled; the numbers along the sides of the electropherogram refer to the molecular weights in kilodaltons of the various polypeptides. (From Ref. 99, used with permission.)

preceding class (157,201). Recently, Bilimoria's laboratory showed that a similar temporal cascade system operates in SF-IPLB-21 cells infected with *S. frugiperda* MNPV (100,202) (Figure 11). It is interesting to note that while inhibition of NPV DNA synthesis with cytosine arabinoside clearly inhibits late protein synthesis in cells infected with *S. frugiperda* MNPV (100,202) and *T. ni* MNPV (200), the same is not true for AcMNPV infections. Both temporal and abundancy controls appear to be operating in the programing of NPV gene expression (203,204).

Although at least seven infected cell-specific proteins including polyhedrin are phosphorylated, the role of phosphorylation in NPV replication is not known (78). Stiles and Wood (77) described nine virus-induced glycoproteins in AcMNPV-infected cells and used the glycosylation inhibitor tunicamycin to show the specificity of the labeled precursor in their experiments. Although there is evidence for slow cleavage (79,199,200), rapid posttranslational cleavage of viral polypeptides in infected cells has not been observed.

### Viral RNA Synthesis

Since NPV infections can be initiated with naked, purified viral DNA (152,205–207), virus-associated enzymes are not needed for the initiation of NPV transcription. Grula et al. (208) suggested that an $\alpha$-amanitin–resistant RNA polymerase was involved in virus-specific RNA synthesis. Because $\alpha$-amanitin inhibits host RNA polymerase II, the result suggests that a viral RNA polymerase may play a role in transcription of the NPV genome. Fuchs et al. (209) showed that viral RNA synthesis was sensitive to $\alpha$-amanitin up to 8 h p.i. With the progress of infection the sensitivity to $\alpha$-amanitin decreased, and by 24 h p.i. viral RNA synthesis was resistant to $\alpha$-amanitin. Thus the shift from early to late viral mRNA synthesis appears to be correlated with appearance of an $\alpha$-amanitin–resistant RNA polymerase. As expected, the bulk of NPV transcription occurs late in the infection cycle (discussed later).

Initial studies on AcMNPV transcription in *S. frugiperda* cells focused on late genes. The late regions that were the most active transcriptionally included the 0–5 map unit (mu), 20–24 mu, and 88–90 mu. These regions code for the polyhedrin, 39 kD, and the 10 kd polypeptides, respectively. The direction of transcription for both polyhedrin and the 10 kD polypeptide was shown to be from left to right on the conventional physical map (187,203,210,211). It was also clear from these studies and the work of Erlandson and Carstens (212), who studied early transcripts, that both late and early transcripts were derived from dispersed regions of the viral genome. Erlandson et al. (213) showed that 12 abundant transcripts were detected by 4 h p.i. There was an increase in the abundancy and number of virus-specific transcripts between 6 and 8 h p.i., coincident with the onset of viral DNA replication. This change in transcription pattern did not occur when cells were infected with a DNA-negative mutant,

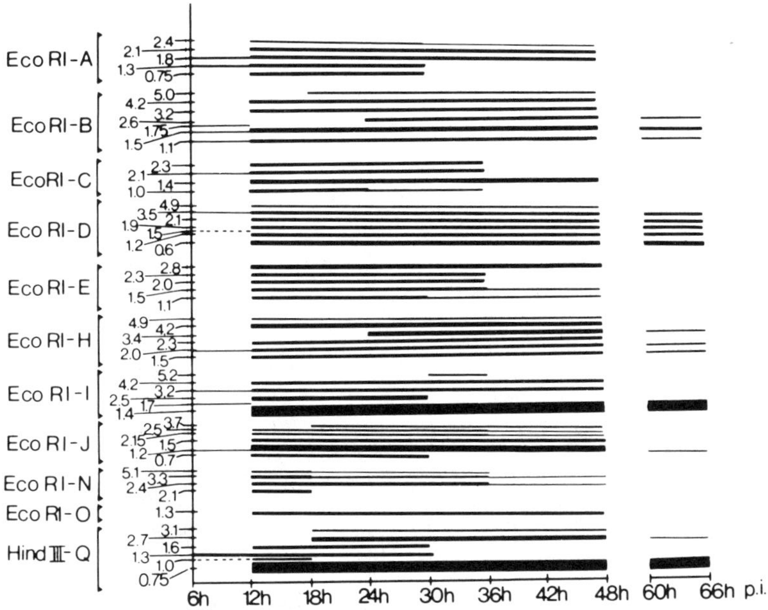


**Figure 9** Schematic diagram depicting extent of hybridization of cytoplasmic poly $(A)^+$ RNA to restriction fragments of DNA from AcMNPV. The thickness of each line represents the intensity of hybridization according to the following arbitrary scheme (from thickest to thinnest lines): extremely strong, very strong, strong, moderate, weak, very weak, and extremely weak (dashed lines). The RNA species hybridizing to the same probe are grouped together, and size estimates (in kb) are indicated on the left. The times of RNA isolation are indicated at the bottom. (From D. Z. Rohel and P. Faulkner, *J. Virol.* 50, 739, 1984, used with permission.)

suggesting that the early transcripts were synthesized prior to the initiation of viral DNA replication. Rohel et al. (203) analyzed the time course of AcMNPV transcripts and showed that most viral RNA was detected 12–24 h p.i. (Figure 9). Polyhedrin and 10 kD polypeptide RNA were the two most abundant late viral RNA transcripts, and they were synthesized at high rates until late in the infection cycle (60 h p.i.). Circumstantial evidence suggested that the abundancy control of polyhedrin and 10 kD polypeptide was at the level of transcription. These workers recognized at least 50 viral RNA transcripts and constructed a transcription map which they correlated with previously published translation maps (Figure 10).

Lübbert and Doerfler (214) mapped 11 early and more than 90 late transcripts of AcMNPV and made the interesting observation that there were overlapping sets of viral RNAs of various lengths and with apparently common 3' or

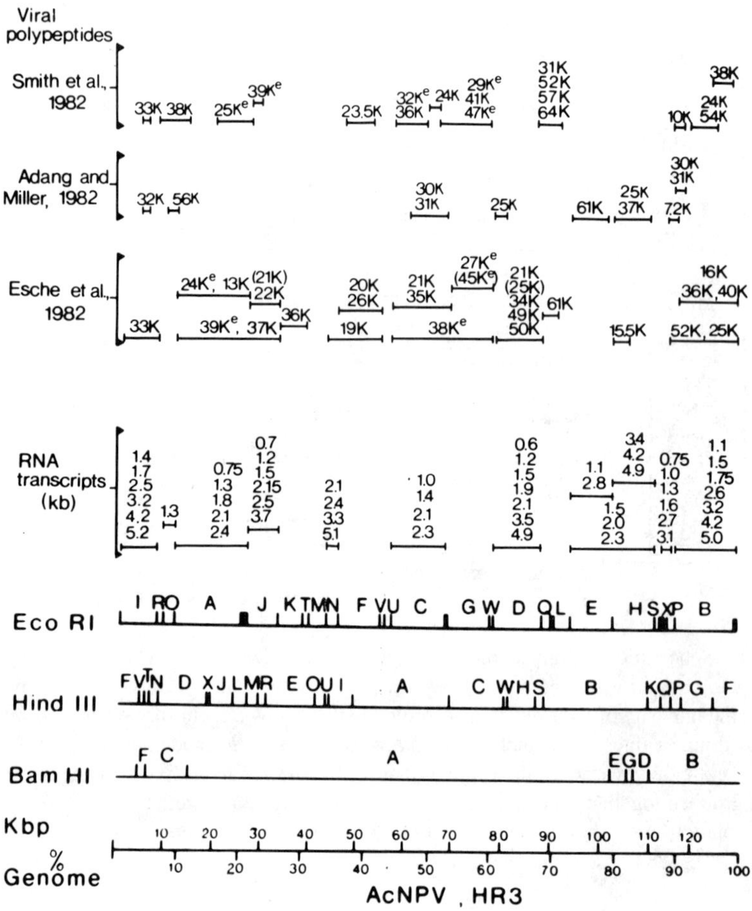


**Figure 10** Transcription map for AcMNPV: linearized physical map of AcMNPV HR3 DNA with the appropriate locations of viral RNA transcripts, which are designated by their estimated sizes (in kb). Previously published translational maps for AcMNPV are shown at the top. The heavy black areas on the physical map of AcMNPV DNA indicate the regions of homology. Kbp, kilobase pairs. (From D. Z. Rohel and P. Faulkner, *J. Virol.* 50, 739, 1984, used with permission.)

common 5' termini. Using high-resolution S1 nuclease mapping, Friesen and Miller (215) located five overlapping RNA transcribed in the same direction and terminating at a common 3' site. Rankin et al. (216) used S1 mapping to confirm the presence of overlapping RNA with common 5' termini in the region coding for the 10 kD polypeptide and showed that three different transcripts (1500,

750, and 2500 bases in length) contained the complete 10 kD protein coding region. Subsequently, Doerfler and his co-workers carried out a detailed investigation of the 81.2–85.0 mu region of the AcMNPV genome and identified at least nine size classes of viral RNA, which ranged in size from 1300 to 4300 bases and exhibited common 3' termini. Detailed restriction maps and nucleotide sequence of this region of the genome revealed several open reading frames (ORFs) on the rightward-transcribed strand. Many of the ORFs and the 5' termini of some of the mapped RNAs were preceded by potential TATA and CAAT signals. The leftward-transcribed strand was devoid of major ORFs. The molecular weights of the five polypeptides derived by in vitro translation of hybridization-elution of RNA from the 81.2 to 85.0 region of viral genome correlated well with values predicted by the ORFs. All RNAs had the same 3' terminus within a sequence of eight nucleotides and the terminus was close to multiple polyadenylation signals. The initiation sites of the nine size classes of RNAs were mapped, and determination of the cap sites of the smaller RNAs (<1800 bases) revealed a multitude of initiation sites (217).

In their study of overlapping transcripts, Friesen and Miller (215) make the very interesting observation that the smallest RNA appeared early but was replaced in time by successively larger RNA initiated further upstream. Primer extension analysis supported the contention that this temporal regulation involved the sequential activation of upstream promoters and the coordinate deactivation of downstream promoters. These workers suggested that transcription from upstream genes may suppress that of downstream genes by transcriptional interference (promoter occlusion) and thus facilitate the expression of different viral functions. In a follow-up study, Miller's group (218) showed that while all the regions of the AcMNPV genome examined displayed overlapping sets of RNAs, not all of these sets manifested a sequential appearance of larger RNAs with the progress of infection. Each overlapping set of RNAs exhibited unique characteristics such as variation in the number of overlapping transcripts, its temporal regulation, and its relative abundance during the course of infection.

Extensive analysis has not revealed any evidence of splicing in NPV transcripts. Rohrmann et al. (219) did not observe intervening sequences in the polyhedrin gene of OpMNPV by R-loop mapping, and this has been confirmed for other polyhedrin genes that have been sequenced. Lübbert and Doerfler (214,220) found no evidence of splicing in the large number of transcripts they studied. While it is still possible that tiny splices may be occurring in some NPV transcripts, this is not borne out by sequencing data on the small number of NPV genes that have been studied in such detail (221,222). The significance of the lack of RNA splicing in AcMNPV infections is not clear. Lübbert and Doerfler (220) proposed that flexibility introduced by overlapping RNA might be a substitute for the versatility afforded by the splicing mechanisms that other animal viruses, but not AcMNPV, had perfected. Other explanations might include the specific inhibition by NPVs of host gene expression via suppression of splicing mechanisms.

Correlation of DNA sequence data and amino acid sequence of polyhedrin suggest that the mRNA for this protein has the following components: (1) a short 5' leader sequence, (2) a 732 nucleotide coding sequence, (3) a 250 nucleotide noncoding region at the 3' end, (4) a poly A tail of an undetermined number of residues (16,219,223). There has been only one report on capping of viral RNA in AcMNPV-infected cells (224).

The study of the mechanisms of the temporal cascade of NPV gene expression is just beginning. Guarino and Summers (196) recently published very interesting results on the regulation of a delayed-early NPV gene by an immediate-early gene product. A transient expression system under the control of the promoter for the 39 kD gene of AcMNPV was expressed in infected cells, but not in uninfected cells, indicating that the 39 kD gene should be classified as delayed-early ($\beta$). Marker rescue experiments indicated that the region of AcMNPV DNA located between 95.0 and 97.5 mu was essential for the activation of the 39 kD promoter. Additional transient assay experiments indicated that the gene for the putative trans-activating factor was an immediate-early ($\alpha$) gene. The nucleotide sequence of this trans-activating gene (IE-1) shows that it encodes a polypeptide of 581 amino acids with a predicted molecular weight of 66,856. In vitro expression of this gene has yielded a single peptide with a molecular weight of 70,000. Quantitative S1 nuclease analysis indicated that IE-1 was maximally synthesized during a 1 h virus adsorption period and that steady-state levels of IE-1 mRNA were maintained during the first 24 h of infection. Since IE-1 activates the transcription of several early AcMNPV genes in addition to the 39 kD gene, it is very likely that IE-1 has a central role in the regulation of $\beta$ gene expression (225).

Several experiments have hinted at transcriptional regulation of late NPV genes by early gene products. Knebel et al. (226) used transient assays to show that the promoter for the late 10 kD gene of AcMNPV was inactive in uninfected *S. frugiperda* cells, but was optimally active in AcMNPV infected cells if the plasmid construct containing the promoter for the 10 kD gene was transfected 18 h after infection with the virus. This finding directly demonstrates that the promoter for the 10 kD gene requires other viral gene products for its activity. The nature of these products is not known. Recently, Guarino and Summers (227) showed that the *hr*5 enhancer is required for expression of a late AcMNPV gene. In addition to *hr*5, two delayed-early ORFs also appear to be required for the activation of this late promoter.

While antimetabolite experiments indicate that $\beta$ and $\gamma$ gene products might be negatively regulating the expression of $\alpha$ and $\beta$ genes, there is no direct evidence of this. There is some evidence that antisense RNA is involved in the inhibition of early gene expression. Friesen and Miller (228) have described two nonoverlapping genes in the 85–88 mu region of the AcMNPV genome, which extend in opposite directions and encode polypeptides with $M_r$ of 35,000 and 94,000. Both genes transcribe unspliced messenger-active RNAs early (2 h) after infection. Transcription of the divergent RNAs was unaffected by cyclohex-imide, indicating that these genes are of the immediate-early class. Late in infection, abundant RNAs are transcribed from promoters located at least 2500 bases upstream from the gene encoding the 35 kD polypeptide. These transcripts completely overlap both the 35 and the 94 kD polypeptide genes but do not appear to code for any proteins. Friesen and Miller suggest that the transcripts may play a role in the suppression of early viral gene expression.

Crawford and Miller (229) characterized an early AcMNPV gene (ETL), located in the EcoRI T fragment (29–30 mu) of the viral genome; this gene codes for a 28 kD polypeptide which directly or indirectly accelerates the expression of two early and numerous late genes.

### Viral DNA Synthesis

Studies on the time course of Ac and SfMNPV DNA replication have been done in *S. frugiperda* cells. Virus-specific DNA synthesis was measured by gradient analysis or DNA–DNA hybridization. Viral DNA synthesis was detectable at 6 h p.i. and the maximum rate was reached at 12 h p.i. After this, the rate of synthesis of viral DNA declined (20,212,230). A viral DNA polymerase with characteristics different from host DNA polymerase has been detected by several laboratories. This polymerase is insensitive to aphidicolin and novobiocin and was inhibited by dTTP and *N*-ethymaleimide, suggesting that this AcMNPV-specific enzyme is a $\gamma$-like DNA polymerase (84,231,232). Kelly (84) reported a thymidine kinase activity in TnMNPV infected *S. frugiperda* cells. This enzyme was virus induced and distinct from the host kinase activity. Thus it is very likely that AcMNPV DNA replication is mediated by a DNA polymerase encoded by the viral genome.

Very little information about the mechanism of replication of the NPV DNA is available. Given the fact that this genome is the largest double-stranded circular DNA genome found in a virus, the study of its replication has the potential of revealing unique mechanisms in DNA replication. Wilson and Miller (233) studied the nature of the DNA-protein complexes assumed by AcMNPV-infected *S. frugiperda* cells. They showed that both parental and progeny viral DNA assumed a chromatinlike structure early in infection. By late times (24 h p.i.), the viral DNA acquired a unique nucleoprotein structure whose basic chromatin-associated proteins had $M_r$ of 15,000 and 39,000. The induction of the 15 kD protein coincided with the appearance of the unique nucleoprotein structure and the cessation of histone synthesis. The role of these two proteins in the replication of NPV DNA is not known.

Cochran and Faulkner (194) had initially suggested that the *hr* sequences in the AcMNPV genome might act as origins for DNA replication. However, while there is now good evidence for the role of these sequences as enhancers of gene expression, their role in the replication of the viral genome remains to be explored.

### Nucleotide Sequences

A considerable amount of information on the sequences of coding regions as well as regulatory elements of NPVs is now available. The polyhedrin gene (Pn gene) was the first NPV gene to be sequenced (222). Transcription of the Pn gene begins at a single start site, and there is a 58 base, 5' leader sequence which is followed by the coding sequence. The $M_r$ of polyhedrin predicted from the coding sequence is 28,872, which is in good agreement with values based on electrophoretic mobility. The striking degree of conservation of the coding regions in the Pn gene is discussed under "Classification." Not unexpectedly, there is also conservation of noncoding, and possibly regulatory, sequences among the NPV. Although putative TATA and CAAT boxes identified upstream of the AcMNPV Pn gene were not conserved in BmNPV, a 12 base sequence (TATAAGGAATTT), located 20 bases upstream of the ATG initiating codon, was conserved in BmNPV, OpMNPV, and OpSNPV (30,181,234).

Lübbert and Doerfler (214,220) sequenced the 5' end of the 10 kD polypeptide (p10) gene of AcMNPV. The transcription initiation sites of two RNAs were each within the same AAA site. Leisy et al. (181) found that the 5' start site of the 10 kD polypeptide gene of OpMNPV is located within the 12 base sequence. These authors emphasize that the sequence is conserved in all late hyperexpressed NPV and granulosis virus (GV) genes examined thus far. Cochran et al. (16) suggested that any consensus sequence arrangement for baculovirus regulatory sequences might be organized with respect to temporal classes. The 12 base sequence appears to be such a sequence for the promoter of the late NPV genes. Similar consensus sequences for early genes may well be a part of the NPV temporal cascade system.

Wilson et al. (235) identified and sequenced the gene for the 6.9 kD arginine-rich polypeptide, which is the core protein, or "basic DNA-binding protein," of viral nucleocapsids in AcMNPV.

### Inhibition of Host Macromolecular Synthesis

The inhibition of host cell protein synthesis as a result of infection with NPV is well established, but the mechanism by which inhibition occurs is not understood (58,78,79). Circumstantial evidence suggests that inhibition of host protein synthesis in cells infected with AcMNPV is mediated via late ICSPs rather than immediate-early or delayed-early ICSPs. Inhibition of host protein synthesis begins at about 10 h p.i. This correlates will with the kinetics of late ICSP synthesis, which begins at 6 h p.i. and increases its rate until 18 h p.i. Kelly and Lescott (200) reported that late and very late ICSPs could inhibit cellular protein synthesis. Liu and Bilimoria (202,236) have shown that host protein synthesis in IPLB-SF-21 cells infected with SfMNPV is essentially similar to that observed for AcMNPV-infected cells. Reinisch and Bilimoria (237) have shown that host protein synthesis is not inhibited in nonpermissive *Bombyx mori* cells infected with AcMNPV. Virus replication in these cells seems to be restricted such that most of the late and the two major very late ICSPs, p10 and polyhedrin, are not synthesized (see below). Gordon and Carstens (156) showed that the temperature-sensitive mutant of AcMNPV, *ts*8, synthesized only early viral proteins at the nonpermissive temperature, and, in contrast to the wild-type virus, did not inhibit host cell protein synthesis. Taken together, the foregoing results rule out the role of viral early proteins in the inhibition of host protein synthesis and support but do not prove the hypothesis of Kelly and Lescott that late and/or very late ICSPs are involved in the inhibition of host protein synthesis.

### Host Specificity

Most NPVs are not species specific, but they do have a narrow host range. Generally, NPVs infect only members of the genus or, in some cases, the family of the original host. Furthermore, there is evidence that the SNPVs are more specific than the MNPVs (25,238). There is a considerable body of evidence indicating that NPVs do not replicate or cause disease in vertebrate organisms (239).

An understanding of the mechanisms of host specificity of the NPVs is of interest from the perspective of differential expression of viral genes; such understanding is also important for developing viral pesticides having broad target specificity and increased virulence. It is a curious fact of baculovirus research that good host range mutants have not been isolated, and this has necessitated the use of semipermissive and nonpermissive cell lines in the study of host specificity.

Studies with vertebrate cell lines have shown that although uptake of virus may occur, there is no convincing evidence of NPV replication in these cells (240,241). The report of Himeno et al. (242) that DNA from BmNPV induces production of PIBs in mammalian amnion cell cultures remains unconfirmed, and at least two laboratories have shown that, contrary to a previous report (243), AcMNPV does not replicate in the hamster ovary cell line CHO-K1 (244,245). Tjia et al. (241) studied the replication of AcMNPV in five human cell lines and a mammalian cell line and showed that even at multiplicities of 100, AcMNPV did not replicate. Tests for the replication or transcription of viral DNA were negative, and there was no evidence for the persistence of viral DNA or fragments of viral DNA in these mammalian cells. AcMNPV appears to enter the nucleus of cell lines derived from frog, turtle, and trout, but there is no synthesis of viral RNA or DNA (246).

NPV infections of semipermissive and nonpermissive insect cell lines show that restriction of infection in alternate hosts can occur at a number of different stages in the replication cycle of the virus.

The first infection of this type was described by Carpenter and Bilimoria (91), who studied the infection of TN-368 cells with SfMNPV. In this infection, the virus induces early cytopathic effect (CPE), including nuclear hypertrophy and the formation of a virogenic stroma characteristic of SfMNPV infection in the permissive IPLB-SF-21 cells but does not produce complete virions. Whereas earlier experiments had suggested that some progeny infectious virus was produced, more recent work strongly suggests that infectious virus is not produced and that the infection is nonpermissive (202). Bilimoria et al. (100) showed that only two infected cell-specific polypeptides (97 and 29 kD) were synthesized in this nonpermissive infection. The effects of reversal of inhibition of ICSP synthesis with cycloheximide and the effect of cytosine arabinoside on the synthesis of these proteins in the permissive IPLB-SF-21 cell line indicates that the 97K ICSP is a delayed-early protein and the 29 kD ICSP is a late protein. Dot-blot hybridization data show that a low level ($<15\%$ of permissive infection) of viral DNA replication does occur (202). While the synthesis of delayed-early and late ICSPs in the absence of immediate-early proteins is somewhat perplexing, it is interesting that the 67 kD ICSP, which is one of the two immediate-early ICSPs in SfMNPV infections of permissive cells, is a structural protein (202). Thus the possibility that the limited expression of a delayed-early and a late ICSP in nonpermissive cells is due to induction by the 67 kD protein in parental virus cannot be ruled out. However, since the synthesis of neither of the two immediate-early ICSPs detected in permissive infections occurs in nonpermissive infection, the primary restriction in this system must occur during expression of the immediate-early genes (236) (Figure 11). Gröner et al. (244) studied AcMNPV infection in the nonpermissive codling moth cell line CP169. They showed that adsorption and engulfment of virus particles occurred, but there was no synthesis of ICSPs or replication of viral DNA. Again, the block in this system seems to be at a stage prior to the initiation of viral gene expression.

In other nonpermissive infection systems, restriction seems to be occurring at a later stage of the NPV growth cycle.

McClintock et al. (247) showed that in infections of nonpermissive gypsy moth cells IPLB-LD-642Y with AcMNPV, early but not late ICSPs were synthesized. The virus produced classic CPE in these cells but failed to produce infectious progeny. Viral DNA replication was initiated from 8 to 12 h p.i. and continued at a maximum rate up to 20 h p.i. The rate of DNA synthesis approximated that observed in permissive TN-368 cells.

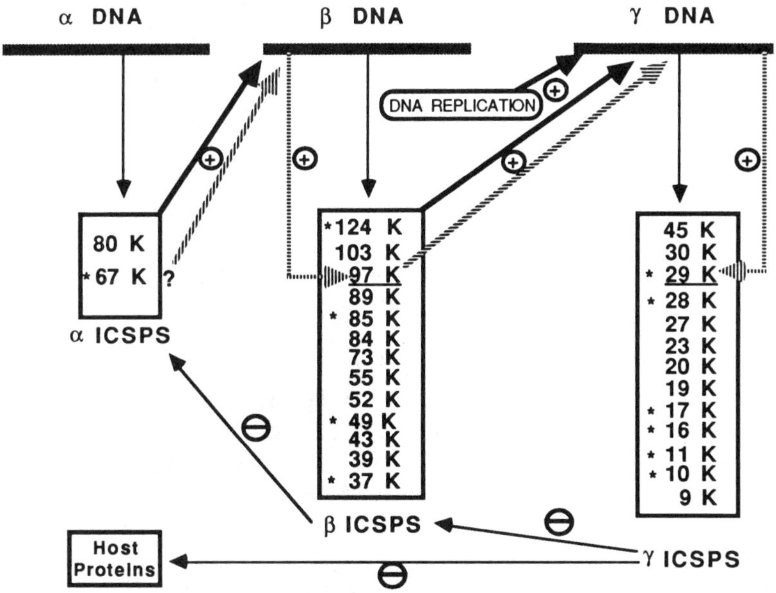


**Figure 11** Cascade regulation model for the synthesis of infected cell-specific polypeptides (ICSPs) in *S. frugiperda* MNPV infections of permissive *Spodoptera frugiperda* cells (solid lines) and nonpermissive *Trichoplusia ni* cells (hatched lines). +, inducing; –, inhibiting; *, structural ICSPs. (From H. S. Liu and S. L. Bilimoria, unpublished data, 1987.)

Bilimoria et al. (100) showed that in AcMNPV infections of *B. mori* BM-5 cells (107,248), the CPE is highly unusual: in addition to developing the nuclear hypertrophy typical of early stage CPE, the cells developed polar terminals and saclike bodies never observed in other baculovirus infections. Recent data acquired by Reinisch and Bilimoria (237) indicate that no infectious virus is produced in these cells; therefore, the infection is nonpermissive. Pulse labeling at late times after infection (17 and 24 h p.i.) with $^{35}$S-methionine and [$^{3}$H] amino acid mixtures indicated that polyhedrin, the 10 kD polypeptide, and 11 other ICSPs were not synthesized in *B. mori* cells but were detected in AcMNPV infections of permissive IPLB-SF-21 cells. An intriguing aspect of this research is that nine ICSPs not found in the permissive cells were detected in *B. mori* cells. The meaning of these dramatic differences in CPE and ICSP synthesis between the permissive and nonpermissive infections is not clear, and it is not known if there is a direct cause-and-effect relationship between the unusual CPE and unique ICSPs observed in the *B. mori* cells.

Rice and Miller (249) have shown that specific AcMNPV transcripts were synthesized in nonpermissive *Drosophila melanogaster* cells, but late viral transcripts were not observed.

Clearly, the preceding three infections of nonpermissive cells with AcMNPV are restricted at apparently different steps of the temporal cascade of NPV gene expression.

Carbonell et al. (250) transfected *D. melanogaster* cells with recombinant plasmids in which the *Escherichia coli* $\beta$-galactosidase gene is under the control of the late polyhedrin promoter of AcMNPV and showed that expression of the bacterial enzyme was less than 0.3% of that in similarly treated IPLB-SF-21 cells, which are permissive for AcMNPV.

It is clear from the foregoing discussion that one or more distinct mechanisms of restriction might be operational in each of the host specificity model systems discussed. The role of enhancers, and the specific motifs within them, in the host specificity of simian virus 40 and other papovaviruses is well established (251). As more information on the specific promoter classes, enhancer sequences, and host-specific factors operational in the temporal and abundance cascade of the NPV replication cycle becomes available, it should be possible to test the role they play in regulating the specificity and tissue tropism of these viruses. At the same time, recent information on characterization of the major surface glycoprotein of AcMNPV (77,111) should provide interesting opportunities for studying cell surface-level mechanisms of NPV specificity.

### Genetic Engineering of NPVs

The impact of recombinant DNA technology on the nature of basic NPV research is clear from the discussion of NPV gene expression. Progress in the application of this technology toward the development of genetically engineered NPV strains is just as brisk. Cochran et al. (16) list four reasons for constructing recombinant baculoviruses: (1) high-efficiency eukaryotic expression vectors, (2) development of improved pest-control agents, (3) models for basic studies on the regulation of gene expression and organization, and (4) models for the regulation of gene expression in invertebrates.

The aspect of NPV genetic engineering that has been the most exploited to date is the potential of the baculovirus vector system to express large amounts of foreign gene products in insect cell cultures and in larvae (9,10,252). Thus far, foreign genes are inserted into the Pn gene region of AcMNPV or BmNPV such that expression is under the control of the Pn promoter. The phenotype of these recombinants would then be $OB^-$ (discussed earlier). Smith et al. (9) inserted the gene for human $\beta$-interferon at a number of locations relative to the transcriptional start site of the Pn gene. They showed that the gene was most efficiently expressed when inserted at a position three nucleotides upstream from the ATG-initiating codon of the Pn gene. The gene was expressed abundantly, and the protein underwent glycosylation and was secreted from the insect cells. Plaque-reduction assays using vesicular stomatitis virus showed that the interferon was biologically active. Miller and co-workers (10) chose to insert the *E. coli* $\beta$-galactosidase gene in developing their expression system which expressed $\beta$-galactosidase as a fusion protein. The expression of this enzyme resulted in the formation of blue plaques in the presence of X-gal(5-bromo-4-chloro-3-indolyl-$\beta$-D-galactopyranoside). The use of $\beta$-galactosidase and blue plaques should be of great value for future genetic work with the NPV. Maeda and colleagues (252) developed an expression vector using BmNPV. This system expresses human $\alpha$-interferon in cell cultures and in silkworms. It should be noted that in each of these systems, the inserted gene was temporally regulated in a fashion similar to polyhedrin expression, with the synthesis first detected 20 h p.i. and increasing rapidly through 42 h p.i. The full potential of the NPV vector system may be estimated by examining the extent of polyhedrin expression in the infected host: greater than 15% of the protein mass found in NPV-infected larvae (29) and 25% of the total protein mass in infected cell cultures (9) is polyhedrin.

The popularity of the baculovirus vector system among molecular biologists is indicated by the fact that more than 150 laboratories are currently utilizing these powerful expression systems to express a variety of genes (253). Advantages of baculovirus vectors include relative safety due to their limited host range and the ability of the insect host cells to effectively carry out posttranslational modifications such as glycosylation (254–257) and proteolytic processing. Smith et al. (258) showed that the *N*-terminal amino acid sequence of human interleukin 2 (IL-2) produced in the baculovirus system was identical to that of natural IL-2, indicating that a mammalian signal peptide was recognized in insect cells; Martin et al. (257) obtained similar results for human glucocerebrosidase. Jeang et al. (259) have shown that the AcMNPV vector can express intron-containing genes, correctly utilizing the splicing signals for the small *t* antigen of simian virus 40. The rod-shaped nucleocapsid of baculoviruses can package at least two genomic lengths (260), suggesting that it may be possible to insert a large number of genes into a single vector.

Recombinant NPVs expressing insecticidal genes will have great potential as highly powerful and specific biological pesticides. Since inclusion body formation is a prerequisite for efficient transmission of NPV in nature, it will be necessary to develop vectors with dual Pn promoters or the insertion of the foreign gene in another location such as the p10 polypeptide gene.

## IN VIVO INFECTION, PATHOGENESIS, AND DIAGNOSIS

To fully understand the molecular biology of NPV virulence and specificity, it is necessary to develop a clear picture of the in vivo infection pathway(s) of these viruses. Several techniques, including electron microscopy, immunochemistry, and virus assay with the use of cell cultures, have been utilized to elucidate the infection pathway.

## Gross Pathology

Larvae infected with NPV do not display the signs and symptoms of infection for several days after ingestion of virus. The symptoms include changes in color and behavior such as loss of appetite and cessation of feeding. Larval growth is retarded and death occurs with liquefaction of the internal tissues and darkening of the cuticle. Rupture of the fragile epidermis often occurs upon death, and this is accompanied by the release of large amounts of OBs (6). Virulent NPV strains may kill larvae by two to five days (261); less virulent strains may take two to three weeks to kill their host (262,263).

OBs have been observed in the nuclei of a large number of tissue types including the hemocytes (264). Benz established a sequence of infection (based on the presence of OBs) for an MNPV infection. OBs appeared in tissues in the following order: (1) fat body, (2) hypodermis, (3) tracheal matrix, (4) muscular sheath, (5) nerve sheath, (6) muscles, (7) ganglia, and (8) pericardial cells. NPV infections of the lepidoptera are generally polyorganotropic (many target cells) (265). On the other hand, NPV infections of mosquitoes (Diptera) and sawflies (Hymenoptera) are strongly monoorganotropic in that only the cells of the midgut epithelium are affected (266–268). The cytopathology of NPV infections in vivo is generally similar to that described for cell culture infections (101,143).

## Pathogenesis

NPV transmission in nature is via OBs which are ingested by larvae during feeding. Other modes of viral entry such as transovum passage and infection through spiracles and by parasitism occur but are not considered important. The foregut and hindgut, which are lined with cuticle, present a formidable barrier to infection. The primary site for virus attachment and entry is considered to be the cells of the midgut epithelium. Ingested NPV OBs are dissolved in the midgut lumen, and PDVs are released and exposed to the midgut juices (6,112) (Figure 12). It is believed that in Lepidoptera, the dissolution process is controlled by the action of the highly alkaline digestive juices (pH 9.5–11.5) and perhaps also by the alkaline protease activity associated with larvae-derived OBs (269,270). It should be noted that this environment is a hostile one for the PDVs in that gut juices have been shown to inactivate released virus (269,271).

There is considerable debate regarding the role of the peritrophic membrane (PM) in the NPV infection process. The PM lines the midgut and is composed primarily of chitin and protein. There is some evidence to support the contention that the PM of Lepidoptera may possess pores which allow PDVs to pass from the lumen to the epithelial side of the PM (272–274). Given the dearth of

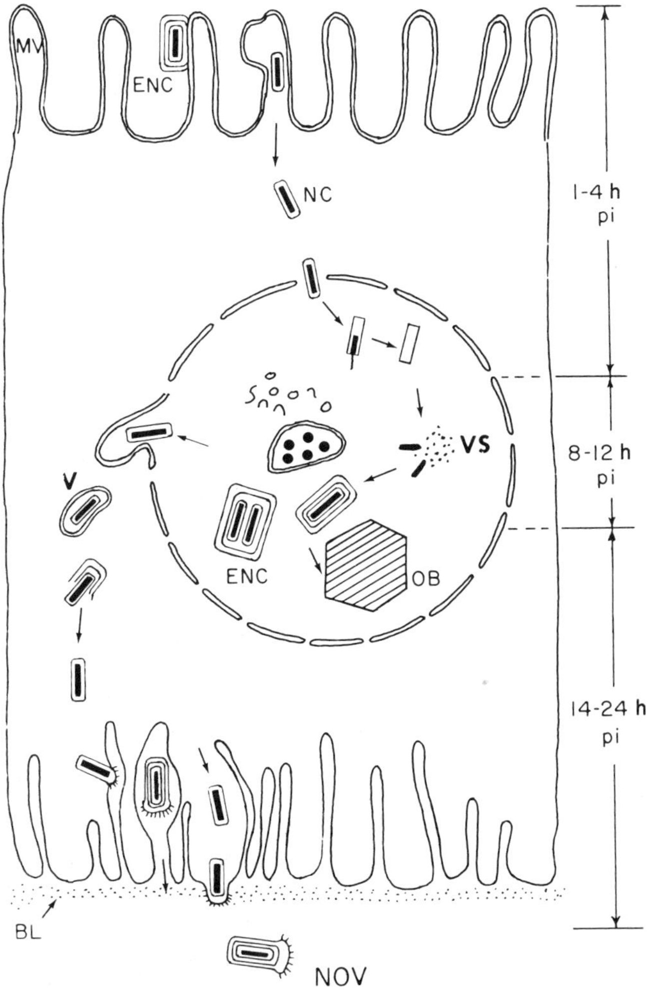


**Figure 12** Replication of *Autographa californica* MNPV in midgut epithelial cells. In phase I (1–4 h p.i.), infection occurs by fusion of ENC (enveloped nucleocapsid) with the microvillus (MV) membrane, entry of nucleocapsids (NC) into the cell and nucleus, and uncoating of the viral genome within the nucleus. Phase II (8–12 h p.i.) represents the replicative stage where virogenic stroma (VS) gives rise to virus NCs, which are enveloped. Other NCs are released into the cytoplasm by means of vesicles (V) originating from the nuclear envelope. In the final phase III (14–24 h p.i.), occlusion bodies (OB), generally virus-free, are produced. In addition, nucleocapsids are released from the vesicles, and they acquire an envelope by budding through the plasma membrane and basal lamina (BL). The budded nonoccluded virus (NOV, i.e., BV) possesses a peplomer structure and is responsible for secondary infection within the host insect hemocoel. (From Ref. 6, used with permission.)

information on the structure and composition of the PM, the relative importance of this structure in NPV infection cannot be assessed at this time (6).

Available data suggest that NPVs attach and enter the midgut cells by membrane fusion, which results in the entry of naked nucleocapsids into the cytoplasm (275). Granados and co-workers (140,276) have shown that the NPV nucleocapsid is uncoated after entering the nucleus. The morphogenesis of NPVs in vivo appears to be similar to that observed in cell culture (6,101) (Figure 12), although it remains to be determined whether the regulation of gene expression in vivo differs substantially from that in cell culture infections. Such studies would be particularly interesting with respect to the observation, in NPV infections of Lepidoptera, that OBs are formed in various other tissues but not in midgut cells (76,112,272). It is generally accepted that the midgut epithelial cells are the primary focus for NPV infection and that the progeny virus from this infection is responsible for secondary infection in the hemocoel (76). However, Granados and Lawler (140), and subsequently other workers (6), have shown that a portion of the inoculum virus bypasses the midgut nuclei, buds into the hemocoel, and is available for the systemic infection of hemocoelic organs.

When nucleocapsids bud through the plasma membrane of midgut cells and thus acquire the viral envelope, the part of the plasma membrane that initially came into contact with the nucleocapsid is covered with spikes, or "peplomers" (101). It has been suggested that this form of BV is responsible for the systemic infection of hemocoelic target cells (6). Whether the susceptible hemocoelic tissues are infected simultaneously or according to some prescribed sequence is not known. Tracheolar cells which surround the susceptible tissue (55), fat body tissue next to the midgut (264), and hemocytes (140) have all been implicated as the principal targets of secondary infection. More research is needed to determine whether any one of these tissues has a preferred role as factories for the production of virus for subsequent infections.

Very little information on the tissue tropism of NPVs in alternate hosts is available. In vivo studies do reveal interesting differences in the tissue tropism of certain NPVs in a number of different hosts. The differential susceptibility of various cell lines from the same host to certain NPVs (discussed earlier) might be a function of the tissues from which the cells have been derived. More research on the molecular basis for such tropism is essential for a clear understanding of the nature of virulence and specificity of NPVs.

## Diagnosis of NPV Infections

Gross pathological changes in NPV infections occur late in infection and are not of significant diagnostic value.

Microscopic examination for nuclear location of PIBs in squashed larvae samples will confirm NPV infection, but this approach cannot discriminate within subgroup A in any reliable fashion. PIBs are highly refractile and "shine" under phase-contrast or dark-field microscopy. Interpretation of stress crystals as PIBs is a potential problem. Several staining methods for the detection of NPV PIBs are discussed by Payne and Kelly (12), but these are of very limited diagnostic value.

Given the recent advances in identification of NPV strains, several potentially specific, quick, and economical diagnostic tools could become available in the very near future. Reinganum (277) has developed a procedure for direct analysis of detergent-disrupted, purified PIBs by SDS-PAGE. The use of silver staining permits sample loads of $4 \times 10^9$ PIBs and allows detection of virion polypeptides as well as polyhedrin in the same lane. Reinganum was able to distinguish eight NPVs using this approach; however, the requirement for purification of PIBs would make this procedure a relatively tedious and time-consuming one. Quant et al. (29) have shown that monoclonal antibodies can distinguish between an SNPV and an MNPV in enzyme-linked immunosorbent assays (ELISA) carried out using extracts of infected insects. Recently, Kalmakoff and his co-workers (278) developed dot-blot hybridization assay for the detection of NPVs. This assay used cloned $^{32}$P-labeled NPV DNA as a probe and was more sensitive than differential staining, phase-contrast microscopy, or indirect solid-phase radioimmunoassay. The assay is specific for NPV and can detect as few as 10 PIBs (150 pg DNA). Samples do not require prior purification or DNA extraction. The last two biochemical approaches have a great deal of potential as rapid, economical tools for the diagnosis of NPV infections. Both methods have the potential to differentiate between strains of NPV and can be used with crude preparations of infected larvae. The extent to which the dot-blot method will register limited DNA replication in nonpermissive infections as false positives remains to be investigated. The development of biotinylated DNA probes should allow wider acceptance of this method since expensive procedures and facilities for the use and disposal of radioactive materials are not required.

## ECOLOGY AND EPIZOOTIOLOGY

Improved techniques in detecting, quantifying, and characterizing NPVs have resulted in much progress on the quantitative description of the ecology and epizootiology of this virus group. This section presents a brief overview of this aspect of NPV research. For a more thorough treatment, the reader is referred to an excellent recent review of this topic by Evans (61).

### Ecology

#### Infectivity and Virus Availability

The modern view of baculovirus ecology as a dynamic interaction between the virus and its host requires a rigorous quantitative approach, which would allow predictive modeling. In this respect, viral infectivity and host specificity, virus availability, and host density are important parameters affecting the ecology of the NPV (61,279).

Infectivity may vary intrinsically at the individual level or it may vary at the population level. Susceptibility of larvae to baculovirus infection decreases with increasing larval age (280–282), and variation at the population level, other than host density effects, is due to the relative proportions of individuals that are susceptible or infected (282–284). Infection in a host population is also related quantitatively to the amount of virus available for ingestion by larvae (61,284).

### Host Density Effects

Host density plays an important role in determining the amount of inoculum needed to initiate and maintain infection in a host population. The NPV that young larvae of a particular generation encounter is the primary inoculum. The quantity and distribution of virus in the primary inoculum is that which survived the periods of nonreplication when no larval stages were present. Primary inoculum is constantly declining as a result of environmental factors such as ultraviolet radiation, rainfall, and chemical conditions on leaf surfaces and in soil. This, together with the low feeding rate and limited dispersal ability of young larvae, precludes a high level of infection if the host populations are sparsely distributed (61,285). With increasing larval age, the direct effect of primary inoculum declines as a result of inoculum depletion and reduced host susceptibility. While these factors are partly compensated for by increased feeding on the part of older larvae, the net effect is a sharp decline (largely due to host density effects) in response to primary inoculum (61).

The effects of host density and distribution on the production and cycling of secondary inoculum are even more important. The large numbers of polyhedra produced in young larvae increase the local concentration of virus by a factor of almost $10^5$, and this has a major influence on events late in a larval generation (280). Greater host density increases the probability of virus–host encounters and leads to density-dependent mortality (61,286,287). The mode of feeding utilized by the host also has an important effect on distribution. Thus in *Neodiprion sertifer,* which is a colonial feeder, the infection is very rapidly transmitted through a population because of secondary cycling of inoculum (61,282).

Interaction between larvae is through secondary cycling of inoculum and is a function of distribution and abundance. Availability of this inoculum is a function of the mechanism of release of virus and varies considerably between insect orders. Thus in the case of NPV infection of sawflies, primarily gut cells are infected. Consequently, PIBs are voided continuously in the feces and also regurgitated. In this manner, secondary inoculum can be disseminated rapidly through the host population well before the peak of the infection (288). On the other hand, in NPV infections of Lepidoptera, the gut is only the initial site of infection, and given the nature of release by rupture of the flaccid larval integument after death, virutally all the PIBs become available over a short period of time following death. This type of secondary inoculum produces very large, localized reservoirs of virus that can become available for ingestion by a large host population (61). The general consensus is that baculoviruses act in a delayed density-dependent manner and induce maximum mortality some time after the host population has reached a peak (283,289,290). The appearance of epizootic virus growth in a population will thus reflect the numbers of susceptible hosts and the amount of primary and secondary inoculum present. Populations with discrete generations will have periods when no hosts are present, and so the production of secondary inoculum will cease. Therefore, the growth of infection will be cyclic with respect to the individual host generations. The situation becomes more complex when the populations have overlapping generations (61,282).

### Specificity

NPVs with broad host specificities have an advantage over monospecific viruses since greater numbers of hosts are available. However, the dosages required for some cross-transmissions can be unrealistically high. The problem of virus identity should also be considered; this is now more amenable to study given the reliability of precise identification tools such as restriction endonucleases (5). Nevertheless, the existence of true cross-infections has been established (291,292) and it has been shown that secondary hosts increase both the spatial and temporal range of the virus (293).

### Transmission

Baculovirus transmission is generally divided into horizontal (spatial) and temporal components. Whereas horizontal transmission is essentially virus dispersal within host generation, the temporal component includes all aspects of virus transmission between generations (61).

In horizontal transmission, virus dispersal can be categorized as primary, secondary, and tertiary. Primary dispersal can be observed most clearly from isolated epicenters of disease in populations that are initially healthy. Secondary dispersal manifests a wavelike form, and there is an accompanying decrease in disease incidence near the epicenter. The latter probably reflects density-dependent response to the decreasing host population (61,286). The tertiary phase of dispersal does not have a clear pattern to the spread of disease: virus is eventually prevalent throughout the host environment (288). It has been suggested that this reflects disease spread from multiple numbers of disease centers generated by discontinuous dispersal by adults or predators (294–296).

Temporal transmission includes vertical (adult to progeny) transmission as well as transmission of virus from one generation to the next through the persistence of virus in the environment (293). It should be emphasized that most cases of vertical transmission cannot be regarded as transovarian transmission due to incorporation of virus in the egg (61,297). It is speculated that vertical transmission by lepidopterous adults in the field results from external contamination of newly emerged females (61). True transovum transmission seems to be limited to species having gut infections in the adult stage and, for the NPV, this is limited to the sawflies (298). Vertical transmission is a potentially important mechanism for the maintenance of virus in a low-density population since, unlike horizontal transmission, it does not require a minimum threshold density (61,299).

The most common means of virus transmission from one generation to another is through the persistence of the agent outside the insect host (293,300). Evans and Entwistle (301) showed that in NPV epizootics in *Gilpinia hercyniae,* virus overwintered on foliage, and there was a direct link between the capacity of spruce foliage to retain virus and primary infection in the following generation. Several investigators have gathered data suggesting that soil is the most stable long-term site of virus retention (279,287,302–307). Evans cautions that since the amount of virus retained in soil (and therefore the amount recycled to the host plant) is very limited, soil would be a significant reservoir only when the host density is high enough to allow virus-host encounters (61).

### Persistence

Virus persistence is the single most important factor in the ecology and epidemiology of baculoviruses (61). Persistence may be biotic or abiotic. In biotic persistence, reservoirs in addition to the adult host would include secondary hosts, predators, and parasitoids. Whether secondary hosts can be utilized would be a function of the host range of the virus. Although predators have been shown to function as dispersal agents for NPVs (e.g., via bird droppings; 296), their role in viral persistence, although quite possible, is not proven. Parasitoids appear to disperse virus by external contamination, and the increased dispersal is thought to increase the probability of viral persistence (61). The main reservoirs for abiotic persistence of NPVs are soil and the host plant. The role of both soil and foliage as reservoirs for NPVs was mentioned earlier. Ultraviolet radiation is the single most important factor in the inactivation of NPVs in the environment (308). Disintegrating cadavers and soil offer a degree of protection from this (61). An interesting aspect of NPV survival on foliage is the adverse effect on the virus of the high pH ($> 9.0$) and high concentrations of divalent cations of dew on cotton leaves (309,310).

## Epizootiology

Evans (61) divides baculovirus epizootiology into three components: enzoosis, sporadic epizoosis, and cyclic epizoosis.

Since density-dependent NPV infection normally occurs above defined host population thresholds, disease is considered to remain enzootic below these thresholds for viruses that have high pathogenicity. For example, Ehler (311) showed that NPV infection of *T. ni* larvae on cotton remained below 5% in the majority of cases, but at times of peak host density, the infection rose to 90% and became epizootic. Viruses of low pathogenicity are more likely to remain enzootic regardless of the host population density.

Sporadic epizootics occur in unstable habitats such as agricultural crops. These epizootics rely on stable reservoirs of virus, as a supply of primary inoculum, and a high host population density and are favored by high virus pathogenicity. *T. ni* NPV is highly pathogenic in the laboratory and is one of the few baculoviruses epizootic in unstable agricultural habitats (61).

Cyclic epizootics of NPV generally are associated with forest habitats. A common feature of such epizootics is a delay in peak virus infection relative to peak host density. There is also a period when the host population density falls below a critical threshold. Thus the disease acts in a delayed density-dependent manner above the threshold density (61,312,313).

## Applications in Pest Control

An important aim of ecological and epizootiological studies on the NPVs, and particularly of mathematical modeling, is to identify the key factors in the population dynamics of the virus and utilize them to optimize the conditions for biopesticidal activity. A very effective example of this is to amplify and extend persistent virus populations by achievement and maintenance of high levels of virus inoculum in the environment. A highly interesting example of this is the control of *Wiseana* sp. by NPV in pastures in New Zealand. Kalmakoff and Crawford (279) showed that infection levels in a natural epizootic were generally higher than those achieved by spraying plots with virus. The relative stability of the pasture habitat resulted in accumulation and efficient persistence of NPV in soil. Ploughing and reseeding the pasture diluted the persistent inoculum to ineffective levels and resulted in the resurgence of the *Wiseana* population. Kalmakoff and Crawford recommended that damaged areas should be oversown and rotational grazing should be increased to enhance virus spread.

Another effective approach to maintaining epizootics is autodissemination (314). The best example of this approach is the control of the *Oryctes rhinoceros* with a subgroup C virus (314–316), although there is some evidence suggesting that this approach will also work (via contamination of adult genitalia) with a number of NPVs (314). Clearly, the chances of success with autodissemination will be highest in controlling insects whose adult stages shed virus in the feces.

## CONCLUSION

Since the mid-1970s much progress has been made in the study of the biology of the NPVs; some of the most dramatic advances have come in the area of NPV replication and molecular biology. The next few years should see the functional mapping of substantial numbers of important genes on the NPV genome. Major challenges will include the elucidation of the detailed mechanism of temporal and abundance regulation of the gene expression cascade. While specific promoter sequences are beginning to be identified, and viral RNA polymerase is implicated in the expression of the late genes, the exact role of the "enhancer" sequences is not known, and little information is available on host factors that might be required for the proper expression of host genes. There is a dearth of information on the mechanism of viral DNA replication, and even less is known about its regulation. The mechanism of virus assembly and its regulation is not understood; for example, how does the virus control the production of BV versus OV?

It should be emphasized that in order to confirm that the replication mechanisms elucidated with the AcMNPV model are generally valid for all NPVs, it is imperative to study a number of other MNPVs, such as SfMNPV, LdMNPV, and OsMNPV, in depth. Similar emphasis on the SNPVs that can now be replicated in cell culture also should pay handsome dividends. Since NPVs are not found in vertebrate hosts, they offer a unique model for gene expression studies and therefore opportunities for fundamental discoveries in eukaryotic molecular biology.

The recent awakening of interest in the possible mechanisms of NPV host specificity is encouraging. Although viable host range mutants of NPVs appear to be very difficult to generate, several laboratories are now looking at differential expression of the NPV genome in alternative cell lines. Much work remains to be done. It is not completely clear if restriction in alternative hosts is mediated by host factors that recognize appropriate promoter classes on the viral genome and whether enhancer or enhancerlike sequences have any role in specificity. The recent elucidation of viral attachment proteins should stimulate research on cellular receptor sites and consequently experiments on cell surface mechanisms of specificity. While some progress has been made in our understanding of the pathogenesis of NPVs, no information is available on the molecular basis of pathogenicity. For example, are there specific viral genes or regulatory sequences that modulate pathogenicity and virulence?

Biotechnological research on the NPVs is off to a good start. We can doubtless expect to see improvements on the existing NPV vector systems and the development of additional vectors based on the p10 and other strongly expressed genes. However, it remains to be seen how this system will compete with mammalian viruses and domestic animals such as sheep and cattle as vectors and hosts of choice for the production of biomedically important proteins. The feasibility of applying basic molecular biological research to the development of genetically engineered NPV for pest control also needs to be assessed. NPVs containing toxin genes and therefore, having a potential "double whammy" effect, will undoubtedly be developed. Will manipulation of key NPV genes yield more virulent strains with the desired host range?

Although a pattern finally is emerging in the ecology and epizootiology of baculoviruses, much more descriptive and mathematical modeling work remains to be done. Future work on the ecology and epizootiology of baculoviruses must aim to improve the quantitative data base on all aspects of virus ecology with particular attention to the study of the ecosystem as a whole. With the development of genetically engineered NPVs, renewed emphasis needs to be placed on the importance and significance of basic ecological and epizootiological studies of these viral systems. Careful studies on the stability of engineered viruses will be important and NPV ecologists will have to determine the degree of stability that is desirable.

Quick and reliable tests for the identification and precise monitoring of NPVs in the field will be essential for ecological and biocontrol studies. The molecular taxonomy of the NPVs should be studied further. A more complete understanding of the taxonomy of the NPVs with emphasis on the phylogenetic relationships between different strains must be developed. The development of a streamlined system of nomenclature and a virus bank for the NPVs also should have high priority.

Thus, beginning with the initial observations of Nysten and through the studies of Glaser and Chapman and Bergold and his contemporaries, modern NPV researchers have utilized the solid foundation laid by the pioneers of baculovirus and insect tissue culture research to make tremendous strides in our understanding of the NPVs. The study of polyhedrosis remains a vibrant field today and should continue to attract some of the best minds in the biological sciences.

## ACKNOWLEDGMENTS

I greatly appreciate the help of Aaron Castro, who assisted with compilation of the bibliography; Priscilla Jones, who cheerfully typed the list of references; Son Le, who efficiently checked the accuracy of the numerical citations; and Victor Montalvo-Lugo and Bob Phillips, who helped with proofing. The assistance of Victor Montalvo-Lugo in printing some of the photographs is also gratefully acknowledged. Portions of the unpublished research described here were supported by SORF grants from the College of Arts and Sciences, Texas Tech University; BRSG (NIH) Awards; and the Fire Ant Research Line Item, Texas.

## REFERENCES

1. Maestri, A. *Frammenti anatomici fisiologici e pathologici sul baco da seta*. Fusi, Pavia, 1856, p. 117.
2. Cornalia, E. Monografia del bombice del gelso, Mem. R. Instit. *Lombardo Sci. Lett.* 6, 3, 1856.
3. Benz, G. A. Introduction: Historical perspectives, in *The Biology of Baculoviruses*, vol. 1, Granados, R. R. and Federici, B. A., Eds., CRC Press, FL, 1986, p. 1.
4. Matthews, R. E. F. Classification and nomenclature of viruses. Fourth report of the international committee on taxonomy of viruses. *Intervirology* 17, 1, 1982.
5. Bilimoria, S. L. Taxonomy and identification of baculoviruses, in *The Biology of Baculoviruses*, vol. 1, Granados, R. D. and Federici, B. A., Eds., CRC Press, Boca Raton, FL, 1986, p. 37.
6. Granados, R. R. and Williams, K. A. In vivo infection and replication of baculoviruses, in *The Biology of Baculoviruses*, vol. 1, Granados, R. R. and Federici, B. A., Eds., CRC Press, Boca Raton, FL, 1986, p. 89.
7. Betz, F. S. Registration of baculoviruses as pesticides, in *The Biology of Baculoviruses*, vol. 2, Granados, R. R. and Federici, B. A., Eds., CRC Press, Boca Raton, FL, 1986, p. 203.
8. Hink, W. F. and Vail, P. V. A plaque assay for titration of alfalfa looper nuclear polyhedrosis virus on a cabbage looper (TN-368) cell line. *J. Invertebr. Pathol.* 22, 168, 1973.
9. Smith, G. E., Summers, M. D., and Fraser, J. J. Production of human beta interferon in insect cells infected with a baculovirus expression vector. *Mol. Cell. Biol.* 3, 2156, 1983.
10. Pennock, G. D., Shoemaker, C., and Miller, L. K. Strong and regulated expression of *Escherichia-coli* beta galactosidase in insect cells with a baculovirus vector. *Mol. Cell. Biol.* 4, 399, 1984.
11. Granados, R. R. and Federici, B. A., Eds. Practical applications for pest control, in *The Biology of Baculoviruses*, 2 vols., CRC Press, Boca Raton, FL, 1986.
12. Payne, C. C. and Kelly, D. C. Identification of insect and mite viruses, in *Microbial Control of Insects and Mites*, 2nd ed., Burgess, D.Ed., Academic Press, New York, 1981, 61.
13. Cibulsky, R. J., Harper, J. D., and Gadauskas, R. T. Biochemical comparison of virion proteins from five nuclear polyhedrosis viruses infecting plusiine larvae (Lepidoptera: Noctuidae). *J. Invertebr. Pathol.* 30, 303, 1977.
14. Summers, M. D. and Smith, G. E. Baculovirus structural polypeptides. *Virology* 84, 790, 1978.
15. Miller, L. K. and Dawes, K. P. Restriction endonuclease analysis to distinguish two closely related nuclear polyhedrosis viruses: *Autographa californica* MNPV and *Trichoplusia ni* MNPV. *Appl. Environ. Microbiol.* 35, 1206, 1978.
16. Cochran, M. A., Brown, S. E., and Knudson, D. L. Organization and expression of the baculovirus genome, in *The Biology of Baculoviruses*, vol. 1, Granados, R. R. and Federici, B. A., Eds., CRC Press, Boca Raton, FL, 1986, p. 239.
17. Knell, J. D. and Summers, M. D. Investigation of genetic heterogeneity in wild

isolates of *Spodoptera frugiperda* nuclear polyhedrosis virus by restriction endonuclease analysis of plaque-purified variants. *Virology* 112, 190, 1981.

18. Kelly, D. C. The DNA contained by nuclear polyhedrosis viruses isolated from four *Spodoptera* sp. (Lepidoptera, Noctuidae): Genome size and homology assessed by DNA reassociation kinetics. *Virology* 76, 468, 1977.
19. Smith, G E. and Summers, M. D. DNA homology among subgroup A, B, and C baculoviruses. *Virology* 123, 393, 1982.
20. Knudson, D. L. and Tinsley, T. W. Replication of a nuclear polyhedrosis virus in a continuous cell line of *Spodoptera frugiperda*: Partial characterization of the viral DNA, comparative DNA-DNA hybridization, and patterns of DNA synthesis. *Virology* 87, 42, 1978.
21. Harrap, K. A., Payne, C. C., and Robertson, J. S. The properties of three baculoviruses from closely related hosts. *Virology* 79, 14, 1977.
22. Kislev, N. DNA homology relationships between *Spodoptera littoralis* nuclearpolyhedrosis virus and other baculoviruses. *Intervirology* 24, 1, 1985.
23. Smith, G. E. and Summers, M. D. Application of a novel radioimmunoassay to identify baculovirus structural proteins that share interspecies antigenic determinants. *J. Virol.* 39, 125, 1981.
24. Bilimoria, S. L. Genomic divergence among single-nucleocapsid nuclear polyhedrosis viruses of plusiine hosts. *Virology* 127, 15, 1983.
25. Harper, J. D. Cross-infectivity of six plusiine nuclear polyhedrosis virus isolates to plusiine hosts. *J. Invertebr. Pathol.* 27, 275, 1977.
26. Vlak, J. M. and Gröner, A. Identification of two nuclear polyhedrosis viruses from the cabbage moth, *Mamestra brassicae* (Lepidoptera: Noctuidae). *J. Invertebr. Pathol.* 35, 269, 1980.
27. Brown, J. A., Evans, H. F., Allen, C. J., and Kelly, D. C. Biological and biochemical investigations on five European isolates of *Mamestra brassicae* nuclear polyhedrosis virus. *Arch. Virol.* 69, 209, 1981.
28. Rohrmann, G. F. Evolution of occluded baculoviruses, in *The Biology of Baculoviruses,* vol. 1, Granados, R. R. and Federici, B. A., Eds., CRC Press, Boca Raton, FL, 1986, p. 203.
29. Quant, R. L., Pearson, M. N., Rohrmann, G. F., and Beaudreau, G. S. Production of polyhedrin monoclonal antibodies for distinguishing two *Orgyia pseudotsugata* baculoviruses. *Appl. Environ. Microbiol.* 48, 732, 1984.
30. Iatrou, K., Ito, K., and Witkiewicz, H. Polyhedrin gene of *Bombyx mori* nuclear polyhedrosis virus. *J. Virol.* 54, 436, 1985.
31. McCarthy, W. J. and Gettig, R. R. Current developments in baculovirus serology, in *The Biology of Baculoviruses,* vol. 1, Granados, R. R. and Federici, B. A., Eds., CRC Press, Boca Raton, FL, 1986, p. 147.
32. Huang, Y. S., Hu, P. C., and Kawanishi, C. Y. Monoclonal antibodies identify conserved epitopes on the polyhedrin of *Heliothis zea* nuclear polyhedrosis virus. *Virology* 143, 380, 1985.
33. Summers, M. D. and Smith, G. E. Comparative studies of baculovirus granulins and polyhedrins. *Intervirology* 6, 168, 1976.
34. Maruniak, J. E. Summers, M. D., Falcon, L. A., and Smith, G. E. *Autographa californica* nuclear polyhedrosis virus structural proteins compared from *in vivo* and *in vitro* sources. *Intervirology* 11, 82, 1979.

35. Kelly, D. C., Brown, D. A., Aynes, M. D., Allen, C. J., and Walker, I. O. Properties of the major nucleocapsids protein of *Heliothis zea* singly enveloped nuclear polyhedrosis virus. *J. Gen. Virol.* 64, 399, 1983.
36. Harrap, K. A. and Payne, C. C. The structural properties and identification of insect viruses, in *Advances in Virus Research*, vol. 25, Lauffer, M. A, Bang, F. B., Maramorosch, K., and Smith, K. M., Eds., Academic Press, New York, 1979, p. 213.
37. Hohmann, A. W. and Faulkner, P. Monoclonal antibodies to baculovirus structural proteins: Determination of specificities by western blot analysis. *Virology* 125, 432, 1983.
38. Bergold, G. H. The nature of nuclear polyhedrosis viruses, in *Insect Pathology, An Advanced Treatise*, vol. 1, Steinhaus, E. A., Ed., Academic Press, New York, 1963, p. 413.
39. Bergold, G. H. The molecular structure of some insect virus inclusion bodies. *J. Ultrastruct. Res.* 8, 360, 1963.
40. Harrap, K. A. The structure of the nuclear polyhedrosis viruses, I. The inclusion body. *Virology* 50, 114, 1972.
41. Minion, F. C., Coons, L. B., and Broome, J. R. Characterization of the polyhedral envelope of the nuclear polyhedrosis virus of *Heliothis virescens*. *J. Invertebr. Pathol.* 34, 303, 1979.
42. Whitt, M. A. and Manning, J. S. A phosphorylated 34-KDa protein and a subpopulation of polyhedrin are thiol linked to the carbohydrate layer surrounding a baculovirus occlusion body. *Virology*, 163, 33, 1988.
43. Summers, M. D. and Arnott, H. J. Ultrastructural studies on inclusion body formation and virus occlusion in nuclear polyhedrosis and granulosis virus infected cells of *Tripchoplusia ni* (Hubner). *J. Ultrastruct. Res.* 28, 462, 1969.
44. MacKinnon, E. A., Henderson, J. F., Stoltz, D. B., and Faulkner, P. Morphogenesis of nuclear polyhedrosis virus under conditions of prolonged passage in vitro. *J. Ultrastruct. Res.* 49, 419, 1974.
45. Federici, B. A. Ultrastructure of baculoviruses, in *The Biology of Baculoviruses*, vol. 1, Granados, R. R., and Federici, B. A., Eds., CRC Press, Boca Raton, FL, 1986, p. 61.
46. Harrap, K. A. The structure of nuclear polyhedrosis viruses. II. The virus particle. *Virology*, 50, 124, 1972.
47. Beaton, C. D. and Filshie, B. K. Comparative ultrastructural studies of insect granulosis and nuclear polyhedrosis viruses. *J. Gen. Virol.* 31, 151, 1976.
48. Burley, S. K., Miller, A., Harrap, K. A., and Kelley, D. C. Structure of the baculovirus nucleocapsid. *Virology* 120, 433, 1982.
49. Teakle, R. E. A nuclear polyhedrosis virus of *Anthela varia* (Lepidoptera: Anthelidae). *J. Invertebr. Pathol.* 14, 18, 1969.
50. Koslov, E. A. and Alexeenko, I. P. Electron microscope investigation of the structure of the nuclear polyhedrosis virus of the silkworm, *Bombyx mori*. *J. Invertebr. Pathol.* 9, 413, 1967.
51. Summers, M. D. and Paschke, J. D. Alkali-liberated granulosis virus of *Trichoplusia ni*. I. Density gradient purification of virus components and some of their *in vitro* chemical and physical properties. *J. Invertebr. Pathol.* 16, 227, 1970.

52. Khosaka, T., Himeno, M., and Onodera, K. Separation and structure of components of nuclear polyhedrosis virus of the silkworm. *J. Virol.* 7, 267, 1971.
53. Stoltz, D. B., Pavan, C., and DaCunha, A. B. Nuclear polyhedrosis virus: A possible example of de novo intracellular membrane morphogenesis. *J. Gen. Virol.* 19, 145, 1973.
54. Hughes, K. M. Fine structure and development of two polyhedrosis viruses. *J. Invertebr. Pathol.* 19, 198, 1972.
55. Adams, J. R., Goodwin, R. H., and Wilcox, T. A. Electron microscope investigation on invasion and replication of insect baculoviruses *in vivo* and *in vitro*. *Biol. Cell.* 28, 261, 1977.
56. Kawamoto, F., Suto, C., Kumada, N., and Kobayashi, M. Cytoplasmic budding of a nuclear polyhedrosis virus and comparative ultrastructural studies of envelopes. *Microbiol. Immunol.* 21, 255, 1977.
57. Kawamoto, F., Kumada, N., and Kobayashi, M. Envelopment of nuclear polyhedrosis virus of the oriental tussock moth, *Euproctis subflava*. *Virology* 77, 867, 1977.
58. Wood, H. A. Protease degradation of *Autographa californica* nuclear polyhedrosis virus proteins. *Virology*, 103, 392, 1980.
59. Bilimoria, S. L. and Arif, B. M. Structural polypeptides of *Choristoneura biennis* entomopoxvirus. *Virology* 104, 253, 1980.
60. Granados, R. R. and Roberts, R. J. Electron microscopy of a pox-like virus infecting an invertebrate host. *Virology* 40, 230, 1970.
61. Evans, H. F. Ecology and epizootiology of baculoviruses, in *The Biology of Baculoviruses*, vol. 2, Granados, R. R. and Federici, B. A., Eds., CRC Press, Boca Raton, FL, 1986, p. 89.
62. Bergold, V. G. Die Isolierung des Polyedervirus und die Natur der Polyeder. *Z. Naturforsch.* 2, b, 122, 1974.
63. Bergold, V. G. Über die kapselvirus-krankheit. *Z. Naturforsch.* 36, 338, 1948.
64. Koslov, E. A., Levitina, T. L., Rudavskii, Y. L., Sogulyaeva, V. M., Sidorova, N. M., and Serebryanyi, S. B. A determination of the molecular weight of the inclusion body protein of the nuclear polyhedrosis virus of the mulberry silkworm *Bombyx mori*. *Biolchimiya*, 38, 1015, 1973.
65. Koslov, E. A., Levitina, T. L., Sogulyaeva, V. M., and Serebryanyi, S. B. Distribution and amino acid composition of tryptic peptides of the inclusion body protein of the mulberry silkworm, *Bombyx mori* nuclear polyhedrosis virus. *Biolchimiya*, 38, 1215, 1973.
66. Smith, G. E., Vlak, J. M., and Summers, M. D. Physical analysis of *Autographa californica* nuclear polyhedrosis virus transcripts for polyhedrin and 10,000 molecular weight protein. *J. Virol.*, 45, 215, 1983.
67. Whitt, M. A. and Manning, J. S. Role of chelating agents monovalent anion and cation in the dissociation of *Autographa californica* nuclear polyhedrosis virus occlusion body. *J. Invertebr. Pathol.* 49, 61, 1987.
68. Kobayashi, M. and Nakagaki, M. Effect of dissolution conditions of polyhedral inclusion bodies in sodium carbonate solution on the yield of nuclear polyhedrosis virus of the silkworm *Bombyx mori* (Lepidoptera: Bombycidae). *Appl. Entomol. Zool.* 19, 280, 1984.

69. Yamafugi, K., Yoshinara, F., and Hirayama, F. Protease and deoxyribonuclease in viral polyhedral crystal. *Enzymologia* 19, 53, 1958.
70. Koslov, W. A., Levitina, T. L., Sidorova, N. M., Radavski, Y. L., and Serebryanyi, S. B. Proteolytic cleavage of polyhedral protein during dissolution of inclusion bodies of the nuclear polyhedrosis viruses of *Bombyx mori* and *Galleria mellonella* under alkaline conditions. *J. Invertebr. Pathol.* 25, 103, 1975.
71. Eppstein, D. A. and Thoma, J. A. Alkaline protease associated with the matrix protein of a virus infecting the cabbage looper. *Biochem. Biophys. Res. Commun.* 62, 478, 1975.
72. Kalmakoff, J., Crawford, A. M., and Moore, S. G. A radioimmunoassay using labeled antibody for polyhedron protein from a nuclear polyhedrosis virus of *Wiseana cervinata. J. Invertebr. Pathol.* 29, 31, 1977.
73. Payne, C. C. and Kalmakoff, J. Alkaline protease associated with virus particles of a nuclear polyhedrosis virus: Assay, purification, and properties. *J. Virol.* 26, 84, 1978.
74. Maruniak, J. E. Baculovirus structural proteins and protein synthesis, in *The Biology of Baculoviruses*, vol. 1, Granados, R. R. and Federici, B. A., Eds., CRC Press, Boca Raton, FL, 1986, p. 129.
75. Volkman, L. E., Summers, M. D., and Hsieh, C. M. Occluded and nonoccluded nuclear polyhedrosis virus grown in *Trichoplusia ni*: Comparative neutralization, comparative infectivity, and in vitro growth studies. *J. Virol.* 19, 820, 1976.
76. Tinsley, T. W. and Harrap, K. A. Viruses of invertebrates, in *Comprehensive Virology,* vol. 12, Fraenkel-Conrat, H. and Wagner, R. R., Eds., Plenum Press, New York, 1978, p. 1.
77. Stiles, B. and Wood, H. A. A study of the glycoproteins of *Autographa californica* nuclear polyhedrosis virus (AcNPV). *Virology* 131, 230, 1983.
78. Maruniak, J. E. and Summers, M. D. *Autographa californica* nuclear polyhedrosis virus phosphoproteins and synthesis of intracellular proteins after virus infection. *Virology* 109, 25, 1981.
79. Dobos, P. and Cochran, M. A. Protein synthesis in cells infected by *Autographa calfornica* nuclear polyhedrosis virus (Ac-NPV): The effect of cytosine arabinoside. *Virology* 103, 446, 1980.
80. Kok, I. P., Chistyakova, A. V., Gudz-Gorban, A. P., and Somomko, A. P. Infectivity and structure of DNA of virus of nuclear polyhedrosis of the silkworm, in *Proc. 13th Int. Congr. Entomol.*, Moscow, 1968, p. 127.
81. Bud, H. M. and Kelly, D. C. The DNA contained by nuclear polyhedrosis viruses isolated from four *Spodoptera* spp. (Lepidoptera, Noctuidae): Genome size and configuration assessed by electron microscopy. *J. Gen. Virol.* 37, 135, 1977.
82. Burgess, S. Molecular weights of lepidopteran baculovirus DNAs: Derivation by electron microscopy. *J. Gen. Virol.* 37, 501, 1977.
83. Summers, M. D. and Anderson, D. L, Characterization of nuclear polyhedrosis virus DNAs. *J. Virol.* 12, 1336, 1973.
84. Kelly, D. C. Baculovirus replication, stimulation of thymidine kinase and DNA polymerase activities in *Spodoptera frugiperda* cells infected with *Trichoplusia ni* nuclear polyhedrosis virus. *J. Gen. Virol.* 52, 313, 1981.

85. Knudson, D. J. and Tinsley, T. W. Replication of a nuclear polyhedrosis virus in a continuous cell culture of *Spodoptera frugiperda*: Purification, assay of infectivity, and growth characteristics of the virus. *J. Virol.* 14, 934, 1974.
86. Dougherty, E. M., Weiner, R. M., Vaughn, J. L., and Reichelderfer, C. F. Physical factors that affect in vitro *Autographa californica* nuclear polyhedrosis virus infection. *Appl Environ. Microbiol.* 41, 1166, 1981.
87. Goodwin, R. H., Vaughn, J. L., Adams, R., Jr., and Louloudes, S. J. The influence of insect cell lines and tissue-culture media on baculovirus polyhedra production, *Mics. Publ. Entomol. Soc. Am.* 9, 66, 1973.
88. Goodwin, R. H. and Adams, J. R. Nutrient factors influencing viral replication in serum free insect cell line culture, in *Invertebrate Systems in Vitro*, Kurstak, E., Maramorosch, K., and Dubendorfer, A., Eds., Elsevier/North-Holland Biomedical Press, Amsterdam, 1980, p. 493.
89. Knudson, E. C. Specificity of baculoviruses, in *Safety Aspects of Baculoviruses as Biological Insecticides: Symposium 13–15 November 1978*, Miltenburger, H. G., Ed., Bundesministerium für Forschung und Technologie, Bonn, 1980, p. 241.
90. Lynn, D. E. and Hink, W. F. Comparison of nuclear polyhedrosis virus replication in five lepidopteran cell lines. *J. Invertebr. Pathol.* 35, 234, 1980.
91. Carpenter, W. M. and Bilimoria, S. L. A semipermissive nuclear polyhedrosis virus infection: Characterization of infection kinetics and morphogenesis. *Virology* 130, 227, 1983.
92. Roberts, P. L. and Wecker, E. Baculovirus replication in *Spodoptera littoralis* cell cultures, in *Safety Aspects of Baculoviruses as Biological Insecticides: Symposium 13–15 November 1978*, Miltenburger, H. H., Ed., Bundesministerium für Forschung und Technologie, Bonn, 1980, p. 135.
93. Wood, H. A., Johnston, L. B., and Burand, J. P. Inhibition of *Autograph californica* nuclear polyhedrosis virus replication in high-density *Trichoplusia ni* cell cultures. *Virology* 119, 245, 1982.
94. Volkman, L. E. and Summers, M. D. Nuclear polyhedrosis virus detection: Relative capabilities of clones developed from *Trichoplusia ni* ovarian cell line TN-368 to serve as indicator cells in a plaque assay. *J. Virol.* 16, 1630, 1975.
95. Hink, W. F. Established insect cell line from the cabbage looper, *Trichoplusia ni*. *Nature* 266, 466, 1970.
96. Gardiner, G. R. and Stockdale, H. Two tissue culture media for production of lepidopteran cells and nuclear polyhedrosis viruses. *J. Invertebr. Pathol.* 25, 363, 1970.
97. McCarthy, W. J., Lambiase, J. T., and Henchal, L. S. The effect of inoculum concentration on the development of the nuclear polyhedrosis virus of *Autographa californica* in TN-368 cells. *J. Invertebr. Pathol.* 36, 48, 1980.
98. Brown, M. and Faulkner, P. Plaque assay of nuclear polyhedrosis viruses in cell culture. *Appl. Environ. Microbiol.* 36, 31, 1978.
99. Volkman, L. E. and Goldsmith, P. A. Generalized immunoassay for *Autographa californica* nuclear polyhedrosis virus infectivity in vitro. *Appl. Environ. Microbiol.* 44, 227, 1982.

100. Bilimoria, S. L., Liu, H. S., and Reinisch, A. J. Gene expression of baculoviruses in semipermissive insect cell lines, in *Fundamental and Applied Aspects of Invertebrate Pathology*, Samson, R. A., Vlak, J. M., and Peters, D., Eds., ICIP, Wageningen, The Netherlands, 1986, p. 51.
101. Volkman, L. E. and Knudson, D. L. In vitro replication of baculoviruses, in *The Biology of Baculoviruses*, vol. 1, Granados, R. R. and Federici, B. A, Eds., CRC Press, Boca Raton, FL, 1986, p. 109.
102. Brown, M. and Faulkner, P. Factors affecting the yield of virus in a cloned cell line of *Trichoplusia ni* infected with a nuclear polyhedrosis virus. *J. Invertebr. Pathol.* 26, 251, 1975.
103. Lynn, D. E. and Hink, W. F. Infection of synchronized TN-368 cell cultures with alfalfa looper nuclear polyhedrosis virus. *J. Invertebr. Pathol.* 32, 1, 1978.
104. Hirumi, H., Hirumi, K., and McIntosh, A. H. Morphogenesis of a nuclear polyhedrosis virus of the alfalfa looper in a continuous cabbage looper cell line. *Ann. N.Y. Acad. Sci.* 226, 302, 1975.
105. Knudson, D. L. and Harrap, K. A. Replication of a nuclear polyhedrosis virus in a continuous cell culture of *Spodoptera frugiperda*: Microscopy study of the sequence of events of the virus infection. *J. Virol.* 17, 254, 1976.
106. Faulkner, P. and Henderson, J. F. Serial passage of a nuclear polyhedrosis disease virus of the cabbage looper *(Trichoplusia ni)* in a continuous tissue culture cell line. *Virology* 50, 920, 1972.
107. Raghow, E. and Grace T. D. Studies on a nuclear polyhedrosis virus in *Bombyx mori* cells in vitro. I. Multiplication kinetics and ultrastructural studies. *J. Ultrastruct. Res.* 47, 384, 1974.
108. Bassemir, V., Miltenburger, H. G., and David, P. Morphogenesis of nuclear polyhedrosis virus from *Autographa californica* in a cell line from *Mamestra brassicae* (cabbage moth). *Cell Tiss. Res.* 228, 587, 1983.
109. Volkman, L. E. and Goldsmith, P. A. Mechanism of neutralization of budded *Autographa californica* nuclear polyhedrosis virus by a monoclonal antibody inhibition of entry by adsorptive endocytosis. *Virology* 143, 195, 1985.
110. Volkman, L. E., Goldsmith, P. A., and Hess, R. T. Alternate pathway of entry of budded *Autographa californica* nuclear polyhedrosis virus fusion at the plasma membrane. *Virology* 148, 288, 1986.
111. Charlton, C. A. and Volkman, L. E. Effect of tunicamycin on the structural proteins and infectivity of budded *Autographa californica* nuclear polyhedrosis virus. *Virology* 154, 214, 1986.
112. Granados, R. R. Infectivity and mode of action of baculoviruses. *Biotechnol. Bioeng.* 22, 1377, 1980.
113. Fraser, M. J. Ultrastructural observations of virion maturation in *Autographa californica* nuclear polyhedrosis virus infected *Spodoptera frugiperda* cell cultures. *J. Ultrastruct. Mol. Struct. Res.* 95, 189. 1986.
114. Volkman, L. E. *Autographa californica* MNPV nucleocapsid assembly: Inhibition by cytochalasin D. *Virology* 163, 547, 1988.
115. Volkman, L. E., Goldsmith, P. A., Hess, R. T., and Faulkner, P. Neutralization of budded *Autographa californica* NPV by a monoclonal antibody: Identification of the target antigen. *Virology* 133, 354, 1984.

116. Summers, M. D. and Volkman, L. E. Comparison of biophysical and morphological properties of occluded and extracellular nonoccluded baculovirus from in vivo and in vitro host system. *J. Virol.* 17, 963, 1976.
117. Chung, K. L., Brown, M., and Faulkner, P. Studies on the morphogenesis of polyhedral inclusion bodies of a baculovirus *Autographa californica* NPV. *J. Gen. Virol.* 46, 335, 1980.
118. Hughes, P. R. and Wood, H. A. In vivo and in vitro bioassay methods for baculoviruses, in *The Biology of Baculoviruses*, vol. 2, Granados, R. R. and Federici, B. A., Eds., CRC Press, Boca Raton, FL, 1986, p. 1.
119. Wood, H. A. An agar overlay plaque assay method for *Autographa californica* nuclear-polyhedrosis virus. *J. Invertebr. Pathol.* 29, 304, 1977.
120. Hink, W. F. and Strauss, E. M. An improved technique for plaque assay of *Autographa californica* nuclear polyhedrosis virus on TN-368 cells. *J. Invertebr. Pathol.* 29, 390, 1977.
121. Brown, M. and Faulkner, P. A plaque assay for nuclear polyhedrosis viruses using a solid overlay. *J. Gen. Virol.* 36, 361, 1977.
122. Knudson, D. L. Plaque assay of baculoviruses employing an agarose-nutrient overlay. *Intervirology* 11, 40, 1979.
123. Gettig, R. R. *Heliothis* spp. Baculoviruses: Genetic Characterization of Geographical Isolates, *In Vitro* Replication and Molecular Cloning, Ph.D. diss., The Pennsylvania State University, University Park, 1983.
124. Smith, G. E., Fraser, M. J., and Summers, M. D. Molecular engineering of the *Autographa californica* nuclear polyhedrosis virus genome: Deletion mutations within the polyhedrin gene. *J. Virol.* 46, 584, 1983.
125. Kurstak, E. and Kurstak, C. Application of the immunoperoxidase technique in virology and cancer virology: Light and electron microscopy, in *Viral Immunodiagnosis*, Kurstak, E. and Morisset, R., Eds., Academic Press, New York, 1974, p. 3.
126. Mims, C. A. Role of persistence in viral pathogenesis, in *Virus Persistence*, Mahy, B. W. J., Minson, A. C., and Darby, G. K., Eds, Cambridge University Press, Cambridge, 1982, p. 1.
127. Burand, J. P., Kawanishi, C. Y., and Huang, Y. S. Persistent baculovirus infection, in *The Biology of Baculoviruses*, vol. 1, Granados, R. R. and Federici, B. A., Eds., CRC Press, Boca Raton, FL, 1986, p. 159.
128. Steinhaus, E. A. Crowding as a possible stress factor in insect disease. *Ecology* 39, 503, 1958.
129. Aruga, H. Induction of virus infections, in *Insect Pathology, An Advanced Treatise*, Steinhaus, E. A., Ed., Academic Press, New York, 1963, p. 499.
130. Longworth, J. F. and Cunningham, J. C. The activation of occult nuclear-polyhedrosis viruses by foreign polyhedra. *J. Invertebr. Pathol.* 10, 361, 1968.
131. Jurkovicova, M. Activation of latent infections in larvae of *Adoxophyes orana* (Lepidoptera: Torticidae) and *Barathra brassicae* (Lepidoptera: Noctuidae) by foreign polyhedra. *J. Invertebr. Pathol.* 34, 213, 1979.
132. McKinley, D. J., Brown, D. A., Payne, C. C., and Harrap, K. A. Cross-infectivity and activation studies with four baculoviruses. *Entomophaga*, 26, 79, 1981.
133. McIntosh, A. H. and Ignoffo, C. M. Replication and infectivity of the single-embedded nuclear polyhedrosis virus, *Baculovirus heliothis* in homologous cell lines. *J. Invertebr. Pathol.* 37, 258, 1981.

134. Ignoffo, C. M., Shapiro, M., and Hink, W. F. Replication and serial passage of infectious *Heliothis* nucleopolyhedrosis virus in an established line of *Heliothis zea* cells. *J. Invertebr. Pathol.* 18, 131, 1971.
135. Sohi, S. S., Percy, J., Arif, B. M., and Cunningham, J. C. Replication and serial passage of singly enveloped baculovirus of *Orgyia leucostigma* in homologous cell lines. *Intervirology* 21, 50, 1984.
136. Granados, R. R., Cawler, K. A., and Burand, J. P. Replication of *Heliothis zea* baculovirus in an insect cell line. *Intervirology*, 15, 71, 1981.
137. Yamada, K. I., Sherman, K. E., and Maramorosch, K. Serial passage of *Heliothis zea* singly embedded nuclear polyhedrosis virus in a homologous cell line. *J. Invertebr. Pathol.* 39, 185, 1982.
138. Yamada, K. I., Sherman, K. E., and Maramorosch, K. In vivo infectivity of early and late passaged *Heliothis zea* polyhedra produced in tissue culture. *Appl. Entomol. Zool.* 16, 504, 1981.
139. McIntosh, A. H., Andrew, P. A., and Ignoffo, C. M. Establishment of two continuous cell lines of *Heliothis virescens* (F.) (Lepidoptera: Noctuidae). *In Vitro*, 17, 649, 1981.
140. Granados, R. R. and Lawler, K. A. In vivo pathway of *Autographa californica* baculovirus invasion and infection. *Virology* 108, 297, 1981.
141. Yamada, K. and Maramorosch, K. Plaque assay of *Heliothis zea* baculovirus employing a mixed agarose overlay. *Arch. Virol.* 67, 187, 1981.
142. McIntosh, A. H., Ignoffo, C. M., and Andrews P. L. In-vitro host range of 5 baculoviruses in lepidopteran cell lines. *Intervirology* 23, 150, 1985.
143. Granados, R. R., Derksen, A. C. G. and Dwyer, K. G. Replication of the *Trichoplusia ni* granulosis and nuclear polyhedrosis viruses in cell cultures. *Virology* 152, 472, 1986.
144. Miller, L. K. The genetics of baculoviruses, in *The Biology of Baculoviruses*, vol. 1, Granados, R. R. and Federici, B. A. Eds., CRC Press, Boca Raton, FL, 1986, p. 217.
145. Ramoska, W. A. and Hink, W. F. Electron microscope examination of two plaque variants from a nuclear polyhedrosis virus of the alfalfa looper, *Autographa californica*. *J. Invertebr. Pathol.* 23, 197, 1974.
146. Potter, K. N., Faulkner, P., MacKinnon, E. A., and Strain, E. A. Selection during serial passage of *Trichoplusia ni* nuclear polyhedrosis virus. *J. Virol.* 18, 1040, 1976.
147. Potter, K. N. and Miller, L. K. Correlating genetic mutations of a baculovirus with the physical map of the DNA genome, in *Animal Virus Genetics,* Fields, B., Janish, R., and Fox, C. F., Eds., Academic Press, New York, 1980, p. 71.
148. Miller, D. W. and Miller, L. K. A virus mutant with an insertion of a *copia*-like transposable element. *Nature* 299, 562, 1982.
149. Fraser, M. J., Smith, G. E., and Summers, M. D. Acquisition of host cell DNA sequences by baculoviruses: Relationship between host DNA insertions and FP mutants of *Autographa californica* and *Galleria mellonella* nuclear polyhedrosis viruses. *J. Virol.* 47, 287, 1983.
150. Fraser, M. J., Brusca, J. S., Smith, G. E., and Summers, M. D. Transposon-mediated mutagenesis of a baculovirus. *Virology* 145, 356, 1985.

151. Beames, B. and Summers, M. D. Comparisons of host cell DNA insertions and altered transcription at the site of insertions in few polyhedra baculovirus mutants. *Virology* 162, 206, 1988.
152. Potter, K. N. and Miller, L. K. Transfection of two invertebrate cell lines with DNA of *Autographa californica* NPV. *J. Invertebr. Pathol.* 36, 431, 1980.
153. Brown, M., Crawford, A. M., and Faulkner, P. Genetic analysis of a baculovirus, *Autographa californica* nuclear polyhedrosis virus. I. Isolation of temperature-sensitive mutants and assortment into complementation groups. *J. Virol.* 31, 190, 1979.
154. Lee, H. H. and Miller, L. K. Isolation, complementation, and initial characterization of temperature sensitive mutants of the baculovirus *Autographa californica* nuclear polyhedrosis virus. *J. Virol.* 31, 240, 1979.
155. Miller, L. K. and Dawes, K. P. Physical map of the DNA genome of *Autographa californica* nuclear polyhedrosis virus. *J. Virol.* 29, 1044, 1979.
156. Gordon, J. D. and Carstens, E. B. Phenotypic characterization and physical mapping of a temperature-sensitive mutant of *Autographa californica* nuclear polyhedrosis virus defective in DNA synthesis. *Virology* 138, 69, 1984.
157. Miller, L. K., Trimarchi, R. E., Browne, D., and Pennock, G. D. A temperature sensitive mutant of the baculovirus *Autographa californica* nuclear polyhedrosis virus defective in an early function required for further gene expression. *Virology* 126, 376, 1983.
158. Duncan, R. and Faulkner, P. Bromodeoxyruidine-induced mutants of *Autographa californica* nuclear polyhedrosis virus defective in occlusion body formation. *J. Gen. Virol.* 62, 369, 1982.
159. Brown, M. and Faulkner, P. A parital genetic map of the baculovirus, *Autographa californica* nuclear polyhedrosis virus, based on recombination studies with ts mutants. *J. Gen. Virol.* 48, 247, 1980.
160. Summers, M. D., Smith, G. E., Knell, J. D., and Burand, J. P. Physical maps of *Autographa californica* and *Rachiplusia ou* nuclear polyhedrosis virus recombinants. *J. Virol.* 34, 693, 1980.
161. Croizier, G., Godse, D., and Vlak, J. Sélection de types virax dans les infections doubles à baculovirus chez les larves de lepidoptera. *C. R. Acad. Sci. Paris* 290, 579, 1980.
162. Croizier, G. and Quiot, J. M. Observation and analysis of two genetic recombinants of baculovirus of lepidoptera, *Autographa californica* Speyer and *Galleria mellonella* L. *Ann. Virol.* 132, 3, 1981.
163. Miller, L. K. Construction of a genetic map of the baculovirus *Autographa californica* nuclear polyhedrosis virus by marker rescue of temperature-sensitive mutants. *J. Virol.* 39, 973, 1981.
164. Brown, M. Faulkner, P., Cochran, M. A., and Chung, K. L. Characterization of two morphology mutants of *Autographa californica* nuclear polyhedrosis virus with large cuboidal inclusion bodies. *J. Gen. Virol.* 50, 309, 1980.
165. Carstens, E. B. Mapping the mutation site of an *Autographa californica* nuclear polyhedrosis virus polyhedron morphology mutant. *J. Virol.* 43, 809, 1982.
166. Carstens, E. B., Krebs, A., and Gallerneault, C. E. Identification of an amino-acid essential to the normal assembly of *Autographa californica* nuclear polyhedrosis virus polyhedra. *J. Virol.* 58, 684, 1986.

167. Duncan, R., Chung, K. L., and Faulkner, P. Analysis of a mutant of *Autographa californica* nuclear polyhedrosis virus with a defect in the morphogenesis of the occlusion body macromolecular lattice. *J. Gen. Virol.* 64, 1531, 1983.
168. Carstens, E. B., Ye, L. B., and Faulkner, P. A point mutation in the polyhedrin gene of baculovirus *Autograph californica* MNPV prevents crystallization of occlusion bodies. *J. Gen. Virol.* 68, 301, 1987.
169. Wood, H. A., Hughes, P. R., Johnston, L. B., and Langridge, W. H. R. Increased virulence of *Autographa californica* nuclear polyhedrosis virus by mutagenesis. *J. Invertebr. Pathol.* 38, 236, 1981.
170. Nathans, D. and Smith, H. O. Restriction endonucleases in the analysis and restructuring of DNA molecules. *Ann. Rev. Biochem.* 44, 273, 1975.
171. Southern, E. M. Measurement of DNA length by gel electrophoresis. *Anal. Biochem.* 100, 319, 1979.
172. Evanson, D. P. Electron microscopy of viral nucleic acids, in *Methods in Virology*, vol. 6, Maramorosch, K. and Koprowski, H., Eds., Academic Press, New York, 1977, p. 220.
173. Donna, K. J. Determination of fragment order through partial digests and multiple enzyme digests, in *Methods in Enzymology*, Grossman, L. and Moldave, K., Eds., Academic Press, New York, 1980, p. 449.
174. Smith H. O. and Bernstiel, M. L. A simple method for DNA restriction site mapping. *Nucleic Acids Res.* 3, 2387, 1976.
175. Sata. S., Hutchinson, C. A., III, and Harris, J. I. A thermostable sequence specific endonuclease from *Thermus aquaticus*. *Proc. Natl. Acad. Sci. U.S.A.* 74, 542, 1977.
176. Legerski, R. J., Hodnett, J. L., and Gray, H. B., Jr. Extracellular nucleases of *Pseudomonas bal*. III. Use of the double-stranded deoxyribonuclease activity as the basis of a convenient method for the mapping of fragments of DNA produced by cleavage with restriction enzymes. *Nucleic Acids Res.* 5, 1445, 1978.
177. Cochran, M. A., Carstens, E. B., Eaton, B. T., and Faulkner, P. Molecular cloning and physical mapping of restriction endonuclease fragments of *Autographa californica* nuclear polyhedrosis virus DNA. *J. Virol.* 41, 940, 1982.
178. Smith, G. E. and Summers, M. D. Restriction maps of five *Autographa californica* MNPV variants, *Trichoplusia ni* MNPV, and *Galleria mellonella* MNPV DNAs with endonucleases *SmaI*, *KpnI*, *BamHI*, *SacI*, *XhoI*, and *EcoRI*. *J. Virol.* 30, 828, 1979.
179. Brown, S. E., Maruniak, J. E., and Knudson, D. L. Physical map of SeMNPV baculovirus DNA: An AcMNPV genomic variant. *Virology* 136, 235, 1984.
180. Lübbert, H., Kruczek, I., Tjia, S., and Doerfler, W. The cloned EcoRI fragments of *Autographa californica* nuclear polyhedrosis virus DNA. *Gene* 16, 343, 1981.
181. Leisy, D. J., Rohrmann, G. F., Nesson, M., and Beaudreau, G. S. Nucleotide sequencing and transcriptional mapping of the *Orgyia-pseudotsugata* multicapsid nuclear polyhedrosis virus P-10 gene. *Virology* 153, 157, 1986.
182. Arif, B. M. and Doerfler, W. Identification and localization of reiterated sequences in the *Choristoneura fumiferana* MNPV genome. *Eur. Mol. Biol. Organ. J.* 3, 525, 1984.

183. Maruniak, J. E., Brown, S. E., and Knudson, D. L. Physical maps of SFMNPV baculovirus DNA and its genomic variants. *Virology* 136, 221, 1984.
184. Wiegers, F. F. and Vlak, J. M. Physical map of the DNA of a *Mamestra brassicae* nuclear polyhedrosis virus variant isolated from *Spodoptera exigua*. *J. Gen. Virol.* 65, 2011, 1984.
185. Knell, J. D. and Summers, M. D. A physical map for the *Heliothis zea* SNPV genome. *J. Gen. Virol.* 65, 445, 1984.
186. Leisy, D. J., Rohrmann, G. F., and Beaudreau, G. S. Conservation of genome organization in two multicapsid nuclear polyhedrosis virus. *J. Virol.* 52, 699, 1984.
187. Adang, M. J. and Miller, L. K. Molecular cloning of DNA complementary to mRNA of the baculovirus *Autographa californica* nuclear polyhedrosis virus: Location and gene products of RNA transcripts found late in infection. *J. Virol.* 44, 782, 1982.
188. Bergsma, D. J., Olive, D. M., Hartzell, S. W., and Subramanian, K. N. Territorial limits and functional anatomy of the simian virus 40 replication origin. *Proc. Natl. Acad. Sci. U.S.A.* 74, 381, 1982.
189. de Villers, J. and Schaffner, N. A small segment of polyoma virus DNA enhances the expression of a cloned β-globin gene over a distance of 1400 base pairs. *Nucleic Acids Res.* 9, 6251, 1984.
190. Given, D., Yee, D., Griem, K., and Kieff, E. DNA of Epstein-Barr virus. V. Direct repeats of the ends of Epstein-Barr virus DNA. *J. Virol.* 30, 852, 1979.
191. Hennessy, K. and Kieff, E. One of two Epstein-Barr virus nuclear antigens contains a glycine-alanine copolymer domain. *Proc. Natl. Acad. Sci. U.S.A.* 80, 3665, 1984.
192. Kitner, C. R. and Sugden, B. The structure of the termini of Epstein-Barr virus. *Cell* 1, 661, 1979.
193. Tamanoi, F. and Stillman, B. W. The origin of adenovirus DNA replication. *Curr. Top. Microbiol. Immunol.* 109, 75, 1983.
194. Cochran, M. A. and Faulkner, P. Location of homologous DNA sequences interspersed at five regions in the baculovirus AcMNPV genome. *J. Virol.* 45, 961, 1983.
195. Guarino, L. A., Gonzalez, M. A., and Summers, M. D. Complete sequence and enhancer function of the homologous DNA regions of *Autographa californica* nuclear polyhedrosis virus. *J. Virol.* 60, 224, 1986.
196. Guarino, L. A. and Summers, M. D. Interspersed homologous DNA of *Autographa californica* nuclear polyhedrosis virus enhances delayed-early gene expression. *J. Virol.* 60, 215, 1986.
197. Liu, A., Qin, J., Rankin, C., Hardin, S. E., and Weaver, R. F. Nucleotide sequence of a portion of the *Autographa californica* nuclear polyhedrosis virus genome containing the Eco-RI site-rich region HR-5 and an open reading frame just 5' of the P-10 gene. *J. Gen. Virol.* 67, 2565; 1986.
198. Arif, B. M., Tjia, S. T., and Doerfler, W. DNA homologies between the genomes of *Choristoneura fumiferana* and *Autographa californica* nuclear polyhedrosis viruses. *Virus Res.* 2, 85, 1985.

199. Carstens, E. B., Tjia, S. T., and Doerfler, W. Infection of *Spodoptera frugiperda* cells with *Autographa californica* nuclear polyhedrosis virus. I. Synthesis of intracellular proteins after virus infection. *Virology* 99, 386, 1979.
200. Kelly, D. C. and Lescott, T. Baculovirus replication: Protein synthesis in *Spodoptera frugiperda* cells infected with *Trichoplusia ni* nuclear polyhedrosis virus. *Microbiologica* 4, 35, 1981.
201. Kelly, D. C. Baculovirus replication. *J. Gen. Virol.* 63, 1, 1982.
202. Liu, H.-S. Host specific expression of the baculovirus, *Spodoptera frugiperda* nuclear polyhedrosis virus, in insect cell lines, Ph.D. diss., Texas Tech University, Lubbock, 1987.
203. Rohel, D. Z., Cochran, M. A., and Faulkner, P. Mapping late mRNA populations of *Autographa californica* nuclear polyhedrosis virus by "criss-cross" DNA-RNA hybridization. *J. Gen. Virol.* 65, 809, 1984.
204. Friesen, P. D. and Miller, L. K. Temporal regulation of baculovirus DNA: Overlapping early and late transcripts. *J. Virol.* 54, 392, 1985.
205. Burand, J. P., Summers, M. D., and Smith, G. E. Transfection with baculovirus DNA. *Virology* 101, 286, 1980.
206. Kelly, D. C. and Wang, X. The infectivity of NPV DNA. *Ann. Virol.* 132E, 247, 1981.
207. Oreshkin, E. N. Investigations of conditions for baculovirus DNA transfection in an insect cell line. *Vopr. Virusol.* 2, 235, 1982.
208. Grula, M. A., Buller, P. L., and Weaver, R. F. Alpha-amanitin-resistant viral RNA synthesis in nuclei isolated from nuclear polyhedrosis virus-infected *Heliothis zea* larvae and *Spodoptera frugiperda* cells. *J. Virol.* 38, 916, 1981.
209. Fuchs, L. Y., Woods, M. S., and Weaver, R. F. Viral transcription during *Autographa californica* nuclear polyhedrosis virus infection: A novel RNA polymerase induced in infected *Spodoptera frugiperda* cells. *J. Virol.* 48, 641, 1983.
210. Rohel, D. Z., Cochran, M. A., and Faulkner, P. Characterization of two abundant mRNAs of *Autographa californica* nuclear polyhedrosis virus present late in infection. *Virology* 124, 357, 1983.
211. Vlak, J. M., Smith G. E., and Summers, M. D. Hybridization selection and in vitro translation of *Autographa californica* nuclear polyhedrosis virus mRNA. *J. Virol.* 40, 762, 1981.
212. Erlandson, M. A. and Carstens, E. B. Mapping early transcription products of *Autographa californica* nuclear polyhedrosis virus. *Virology* 126, 398, 1983.
213. Erlandson, M. A., Gordon, J., and Carstens, E. B. Size and map locations of early transcription products on the *Autographa californica* nuclear polyhedrosis virus genome. *Virology* 142, 12, 1985.
214. Lübbert, H. and Doerfler, W. Transcription of overlapping sets of RNA from the genome of *Autographa californica* nuclear polyhedrosis virus: A novel method for mapping RNA. *J. Virol.* 52, 255, 1984.
215. Friesen, P. D. and Miller, L. K. Temporal regulation of baculovirus RNA overlapping early and late transcripts. *J. Virol.* 54, 392, 1985.
216. Rankin, C., Ladin, B. F., and Weaver, R. F. Physical mapping of temporally regulated overlapping transcripts in the region of the 10-kilodalton protein gene in *Autographa californica* nuclear polyhedrosis virus. *J. Virol.* 57, 18, 1985.

217. Oellig, C., Happ, B., Mueller, T., and Doerfler, W. Overlapping sets of viral RNA species reflect the array of polypeptides in the Eco-RIJ and N fragments, map positions 81.2 to 85.0 of the *Autographa californica* nuclear polyhedrosis virus genome. *J. Virol.* 61, 3048, 1987.
218. Mainprize, T. H., Lee, K., and Miller, L. K. Variation in the temporal expression of overlapping baculovirus transcripts. *Virus Res.* 6, 85, 1986.
219. Rohrmann, G. F., Leisy, D. J., Chow, K. C., Pearson, G. D., and Beaudreau, G. S. Identification, cloning, and R-loop mapping of the polyhedrin gene from the multicapsid nuclear polyhedrosis virus of *Orgyia pseudotsugata*. *Virology* 121, 51, 1982.
220. Lübbert, H. and Doerfler, W. Mapping of early and late transcripts encoded by the *Autographa californica* nuclear polyhedrosis virus genome: Is viral RNA spliced? *J. Virol.* 50, 497, 1984.
221. Howard, S. C., Ayres, M. D., and Possee, R. D. Mapping the 5' and 3' ends of *Autographa californica* nuclear polyhedrosis virus polyhedrin messenger RNA. *Virus Res.* 5, 109, 1986.
222. Hooft van Iddekinge, B. J., Smith, G. E., and Summers, M. D. Nucleotide sequences of the polyhedrin gene of *Autographa californica* nuclear polyhedrosis virus. *Virology* 131, 461, 1983.
223. Kozlov, E. A., Levitina, T. L., Gusak, N. M., and Serebryanyi, S. B. Comparison of the amino acid sequence of inclusion body proteins of nuclear polyhedrosis viruses *Bombyx mori, Porthetria dispar* and *Galleria mellonella*. *Bioorg. Chem.* 7, 1008, 1981.
224. Jun-Chuang, Q. and Weaver, R. F. Capping of viral RNA in cultured *Spodoptera frugiperda* cells infected with *Autographa californica* nuclear polyhedrosis virus. *J. Virol.* 43, 234, 1982.
225. Guarino, L. A. and Summers, M. D. Nucleotide sequence and temporal expression of a baculovirus regulatory gene. *J. Virol.* 61, 2091, 1987.
226. Knebel, D., Lübbert, H., and Doerfler, W. The promoter of the late P-10 gene in the insect nuclear polyhedrosis virus *Autographa californica:* Activation by viral gene products and sensitivity to DNA methylation. *EMBO J.* 4, 1301, 1985.
227. Guarino, L. A. and Summers, M. D. Functional mapping of *Autographa californica* nuclear polyhedrosis virus genes required for late gene expression. *J. Virol.* 62, 463, 1988.
228. Friesen, P. D. and Miller, L. K. Divergent transcription of early 35 and 94-kilodalton protein genes encoded by the *Hind*-D-III K genome fragment of the baculovirus *Autographa californica* nuclear polyhedrosis virus. *J. Virol.* 61, 2264, 1987.
229. Crawford, A. M. and Miller, L. K. Characterization of an early gene accelerating expression of late genes of the baculovirus *Autographa californica* nuclear polyhedrosis virus. *J. Virol.* 62, 2773, 1988.
230. Tjia, S. T., Carstens, E. B., and Doerfler, W. Infection of *Spodoptera frugiperda* cells with *Autographa californica* nuclear polyhedrosis virus. II. The viral DNA and kinetics of its replication. *Virology* 99, 399, 1979.
231. Miller, L. K., Jewell, J. E., and Browne, D. Baculovirus induction of a DNA polymerase. *J. Virol.* 40, 305, 1981.

232. Flore, P. H., Burand, J. P., Gettig, R. R., and Wood, H. A. Characterization of DNA polymerase activity in *Trichoplusia ni* cells infected with *Autographa californica* nuclear polyhedrosis virus. *J. Gen. Virol.* 68, 2025, 1987.
233. Wilson, M. E. and Miller, L. K. Changes in the nucleoprotein complexes of a baculovirus DNA during infection. *Virology* 151, 315, 1986.
234. Leisy, D., Rohrmann, G., and Beaudreau, G. The nucleotide sequence of the polyhedrin gene region from the multicapsid baculovirus of *Orgyia pseudotsugata Virology* 153, 280, 1986.
235. Wilson, M. E., Mainprize, T. H., Friesen, P. D., and Miller, L. K. Location transcription and sequence of a baculovirus gene encoding a small arginine-rich polypeptide. *J. Virol.* 61, 661, 1987.
236. Liu, H. A. and Bilimoria, S. L. Unpublished data, 1987
237. Reinisch, A. J. and Bilimoria, S. L. Unpublished data, 1987.
238. Vail, P. V. and Jay, D. L. Pathology of a nuclear polyhedrosis virus of the alfalfa looper in alternate hosts. *J. Invertebr. Pathol.* 21, 198, 1973.
239. Gröner, A. Specificity and safety of baculoviruses, in *The Biology of Baculoviruses*, vol. 1, Granados, R. R., and Federici, B. A., Eds., CRC Press, Boca Raton, FL, 1986, p. 177.
240. McIntosh, A. H., Maramorosch, K., and Riscoe, R. *Autographa californica* nuclear polyhedrosis virus (NPV) in a vertebrate cell line: Localization by electron microscopy. *J. N. Y. Entomol. Soc.* 87, 55, 1979.
241. Tjia, S. T., Meyer zu Altenchildesche, G., and Doerfler, W. *Autographa californica* nuclear polyhedrosis virus (AcNPV) DNA does not persist in mass cultures of mammalian cells. *Virology* 125, 107, 1983.
242. Himeno, H., Sakai, F., Onodera, K., Nakni, H., Fukada, T., and Kawade, Y. Formation of nuclear polyhedral bodies and nuclear polyhedrosis virus of silk worm in mammalian cells infected with viral DNA. *Virology* 33, 507, 1967.
243. McIntosh, A. H. and Shamy, R. Biological studies of a baculovirus in a mammalian cell line. *Intervirology* 13, 331, 1980.
244. Gröner, A., Granados, R. R., and Burand, J. P. Interaction of *Autographa californica* nuclear polyhedrosis virus with 2 nonpermissive cell lines. *Intervirology* 21, 203, 1984.
245. Volkman, L. E., and Goldsmith, P. A. In vitro survey of *Autographa californica* nuclear polyhedrosis virus interaction with nontarget vertebrate host cells. *Appl. Environ. Microbiol.* 45, 1085, 1983.
246. Brusca, J., Summers, M., Couch, J., and Courtney, L. *Autographa californica* nuclear polyhedrosis virus efficiently enters but does not replicate in poikilothermic vertebrate cells. *Intervirology* 26, 207, 1986.
247. McClintock, J. T., Dougherty, E. M., and Weiner, R. M. Semipermissive replication of a nuclear polyhedrosis virus of *Autographa californica* in a gypsy moth cell line. *J. Virol.* 57, 197, 1985.
248. Grace, T. D. C. Establishment of a line of cells from the silkworm *Bombyx mori*. *Nature* 216, 613, 1967.
249. Rice, W. C. and Miller, L. K. Baculovirus transcription in the presence of inhibitors and in nonpermissive *Drosophila* cells. *Virus Res.* 6, 155, 1986.
250. Carbonell, L. E., Klowden, M. J., and Miller, L. K. Baculovirus-mediated expression of bacterial genes in dipteran and mammalian cells. *J. Virol.* 45, 153, 1985.

251. Maniatis, T., Goodbourn, S., and Fishcher, J. A. Regulation of inducible and tissue-specific gene expression. *Science* 236, 1237, 1987.
252. Maeda, S., Kawai, T., Obinata, M., Fujiwara, H., Horiuchi, T., Saeki, Y., Sato, Y. and Furusawa, M. Production of human $\alpha$-interferon in silkworm using a baculovirus vector. *Nature* 315, 592, 1985.
253. Jasny, B. R. Insect viruses invade biotechnology. *Science* 238, 1653, 1987.
254. Possee, R. D. Cell-surface expression of influenza virus hemagglutinin in insect cells using a baculovirus vector. *Virus Res.* 5, 43, 1986.
255. Kuroda, K., Hauser, C., Rott, R., Klenik, H. D., and Doerfler, W. Expression of the influenza virus hemagglutinin in insect cells by a baculovirus vector. *EMBO J.* 5, 1359, 1986.
256. Matsura, Y., Possee, R. D., Overton, H. A., and Bishop, D. H. L. Baculovirus expression vectors: The requirements for high level expression of proteins including glycoproteins. *J. Gen Virol.* 68, 1233, 1987.
257. Martin, B. M., Tsuji, S., Lamarca, M. E., Maysak, K., Eliason, W., and Ginns, E. I. Glycosylation and processing of high levels of active human glucocerebrosidase in invertebrate cells using a baculovirus expression vector. *DNA* 7, 99, 1988.
258. Smith, G. E., Ju, G., Ericson, B. L., Moschera, J., Lahm, H. W., Chizzoni, T. R., and Summers, M. D. Modification and secretion of human interleukin-2 produced in insect cells by a baculovirus expression vector. *Proc. Natl. Acad. Sci. U.S.A.* 82, 8404, 1985.
259. Jeang, K. T., Holmgren-Konig, M., and Khoury, G. A baculovirus vector can express intron-containing genes. *J. Virol.* 61, 1761, 1987.
260. Skuratovskaya, I. N., Strokovskaya, L. I., Zherebtsova, E. N., and Gudz-Gorban, A. P. Properties of the nuclear polyhedrosis virus of the greater wax moth: Oligomeric circular DNA and the characteristics of the genome. *Virology* 120, 465, 1982.
261. Ignoffo, C. M. Effects of age on morality of *Heliothis zea* and *Heliothis virescens* larvae exposed to a nuclear-polyhedrosis virus. *J. Invertebr. Pathol.* 8, 279, 1966.
262. Huger, A., Granulosis of insects, in *Insect Pathology: An Advanced Treatise*, vol. 1, Steinhaus, E. A., Ed., Academic Press, New York, 1963, chap. 16.
263. Smirnoff, W. A. Observations on the effect of virus infection on insect behavior. *J. Invertebr. Pathol.* 7, 387, 1965.
264. Benz, G. A. Nuclear polyhedrosis of *Malacosoma alpicola* (Staudinger). *J. Insect. Pathol.* 5, 215, 1963.
265. Tanada, Y., Hess, R. T., and Omi, E. M. Unique virus morphogenesis and cytopathology of a baculovirus (hypertrophy strain) in larva of the armyworm, *Pseudaletia unipuncta*, *J. Invertebr. Pathol.* 40, 197, 1982.
266. Federici, B. A. Mosquito baculovirus: Sequence of morphogenesis and ultrastructure of the virion. *Virology* 100, 1, 1980.
267. Bird, F. T. The use of a virus disease in the biological control of the European pine sawfly, *Neodiprion sertifer* (beoffr.). *Can. Entomol.* 85, 437, 1953.
268. Stiles, B., Dunn, P. E., and Paschke, J. D. Histopathology of a nuclear polyhedrosis infection in *Aedes epactius* with observations in four additional mosquito species. *J. Invertebr. Pathol.* 41, 191, 1983.

269. Pritchett, D. W., Young, S. Y., and Yearian, W. C. Dissolution of *Autographa californica* nuclear polyhedrosis virus polyhedra by the digestive fluid of *Trichoplusia ni* (Lepidoptera: Noctuidae) larvae. *J. Invertebr. Pathol.* 39, 354, 1982.
270. Eppstein, D. A., Thoma, J. A., Scott, H. A., and Young, S. Y. Degradation of matrix protein from a nuclear-polyhedrosis virus of *Trichoplusia ni* by endogenous protease. *Virology* 67, 591, 1975.
271. Aizawa, K. Antiviral substances in the gut juice of the silkworm, *Bombyx mori* (Linnaeus). *J. Insect Pathol.* 4, 72, 1962.
272. Paschke, J. D. and Summers, M. D. Early events in the infection of the arthropod gut by pathogenic insect viruses, in *Invertebrate Immunity*, Maramorosch, K. and Shope, R. E., Eds., Academic Press, New York, 1975, p. 75.
273. Adang, M. J. and Spence, K. G. Surface morphology of peritrophic membrane formation in the cabbage looper, *Trichoplusia ni*. *Cell Tiss. Res.* 218, 141, 1981.
274. Adang, M. J. and Spence, K. D. Permeability of the peritrophic membrane of the douglas fir tussock moth *(Orgyia pseudotsugata)*, *Comp. Biochem. Physiol. A.* 74, 233, 1983.
275. Tanada, Y., Hess, R. T., and Omi, E. M. The movement and invasion of an insect baculovirus in tracheal cells of the armyworm *Pseudaletia unipuncta*. *J. Invertebr. Pathol.* 44, 198, 1984.
276. Granados, R. R. Early events in the infection of *Heliothis zea* midgut cells by a baculovirus. *Virology* 90, 170, 1978.
277. Reinganum, C. Polyacrylamide gel electrophoresis of baculovirus proteins: a simplified and sensitive modification for differentiating between isolates. *J. Invertebr. Pathol.* 44, 134, 1984.
278. Ward, V. K., Fleming, S. B., and Kalmakoff, J. Comparison of a DNA-DNA Dot-BLOT hybridization assay with light microscopy and radioimmunoassay for the detection of a nuclear polyhedrosis virus. *J. Virol. Methods* 15, 65, 1987.
279. Kalmakoff, J. and Crawford, A. M. Enzootic virus control of *Wiseana* spp. in a pasture environment, in *Microbial and Viral Pesticides*, Kurstak, E., Ed., Marcel Dekker, New York, 1982, p. 435.
280. Evans, H. F. Quantitative assessment of the relationships between dosage and response of the nuclear polyhedrosis virus of *Mamestra brassicae*. *J. Invertebr. Pathol.* 37, 101, 1981.
281. Allen, G. E. and Ignoffo, C. N. The nucleopolyhedrosis of *Heliothis:* Quantitative in vivo estimates of virulence. *J. Invertebr. Pathol.* 13, 378, 1969.
282. Young, S. Y., Lack, J. R., and Yearian, W. C. Transmission of a nuclear polyhedrosis virus in *Anticarsia gemmatalis* (Lepidoptera: Noctuidae) larval populations on caged soybeans. *Environ. Entomol.* 16, 691, 1987.
283. Anderson, R. M. and May, R. M. The population dynamics of microparasites and their invertebrate hosts. *Phil. Trans. R. Soc.* 291, 451, 1981.
284. Young, S. Y. and Yearian, W. C. Secondary transmission of nuclear polyhedrosis virus by *Pseudoplusia includens* and *Anticarsia gemmatalis* larvae on semisynthetic diet. *J. Invertebr. Pathol.* 51, 133, 1988.
285. Harper, J. D. Food consumption by cabbage loopers infected with nuclear polyhedrosis virus. *J. Invertebr. Pathol.* 21, 191, 1973.

286. Woods, S. A. and Elkinton, J. S. Bimodal patterns of mortality from nuclear polyhedrosis virus in gypsy moth *Lymantria dispar* populations. *J. Invertebr. Pathol.* 50, 151, 1987.
287. Fleming, S. B., Kalmakoff, J., Archibald, R. D., and Stewart, K. M. Density-dependent virus mortality in populations of *Wiseana* (Lepidoptera: hepialidae). *J. Invertebr. Pathol.* 48, 193, 1986.
288. Entwistle, P. F., Adams, P. H. W., Evans, H. F., and Rivers, C. F. Epizootiology of a nuclear polyhedrosis virus (Baculoviridae) in European spruce sawfly *(Gilpinia hercyniae)*: Spread of disease from small epicentres in comparison with spread of baculovirus disease in other hosts. *J. Appl. Ecol.* 20, 473, 1983.
289. Tanada, Y. Ecology of insect viruses, in *Perspectives in Forest Entomology*, Anderson, J. F. and Kaya, H. K., Eds., Academic Press, New York, 1976, p. 265.
290. Donne, C. C. Ecology of pathogens of the gypsy moth, in *Perspectives in Forest Entomology,* Anderson, J. F. and Kaya, H. K., Eds., Academic press, New York, 1976, p. 285.
291. Tompkins, G. J., Vaughn, J. L., Adams, J. R., and Reichelderfer, C. F. Effects of propagating *Autographa californica* nuclear polyhedrosis virus and its *Trichoplusia ni* variant in different hosts. *Environ. Entomol.* 10, 801, 1981.
292. Shapiro, M. In vivo mass production of insect viruses, in *Microbial and Viral Pesticides,* Kurstak, E., Ed., Marcel Dekker, New York, 1982, p. 463.
293. Evans, H. F. and Harrap, K. A. Persistence of insect viruses, in *Virus Persistence,* Mahy, B. W. J., Minson, A. C., and Darby, G., Eds., Cambridge University Press, Cambridge, 1982, p. 57.
294. Bird, F. T. Transmission of some insect viruses with particular reference to ovarial transmission and its importance in the development of epizootics. *J. Insect Pathol.* 3, 352, 1961.
295. Neilson, M. M., Martineau, R., and Rose, A. H. Diprion hercyniae (Hartig.), European spruce sawfly (Hymenoptera: Diprionidae): Biological control programmes in Canada 1959–1968. *Commonwealth Inst. Biol. Control Tech Bull.* 4, 136, 1971.
296. Entwistle, P. F., Adams, P. H. W., and Evans, H. F. Epizootiology of a nuclear-polyhedrosis virus in European spruce sawfly *(Gilpinia hercyniae)*: The status of birds as dispersal agents of the virus during the larval season. *J. Invertevr. Pathol.* 29, 354, 1977.
297. Martignoni, M. E. and Milstead, J. E. Trans-ovum transmission of the nuclear polyhedrosis virus of *Coliac enrytheme* Biosduval through contamination of the female genitalia. *J. Insect Pathol.* 4, 113, 1962.
298. Cunningham, J. C. and Entwistle, P. F. Control of sawflies by baculoviruses, in *Microbial Control of Pests and Plant Diseases,* Burgess, H. D., Ed., Academic Press London, 1981, p. 379.
299. Tanada, Y. and Omi, E. M. Epizootiology of virus disease in three lepidopterous insect species on alfalfa. *Res. Popul. Ecol.* 16, 59, 1974.
300. Elgee, E. Persistence of a virus of the white-marked tussock moth on balsam fir foliage. *Bi-mon. Res. Notes* 31, 33, 1975.
301. Evans, H. F. and Entwistle, P. F. Epizootiology of the nuclear polyhedrosis virus

of European spruce sawfly with emphasis on persistence of virus outside the host, in *Microbial and Viral Pesticides,* Kurstak, E., Ed., Marcel Dekker, New York, 1982, p. 449.
302. Thompson, C. G. and Scott, D. W. Production and persistence of the nuclear polyhedrosis virus of the Douglas-fur tussock moth, *Orgyia pseudotsugata* (Lepidotera: Lymantriidae), in the forest ecosystem. *J. Invertebr. Pathol.* 33, 57, 1979.
303. Jaques, R. P. Persistence, accumulation, and denaturation of nuclear polyhedrosis virus and granulosis viruses, in *Baculoviruses for Insect Pest Control: Safety Considerations*, Summers, M., Engler, R., Falcon, L. A., and Vail, P., Eds., American Society for Microbiology, Washington, D. C., 1975, p. 90.
304. Thompson, C. G., Scott, D. W., and Wickman, B. E. Long term persistence of the nuclear polyhedrosis virus of the Douglas-fur tussock moth, *Orgyia pseudotsugata* (Lepidoptera: Lymantriidae) in forest soil. *Environ. Entomol.* 10, 254, 1981.
305. Weseloh, R. M. and Andreadis, T. G. Laboratory assessment of forest microhabitat substrates as sources of the gypsy moth nuclear polyhedrosis virus. *J. Invertebr. Pathol.* 48, 27, 1986.
306. Young, S. Y. and Yearian, W. C. Movement of a nuclear polyhedrosis virus from soil to soybean and transmission in *Anticarsia gemmatalis* (Lepidoptera: Noctuidae) populations on soybean. *Environ. Etomol.* 15, 573, 1986.
307. Olofsson, E. Dispersal of the nuclear polyhedrosis virus of *Neodiprion sertifer* from soil to pine foliage with dust. *Entomol. Exp. Appl.* 46, 181, 1988.
308. McLeod, D. J., Yearian, W. C., and Young, S. Y. Inactivation of *Baculovirus heliothis* by ultraviolet irradiation, dew, and temperature. *J. Invertebr. Pathol.* 30, 237, 1977.
309. Young, S. Y., Yearian, W. C., and Kim, K. S. Effect of dew from cotton and soybean foliage on activity of *Heliothis* nuclear polyhedrosis virus. *J. Invertebr. Pathol.* 30, 237, 1977.
310. Elleman, C. J. and Entwistle, P. F. Inactivation of a nuclear polyhedrosis virus on cotton by the substances produced by the cotton leaf surface glands. *Ann. Appl. Biol.* 106, 83, 1985.
311. Ehler, L. E. Natural enemies of cabbage looper on cotton in the San Joaquin Valley. *Hilgardia* 45, 73, 1977.
312. Bird, I. T. and Elgee, D. E. A virus disease and introduced parasites as factors controlling the European spruce sawfly, *Diprion hercyniae* (Htg) in central New Brunswick. *Can. Entomol.* 89, 371, 1957.
313. Evans, H. F. and Entwistle, P. F. The development of infection during a virus epizootic in spruce sawfly populations in mid-Wales, in *Proc. 1st Int. Colloq. Invertebr. Pathol.*, Kingston, Canada, 1976, p. 350.
314. Ignoffo, C. M. Strategies to increase the use of entomopathogens. *J. Invertebr. Pathol.* 31, 1, 1978.
315. Bedford, G. O. Biology, ecology, and control of palm rhinoceros beetles. *Ann. Rev. Entomol.* 25, 309, 1980.
316. Zelazny, B. Transmission of a baculovirus in populations of *Orcytes rhinoceros*. *J. Invertebr. Pathol.* 27, 221, 1976.

# 2

# Baculoviridae: Subgroup B

## Comparative Aspects of Granulosis Viruses

NORMAN E. CROOK

*Horticulture Research International, West Sussex, England*

## INTRODUCTION

The granulosis viruses (GVs) (Baculoviridae: subgroup B) are a large group of occluded viruses isolated from larvae of insect species all within the order Lepidoptera. They are closely related to the nuclear polyhedrosis viruses (NPVs) but are easily distinguished on the basis that GVs contain a single virion within each capsular occlusion body (OB), whereas NPVs contain many virions within a much larger OB, which is usually polyhedral in shape. Baculoviruses have been isolated only from arthropods and are unusual among invertebrate viruses in having no known related vertebrate viruses.

Diseases caused by GVs were first described by Paillot (1,2) in larvae of *Pieris brassicae* and *Agrotis segetum*. Steinhaus (3) first applied the name "granulosis" to the disease when he examined cells of infected *Peridroma margaritosa* and observed minute granules under the light microscope. Confirmation that these diseases were caused by viruses was obtained by Bergold (4), who examined diseased *Choristoneura murinana* by electron microscopy and showed both the rod-shaped virions and capsular OBs. Since that time, the number of insect species from which GVs have been isolated has steadily increased and is currently around 150 (5,6).

One of the main reasons for studying these viruses has been their considerable potential for use as pest control agents. Their narrow host range, high infectivity, and high mortality in the host larvae are ideal characteristics for a microbial pesticide. The very high levels of virus obtained from diseased larvae allow it to be produced economically and the occluded nature of the virus makes it very stable during long-term storage and conventional spray application techniques. Moreover, in view of the very narrow host range of baculoviruses and the absence of any deleterious effects in extensive safety studies (7), it appears that the environmental risk of using these viruses, especially in terms of any hazard to vertebrates, is extremely small.

Since the 1970s, extensive field trials have been carried out in several countries by many workers using the GVs from *Pieris rapae* (PrGV) and *Cydia pomonella* (CpGV) [reviewed by Payne (8,9)]. More limited trials have been conducted with GVs from several other species, e.g., *Phthorimaea operculella* (10), *A. segetum* (AsGV) (11), *Plodia interpunctella* (PiGV) (12–14), *Lacanobia oleracea* (LoGV) (15), *Homona magnanima*, *Adoxophyes orana* (16), and *C. murinana* (17). (See Table 1 for list of abbreviations for viruses.) Although many of these trials have produced very encouraging results, only CpGV has thus far been produced commercially. Even in this case, although at least five companies in Europe and North America became involved in its development and production at varying times during the 1980s, the virus has not been widely available commercially. In 1989 it finally received full registration in West Germany and Switzerland, and it is hoped this will encourage further commercial development.

In countries outside Europe and North America the relative costs of biological and chemical pesticides more frequently favor the former, and since safety testing and registration of biological insecticides are frequently easier, several baculoviruses are widely used. Among the GVs, PrGV is probably the most extensively used with over 5000 ha per annum of cabbage being treated in China (18); in this instance extensive safety testing was carried out prior to large-scale use.

Despite limited commercial development to date, the continuing trend toward reduced chemical treatment of crops, the increasing safety standards for chemical pesticides together with increasing costs of development of new products, and the development of resistance in insects provide a favorable climate for the continuing study of viral pesticides. Further details of the current status of the insecticidal use of baculoviruses worldwide have recently been provided by Huber (19) and Cunningham (20).

Although many GVs have now been isolated, relatively few have been studied in detail and most of those that have been well characterized are either among those previously mentioned which have been used in field trials or those isolated from other major lepidopterous pests such as *Heliothis* spp., *Spodoptera* spp., and *Trichoplusia* spp. This chapter therefore deals with work relating to this relatively small number of GVs.

## CLASSIFICATION

Although subgrouping of the Baculoviridae into NPVs, GVs, and nonoccluded viruses is straightforward, there has been no attempt at any sort of classification of the 150 or so viruses within the granulosis subgroup. Most of these viruses have been identified only on the basis of their morphology, which is very similar

**Table 1** Abbreviations Used for Some Granulosis Viruses

| Host species | Abbreviation |
| --- | --- |
| *Agrotis segetum* | AsGV |
| *Cadra cautella* | CcGV |
| *Cydia pomonella* | CpGV |
| *Estigmene acrea* | EaGV |
| *Heliothis armigera* | HaGV |
| *Lacanobia oleracea* | LoGV |
| *Pieris brassicae* | PbGV |
| *Pieris (Artogeia) rapae* | PrGV, ArGV |
| *Plodia interpunctella* | PiGv |
| *Pseudaletia unipuncta* | PuGV |
| *Scotogramma trifollii* | StGv |
| *Spodoptera frugiperda* | SfGV |
| *Trichoplusia ni* | TnGV |

for all GVs. Limited numbers of GVs have been compared by more discriminating techniques such as serology, polypeptide analysis, DNA hybridization, and restriction enzyme analysis. These studies have not revealed any particular similarities and dissimilarities within the subgroup. However, where a large number of isolates of a single virus (PrGV) have been examined (21), it has been possible to arrange them into subtypes on the basis of their restriction profiles. These subtypes correlated closely with the geographic origins of the isolates, depending on whether they came from North America, Europe, Asia, or Australia.

Since most baculoviruses have been classified solely on the basis of their morphology and the host species from which they were isolated, they have been named on the basis of this information. This is informative in one respect, yet it suffers from a number of disadvantages, in particular, (1) the name of a virus can change from time to time and be different in different countries, depending on the views of insect taxonomists in classifying the host species, (2) the same virus may be isolated from more than one host species and thus have more than one name, and (3) more than one virus may be isolated from the same species and thus quite different viruses have the same name. For example, GVs isolated from *Pieris brassicae*, *Artogeia rapae*, and *A. napi* have been shown to be variants of the same virus (21,22). Moreover, some authorities consider *Artogeia* to be merely a subgenus within the genus *Pieris*. Thus a variety of different names are commonly used for the same virus. On the other hand, two isolates of *Pseudaletia unipuncta* (PuGV) have been shown to be unrelated (23). Naming of virus isolates must therefore not be confused with identification of the virus.

Undoubtedly the best method of identification at present is by restriction enzyme analysis of the viral DNA, since it not only readily distinguishes different viruses but also discriminates between different isolates of the same virus and even between different genotypes present in a single isolate (24). Restriction profiles using at least one enzyme have been published for most of the more commonly studied GVs (see Table 2 below) and thus provide a basis by which to compare any new isolates of viruses from these species.

# STRUCTURE AND PHYSICOCHEMICAL PROPERTIES

## Morphology

Granulosis viruses show relatively little variation in either shape or size. The OBs, or capsules, are generally ovocylindrical and range in size from 300 to 500 nm in length by 120 to 300 nm in width (25,26). Electron microscopy of purified virus preparations or virus-infected tissue (Figure 1) shows that GVs are structurally complex with a rod-shaped nucleocapsid surrounded by an envelope and embedded within a crystalline protein matrix referred to as granulin. An outer envelope, or epicapsular layer, has been observed on several GVs (27–29). It appears to contain protein components distinct from granulin (30) and may be similar to the polyhedrin membrane in NPVs, which is reported to be composed of carbohydrate (31) and a phosphorylated protein distinct from polyhedrin (32). A thorough and detailed account of the ultrastructure of GVs and other baculoviruses can be found in a recent review by Federici (33). Unusual morphological forms have been noted for a number of GVs. Arnott and Smith (34) observed rectangular and greatly elongated OBs for PiGV and also reported capsules containing two virus particles as well as virus particles containing up to nine nucleocapsids. The incidence of these unusual forms was described as rare, but similar forms have since been observed for several other GVs including *Archips argyrospila* GV (35), which had enlarged rugose capsules and virus particles containing up to four nucleocapsids, *Plutella xylostella* GV (36), LoGV (15, 37), where multiple nucleocapsid forms occurred in 1–2% of capsules, and CpGV, where angular and large cuboidal OBs have been observed (38) and multiple nucleocapsid forms commonly occurred in midgut tissues with up to 22 nucleocapsids within a single envelope (39). Long virus particles were occasionally found in preparations of GV from six different species examined by Ackermann and Smirnoff (26) and extremely long and sometimes highly branched rods have been found associated with GVs in some noctuid larvae (40), in *P. rapae* (40), and in *Cryptophlebia leucotreta* (41), but the nature and origin of these rods have not been established. Hunter and Hoffmann (42) compared *Cadra cautella* GV (CcGV) from the homologous host and also from an alternative host, *P. interpunctella*. Many aberrant forms were observed in the virus from *P. in-*

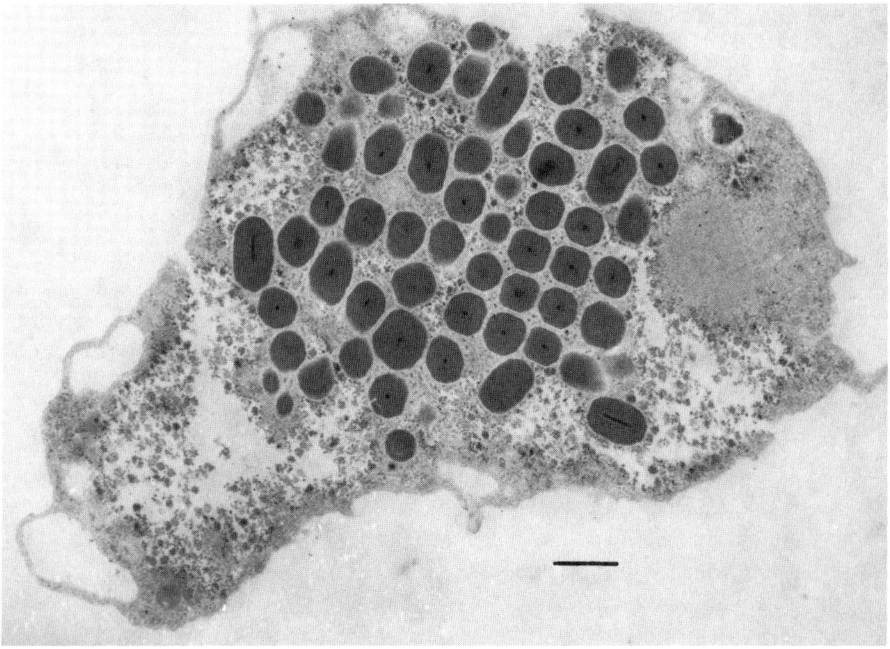

**Figure 1** Electron micrograph of section through CpGV-infected *C. pomonella* cell cultured in vitro. Bar = 500 nm. (Courtesy of Dr. D. Winstanley.)

*terpunctella* larvae, but when this virus was passaged again through *C. cautella* the morphology returned to normal. This suggests that morphological aberrations may be host or cell mediated rather than caused by any mutation in the virus.

## Structural Components and Protein Characterization

### Isolation

The major structural components of GVs can be isolated by successive treatments of OBs with dissolution agents. In general, OBs are dissolved by exposure to high pH, e.g., sodium carbonate solution, and the released virus particles are banded on sucrose gradients. Treatment of virus particles with the non-ionic detergent Nonidet P40 (NP40), usually at 1% concentration, removes the viral envelope leaving intact nucleocapsids which can be purified on sucrose gradients. Finally, the core proteins and DNA can be extracted by incubation with 0.6–1.0 *M* sodium chloride at pH 8.5, leaving only tubular capsids. These procedures have been well established for many years and have been reviewed by Tweeten et al. (43).

## Occlusion Body Proteins

The matrix protein of the OB, called granulin, has been well characterized both biochemically and by cloning and sequencing of the gene from some GVs. Early estimates of the molecular weight of the granulin from several GVs by polyacrylamide gel electrophoresis varied from 26,300 to 29,100 (43) but more recent sequencing data for the granulins from *Trichoplusia ni* GV (TnGV) (44), *P. brassicae* GV (PbGV) (45), and CpGV (46) gave molecular weights of 29,286, 29,318, and 28,188, suggesting that there may be very little variation in the size of these proteins. Low levels of phosphorylation have been detected in the granulins of TnGV (47) and PiGV (48), but whether this occurs at specific residues in the molecule is unknown. There is no evidence for glycosylation of granulins.

The OBs of several GVs have been shown to contain proteins that enhance baculovirus infectivity (49,50). GVs from *P. unipuncta*, *Scotogramma trifolii* (StGV), and *Spodoptera exigua* enchanced GV and NPV infections of *P. unipuncta*, *T. ni*, and *S. exigua* (49). The enhancing factor (referred to as synergistic factor) from PuGV had an $M_r$ of 126,000 and contained phospholipid, the removal of which with phospholipase C caused complete loss of activity (51). The protein appeared to facilitate the attachment of virus particles to the plasma membrane of midgut cells. A protein present in the granulin fraction of TnGV has been shown to enhance the infectivity of *Autographa californica* multiple nucleocapsid NPV (AcMNPV) (50). This same fraction also caused the loss of three glycoproteins from the peritrophic membrane of *T. ni* larvae, making it very fragile, though the significance of this in relation to increased larval infection was unknown. Preliminary studies indicated that the enhancing factor was a protein of $M_r$ 101,000. Isolation of genes coding for these enhancing factors and insertion of such a gene into the genome of baculoviruses lacking this factor (so that it is expressed and incorporated into the OB) may provide a means of increasing the infectivity of some viruses.

A third protein apparently associated with the OB matrix is an alkaline protease which causes rapid breakdown of the granulin when it is dissolved in sodium carbonate (15,47,52–54). It appears likely that this protease aids dissolution of the GV capsules, and release of virus particles, in the alkaline conditions of the larval midgut. Tweeten et al. purified the protease from PiGV by a combination of ion exhange and gel filtration chromatography (52) and showed it to have an $M_r$ of 14,000 and a pH optimum of 10.5. In contrast, Langridge and Balter found a much lower pH optimum for *Estigmene acrea* GV (EaGV) of pH 8.2 and only very low activity above pH 8.7 (54). Although the protease appears to be an integral part of the OB, two pieces of evidence suggest that it is acquired from the insect host rather than being a virus-coded protein. Nagata and Tanada (53) were able to purify PuGV with greatly reduced protease activity by removing the digestive tracts from infected larvae prior to virus purification, indicating that the protease may be acquired from the digestive juices during virus purification. Moreover, it was found that a baculovirus produced in cultured cells did not contain a protease, but when the same virus was produced in vivo the protease was present (55).

Virus Particle Proteins

The structural polypeptides present in virus particles have been analyzed by SDS polyacrylamide gel electrophoresis for a number of GVs including isolates from *Spodoptera frugiperda* (SfGV), *T. ni* (56,57), *P. brassicae* (22,58), *P. rapae* (22), *P. unipuncta* (59), *P. interpunctella* (48,57), *H. armigera* (57), *L. oleracea* (15), *C. pomonella* (60), *Hyphantria cunea*, *Diacrisia virginica* (61), and three *Choristoneura* spp. (62). Between 12 and 35 polypeptides have been detected varying in $M_r$ from 12,000 to 200,000. For several of these viruses, polypeptide profiles have also been obtained from nucleocapsids and show the presence of up to eight bands varying in $M_r$ from 12,000 to 73,000 (22,48,58,59,63). A direct comparison between virus particles obtained from six GVs is shown in Figure 2, from which it is clear that there is very little similarity between profiles obtained from different viruses. However, where variants of the same virus have been compared, such as with *Pieris* GVs, *Choristoneura* GVs, CpGV, and LoGV, the profiles have shown only a few differences. In PbGV and PrGV there were small differences in the $M_r$ of three of the polypeptides, all of which appeared to occur in the virus particle envelope since they were not present in the nucleocapsids (22). GVs from *Choristoneura fumiferana*, *C. occidentalis*, and *C. retiana* showed differences only between two of the larger major polypeptides ($M_r$ 162,000–182,000) and in some of the minor polypeptides ranging in $M_r$ from 116,000 to 164,000 (62). There were no significant differences noted between two variants of LoGV (15) but with CpGV, one variant, CpGV-R, lacked two low molecular weight polypeptides present in CpGV-M and showed a small increase in the molecular weight of a third polypeptide (60).

Despite the great variation between the structural proteins of different GVs, two major virus particle proteins appear to be present in many GVs. One of these is usually in the region of 12,000–14,000 $M_r$. This protein (VP12) has been isolated from PiGV and is extremely basic due to its high content of arginine (27%) and histidine (12%) (63). The protein occurs in the nucleocapsid core and appears to bind to the virus DNA (63–65). The second major protein common to most GVs has an $M_r$ of about 30,000–32,000 and is also present in the nucleocapsid. This protein is the major component of empty capsids of TnGV (56) and PiGV (63) and is presumably the major structural protein of other GV capsids.

Also associated with purified capsids of PiGV is a protein kinase activity

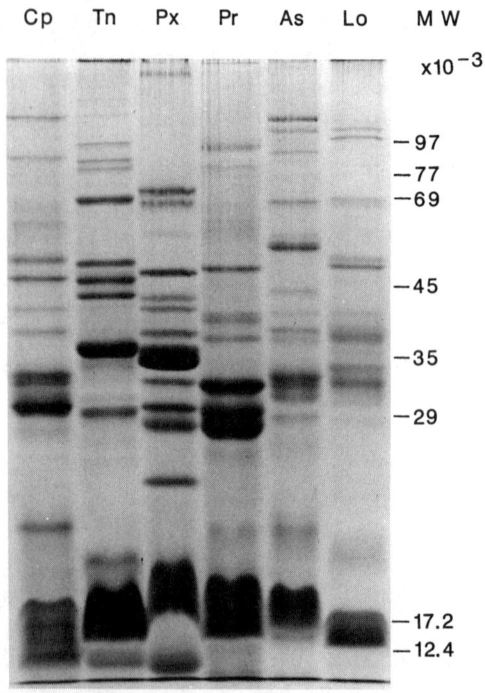


**Figure 2** Electrophoresis on 11% polyacrylamide gel of virus particle proteins obtained from CpGV, TnGV, PxGV, PrGV, AsGV, and LoGV. The position and size of molecular weight markers are indicated.

which has been shown to catalyze phosphorylation of VP12 (66). Wilson and Consigli (65) proposed that this enzyme became activated when the nucleocapsid aligned with the nuclear pore, thus phosphorylating VP12 and causing decondensation of the DNA and its release into the cell nucleus to initiate infection.

## Serology

Serological techniques have been used for the detection and identification of GVs, for studying the relationship between different GVs and between their structural components, and for detecting viral replication in cell cultures (discussed later). Techniques used have included immunodiffusion, immunoelectrophoresis, radioimmunoassay, enzyme-linked immunosorbent assay, and radioimmune protein blotting.

Immunodiffusion provides a simple means of studying the relationship between different viruses but the results can be difficult to interpret and can vary, depending on the pretreatment of the immunogen and the antigen. Studies using TnGV (67) and PrGV (22) showed that different precipitin line patterns were obtained, depending on whether the alkaline protease, present in OBs, had been inactivated, but no such differences were detected with PuGV (68). However, differences were found in immunodiffusion with PuGV using antisera produced against either whole capsules or dissolved capsule proteins. Antiserum to the former had a narrower specificity and gave sharper precipitin lines than antiserum to the latter (68). Purified granulin from TnGV has been shown, by immunodiffusion, to have at least three different antigenic determinants (67), of which only one cross-reacted with other granulins and polyhedrins. Cross-reactivity, though not complete identity, has also been shown for CpGV and PbGV using antiserum to CpGV (69).

One- and two-dimensional immunoelectrophoresis has been used to compare a number of GVs (59,68,70). The single-dimension technique was used to confirm the identity of progeny virus obtained from several insect species infected with PuGV (68). This method allows two samples to be run side by side, making it easy to compare different isolates. Two-dimensional electrophoresis was used in one study (70) to compare GVs from *Ephestia cautella*, *Spodoptera littoralis*, *P. brassicae*, *C. pomonella*, and two strains of *P. interpunctella* using both homologous and heterologous antisera. A considerable amount of cross-reactivity was observed but the patterns obtained with different viruses were distinct and even the two strains of PiGV could be distinguished. It was also found that peaks of precipitation were more clearly defined when viral antigens were succinilated. The technique has also been used to compare the capsule proteins of two isolates of PuGV (71) and to analyze the virion envelope from one of these strains (59).

Summers and Hoops (67) used a radioimmunoassay (RIA) to quantify antigenic relationships between OB proteins from TnGV, SfGV, and a number of NPVs. In a competitive inhibition RIA, they found that the amount of purified OB protein required to produce 50% inhibition of binding of $^{125}$I-TnGV granulin to homologous antisera was only slightly greater for SfGV (9.5 ng) than for TnGV (7 ng) but was much greater for all the NPVs used, varying from 40 ng (AcMNPV) up to 4 $\mu$g [*Heliothis armigera* MNPV (HaMNPV)]. This indicated a much closer relationship between the two granulins than between TnGV granulin and any of the polyhedrins.

A detailed comparison of the antigenic cross-reactivity of virus particle polypeptides from GVs and NPVs was carried out by Smith and Summers (57) using a protein-blot RIA. Purified virus particles were electrophoresed in polyacrylamide gels, blotted onto nitrocellulose, and allowed to react with antisera to virus particles from TnGV and various NPVs. Antibody binding was detected with $^{125}$I-labeled protein A. The TnGV antisera reacted with seven polypeptide bands from TnGV, six from SfGV, two from *H. armigera* GV (HaGV), one from PiGV, and also one from each of AcMNPV, *Rachiplusia ou* MNPV, and HaMNPV; there was no reaction with seven other MNPVs or three SNPVs. However, the only cross-reactions between antisera to NPVs and GV polypeptides were obtained with single nucleocapsid NPV (SNPV) antisera. Antisera to *Heliothis zea* SNPV reacted with SfGV (two bands) and HaGV (one band) and antisera to *T. ni* SNPV (TnSNPV) reacted with HaGV and PiGV (one band each).

A similar analysis was used to compare three isolates of CpGV but no serological differences between isolates were observed (60).

Crook and Payne studied the cross-reactivity of different GVs using three different enzyme-linked immunosorbent assay (ELISA) methods (72). The double antibody sandwich ELISA gave the highest specificity and very low levels of cross-reactivity were observed for heterologous viruses using antisera to PbGV and AsGV capsules with dissolved OBs of PbGV, AsGV, CpGV, and *S. littoralis* NPV as antigen. In the direct ELISA, rather higher levels of cross-reaction were observed; in the indirect method, there was no discrimination at all between the four viruses with any of the antisera unless they were used at low concentrations (1 μg IgG/ml). The double antibody sandwich ELISA was also shown to be the best method for detecting virus in crude larval homogenates (72) and has been used as a rapid means of quantifying CpGV in commercial production batches (73).

## MOLECULAR BIOLOGY AND GENETICS

### Physical Characterization and Restriction Enzyme Analysis of DNA

Early studies on the size, conformation, and GC content of the nucleic acid of GVs made use of a variety of techniques including chemical analysis, melting point, electron microscopy, reassociation kinetics, and centrifugation in neutral or alkaline sucrose gradients or in cesium chloride containing ethidium bromide or propidium iodide. These studies, reviewed by Harrap and Payne (74), showed that the GV genome was a very large molecule of predominantly circular double-stranded DNA containing both supercoiled and open circular forms and with a GC content varying from 34.8 to 50%. Estimates of the molecular size of the DNA using these methods varied considerably even for the same virus isolate and it is only since restriction enzymes have been used to analyze these DNAs and especially where physical maps have been produced that reliable and accurate determination of molecular sizes has been achieved. The digested DNA is electrophoresed in an agarose gel alongside sequenced DNA fragments to provide suitable size standards. Total genome size is obtained by summing the sizes of all the fragments. Although the size of most individual fragments can be calculated with considerable accuracy, errors in the overall size estimate can still occur because (1) very small fragments are not being detected, (2) there is difficulty in estimating sizes of very large fragments, and (3) several fragments migrating very close together can be difficult to interpret. Analysis of several profiles using different enzymes or double digests helps to reduce these errors, but the most reliable genome size estimates are obtained when a physical map is produced using several enzymes.

To date, restriction profiles for GVs isolated from about 12 host species have been published (Table 2). With the exception of EaGV, all these reports have used at least EcoRI. For the purpose of identification of the virus being used, it is highly desirable that all studies have at least one enzyme in common and EcoRI is clearly the enzyme of choice, not only because it is very readily and cheaply available but also because with all GVs examined thus far, it provides a suitable number (between 12 and 27) and size range of fragments to make accurate analysis and comparison with other viruses very straightforward. BamHI (or BstI) and HindIII have also commonly been used for similar reasons, although CpGV is very unusual in having only two HindIII sites. Where used, KpnI has also given a good range of fragments, whereas SalI has given rather a lot of smaller fragments. Enzymes with a 6 base recognition sequence comprising only Gs and Cs (ApaI, BglI, SacII, SmaI) have provided very few sites, as would be anticipated in view of the low GC content of most GVs. For example, PbGV, with a GC content of 37% (58), has no ApaI site, one SmaI site, and two BglI sites (21); a related isolate from *P. rapae* with similar sites for these three enzymes (21) has five SacII sites (78).

Restriction enzyme analysis has shown that different isolates of GV from the same host species are frequently distinct genotypes. Small differences have been shown between two isolates each from *L. oleracea* (15) and *S. trifolii* (23) and three isolates from *C. leucotreta* (86). Fifteen isolates of GV from *P. brassicae* and *A. rapae*, collected in four continents, have been compared (21) and only three, all from one area of Queensland, were indistinguishable. Considerable variability was found between some of these isolates with less than 70% similarity in restriction profiles (22). The degree of dissimilarity correlated closely with the separation of geographic locations from where the isolates were obtained. In addition to this regional variation, a number of the individual isolates comprised a mixture of variants (21,78) and up to eight different genotypes have been purified, by in vivo cloning, from a single isolate (24), suggesting that the total number of genotypes of this virus is very large and could well exceed a hundred.

By contrast, a number of isolates of CpGV from several countries were all found to be indistinguishable when analyzed with EcoRI (83). This may, however, be simply because a single strain of CpGV has been used in all these countries as a pesticide and has then been reisolated. Nevertheless, two isolates, one from Russia (CpGV-R) and one from England (CpGV-E), did show small differences

**Table 2** GVs Analyzed Using Restriction Enzymes

| Host insect species | Number of isolates | Enzymes used[a] | Genome size[b] | Reference |
|---|---|---|---|---|
| *Trichoplusia ni* | 1 | E | 174.8 | 75 |
| | 1 | E | 176.5 | 76 |
| *Spodoptera frugiperda* | 1 | E | 121.7 | 75 |
| | 1 | E | 125.3 | 76 |
| *Plodia interpunctella* | 1 | E,B,H | 110.6 | 77 |
| | 1 | E | 103.9 | 76 |
| *Pieris* spp. | 1 | E,B,H. | 117.5 | 77 |
| | 2 | E,B.H | | 22 |
| | 15 | E,B,H,X,Sm,P,Bg | | 21 |
| | 1 | E,B,H,K,Sa,Sc | 106.5 | 78 |
| | 4 | E,B,H,K | 109.4–110.1 | 79 |
| | 1 (8 genotypes) | E,K,H. | | 24 |

| | | | | |
|---|---|---|---|---|
| *Heliothis armigera* | 1 | E | 165.9 | 76 |
| | 1 | E | 136.2 | 80 |
| | 1 | E[c] | 136.4 | 81 |
| *Lacanobia oleracea* | 2 | E | 125.2 | 15 |
| *Estigmene acrea* | 1 | B,H | 114.6 | 82 |
| *Cydia pomonella* | 1 | E | 126.9 | 80 |
| | 3 | E,B,H,K,X,Sl | 118.1–120.0 | 60 |
| | 3 | E,B,H,X,Sm,A | 123.6–126.8 | 83 |
| *Phthorimaea operculella* | 1 | E | 107.6 | 80 |
| *Argotis segetum* | 1 | E | 132.1 | 84 |
| | 1 | E,B[c] | 114 | 85 |
| *Pseudaletia unipuncta* | 2 | E,B,H,Sl | 105.5, 133.9 | 23 |
| (*Scotogramma trifolii*)[d] | 1[e] | E,B,H,Sl | 130.7 | 23 |
| *Choristoneura* spp. | 3 | E,B,H,K | 107 | 62 |

[a]Abbreviations: E, EcoRI; B, BamHI or BstI; H, HindIII; K, KpnI; X, XhoI; P, PstI; Bg, BglI; Sm, SmaI; Sa, SacII; Sc, ScaI; Sl, SalI; A, ApaI.
[b]Where given in $M_r$, converted to kbp using $M_r$ $1 \times 10^6 = 1.575$ kbp.
[c]Fragment sizes given but gel profiles not illustrated.
[d]The virus was obtained from *P.unipuncta* following infection with a GV isolated from *S. trifolii*.
[e]A second isolate referred to but not shown.

from the original Mexican strain (60,83). CpGV-E was shown to be a mixture of two genotypes, one of which (CpGV-E1) was indistinguishable from CpGV-M while the other (CpGV-E2) had small differences in the EcoRI and BamHI profiles.

Only one insect species, *P. unipuncta*, has thus far been found to be the natural host for two GVs that show no similarity in their restriction profiles (23).

## Genome Maps and Phenotypic Variation

More detailed characterization of genomes requires construction of physical maps to identify the position of restriction fragments along the DNA molecule. Maps have been produced for CpGV (83), AsGV (85), and PrGV (24,78,79).

Crook et al. (83) mapped the three genotypes of CpGV with five enzymes, giving a total of 36 (CpGV-M and CpGV-R) or 37 sites (CpGV-E2). As expected from the restriction profiles, differences between the three maps were small. CpGV-R contained all the sites mapped for CpGV-M but had a deletion of 2.4 kbp somewhere between 50 and 65 kbp from the zero point (the granulin gene). All the differences found in GpGV-E2 relative to CpGV-M occurred between 12 and 34 kbp from the zero point and consisted of a shift in position of a BamHI site, an additional EcoRI site, and in total about 1 kbp more DNA. It would be especially interesting to identify any phenotypic differences between these isolates since the genotypic differences appear to be restricted to small areas of the genome, though undetected minor differences may occur elsewhere. Infectivity differences of one to two orders of magnitude were reported by Harvey and Volkman (60) between CpGV-M and CpGV-R but Crook et al. (83) failed to find any significant difference between these isolates. A comparison of CpGV polypeptides by Harvey and Volkman (60) showed CpGV-M contained two polypeptides, of $M_r$ 18,000 and 14,000, which are absent in CpGV-R. The only other difference between the remaining 22 structural polypeptides detected in the two isolates was that one of $M_r$ 90,000 in CpGV-M increased to 92,000 in CpGV-R. The absence of the two small polypeptides in CpGV-R may be due to the 2.4 kbp deletion in the genome. The function of these proteins, and any other gene products coded for in this region which is deleted in CpGV-R, is unknown but clearly must be nonessential for the survival of the virus. Harvey and Volkman (60) suggested that one of these polypeptides may be involved in modifying or clipping the 92,000 $M_r$ polypeptide into one of $M_r$ 90,000, explaining why this polypeptide is larger in CpGV-R. Alternatively, the increased size of the larger polypeptide may be due to an additional difference in the genome—the increase in $M_r$ of 2000 would require an insertion of about 50 bp of DNA, which could go undetected in a comparison of restriction profiles.

Although Qiu et al. (85) constructed a simple map for AsGV, very little molecular biology has been done on the virus. The genome was mapped for

BamHI with a total of 13 sites (though the sizes of only 12 BamHI fragments were listed) but the granulin gene was not located. Since no restriction profiles were shown, it is difficult to tell whether this Chinese isolate was similar to the isolate from Denmark, an EcoRI profile of which was provided by Allaway and Payne (84).

In contrast, GV isolates from *P. rapae* and *P. brassicae* have been the most intensively studied GVs and maps of about 20 variants have been produced (24,78,79). Variation in the form of insertions and/or deletions and restriction site changes occurs at many points throughout the genome, although in a comparison of four isolates, Smith and Crook (79) found that 13 of a total of 17 differences detected occurred in a region occupying only 30% of the genome, between 22 and 55 kbp from the zero point (granulin gene). A more extensive comparison of seven isolates seen in Figure 3 shows that even with the less closely related isolates (ArGV5, 6, and 9) certain regions accounting for less than half the genome, i.e., 17–51, 78–80, and 92–108 kbp along the map, contain over 85% of the differences. Different genotypes from a single isolate have also been mapped (Figure 4) and again all the variation apart from the additional EcoRI site at 58 kbp occurred in these same regions. It was suggested (24) that this genotypic diversity may have arisen by recombination between a smaller number of variants, possibly in this particular case even within the single larva from which all the genotypes were isolated.

None of the variation found in PrGV caused any significant effect on its infectivity for *P. rapae* but did cause a dramatic change in infectivity for *P. brassicae* (21). Two isolates were moderately infectious for *P. brassicae* with $LD_{50}$ values of $10^{5.8}$ and $10^{5.9}$ for fourth instar larvae. All other isolates caused few or no deaths in *P. brassicae* at doses up to 1000-fold greater. Only a single map difference correlated in all genotypes with the difference in infectivity; i.e., the additional HindIII site at 22 kbp was present in all genotypes noninfectious for *P. brassicae*, but there were no grounds to suggest that this change was biologically significant.

## Homology Between Virus Genomes

A precise measure of homology can be obtained only by a comparison of sequences such as has been done with a few granulin genes (discussed later). Homology between closely related isolates has been estimated from the percentage of restriction fragments which comigrate in agarose gels using a formula devised by Upholt (87). However, this formula takes into account variation due to random point mutations only and does not allow for changes due to insertions and deletions, which are a frequent source of variation in GVs (24,79,83). Nevertheless, viruses that have recognizably similar restriction profiles will have a very high degree of homology, generally well over 90%. Most GV isolates

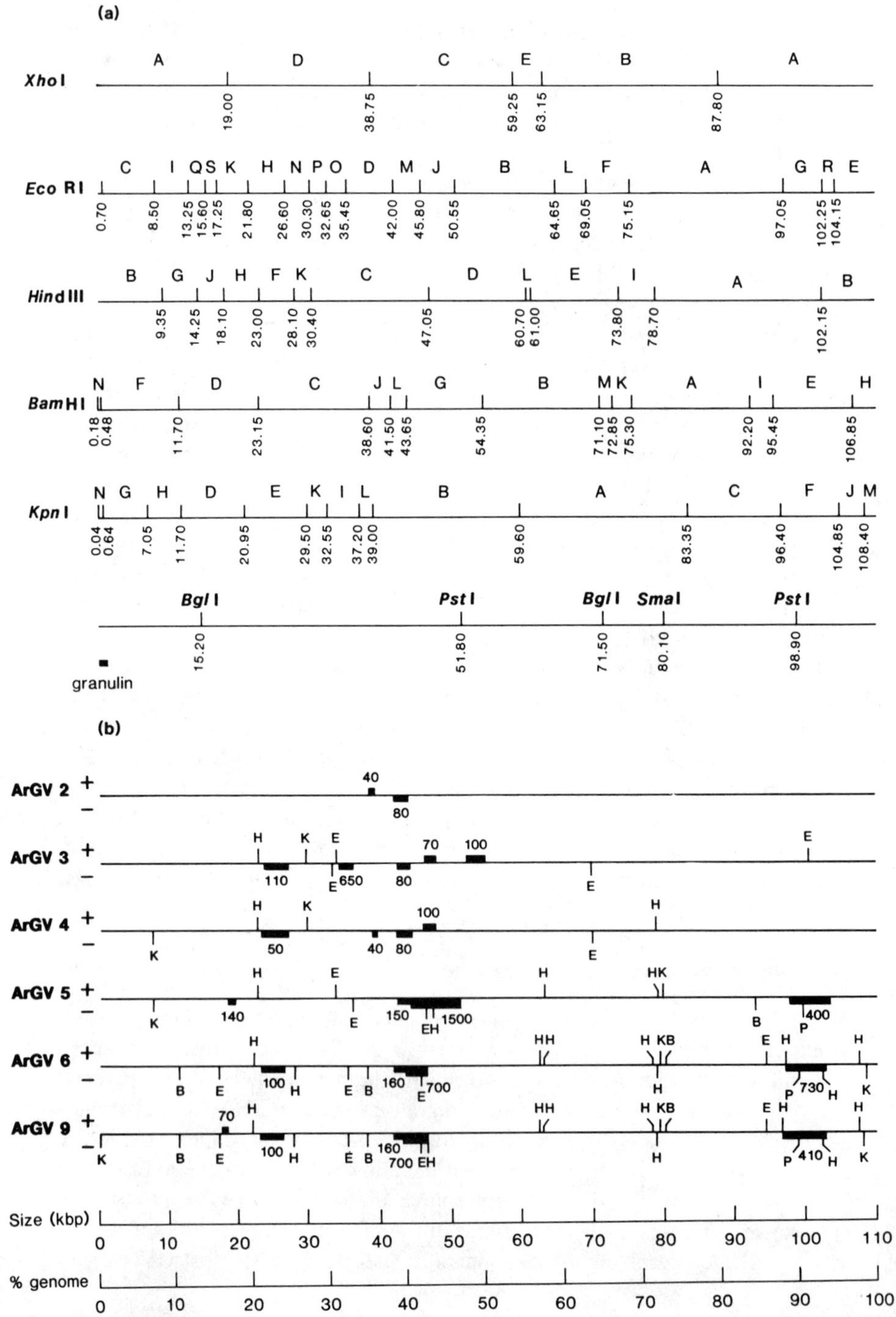
(a)
Xho I
A D C E B A
19.00 38.75 59.25 63.15 87.80
Eco RI
C I Q S K H N P O D M J B L F A G R E
0.70 8.50 13.25 15.60 17.25 21.80 26.60 30.30 32.65 35.45 42.00 45.80 50.55 64.65 69.05 75.15 97.05 102.25 104.15
Hind III
B G J H F K C D L E I A B
9.35 14.25 18.10 23.00 28.10 30.40 47.05 60.70 61.00 73.80 78.70 102.15
Bam HI
N F D C J L G B M K A I E H
0.18 0.48 11.70 23.15 38.60 41.50 43.65 54.35 71.10 72.85 75.30 92.20 95.45 106.85
Kpn I
N G H D E K I L B A C F J M
0.04 0.64 7.05 11.70 20.95 29.50 32.55 37.20 39.00 59.60 83.35 96.40 104.85 108.40
Bgl I Pst I Bgl I Sma I Pst I
15.20 51.80 71.50 80.10 98.90
granulin
(b)
ArGV 2
40 80
ArGV 3
H K E 70 100 E
110 E 650 80 E
ArGV 4
H K 100 H
K 50 40 80 E
ArGV 5
H E H HK
K 140 E 150 EH 1500 B 400 P
ArGV 6
H HH H KB E H H
B E 100 H E B 160 E 700 H P 730 H K
ArGV 9
70 H HH H KB E H H
K B E 100 H E B 160 700 EH H P 410 H K
Size (kbp)
0 10 20 30 40 50 60 70 80 90 100 110
% genome
0 10 20 30 40 50 60 70 80 90 100

obtained from the same host species have had restriction profiles showing many comigrating fragments. Around 90% of fragments comigrated for different isolates of GVs obtained from *L. oleracea* (15), *C. pomonella* (60,83), *S. trifolii* (23), and *C. leucotreta* (88), although greater variation was found between GV isolates from *P. rapae* with some isolates having less than 70% comigrating fragments (21,22).

Homology can also be estimated from the degree of hybridization that occurs between different DNAs, particularly under differing conditions of temperature and formamide concentration. Different isolates of CpGV (60) and StGV (23) examined by Harvey and colleagues showed complete homology by Southern blot hybridization under stringent conditions (65°C, no formamide). StGV also hybridized weakly to one isolate of PuGV from Oregon (PuGVO) but not to another isolate from Hawaii (PuGVH) (23). Conversely, PuGVO, but not PuGVH, hybridized weakly to StGV. Most interestingly, the two isolates of PuGV did not cross-hybridize at all, even under conditions of lower stringency (58°C, no formamide), showing that these two isolates are different viruses (23). However, Smith (89) found that at least a few and sometimes most EcoRI fragments from 10 different GVs hybridized to PrGV DNA under conditions of low stringency (37°C, 35% formamide). No hybridization of AcMNPV DNA was detected under these conditions. The PrGV EcoRI-E fragment containing most of the granulin gene hybridized to one or more EcoRI or BamHI fragments from all of the 10 GVs but not to AcMNPV (89), indicating that there is a greater homology between many granulin genes than between granulin and polyhedrin genes. At still lower stringency (42°C, 20% formamide), Dwyer and Granados (78) detected hybridization between 7 AcMPNV EcoRI fragments (A, D, E, F, I, K, and M) and PrGV but another 7 AcMNPV EcoRI fragments (B, G, H, J, L, P, and Q) gave no detectable hybridization. Thus the homologous fragments included EcoRI-I containing the polyhedrin gene but not EcoRI-P and J containing regions coding, respectively, for the 10 kD early and 39 kD late nonstructural polypeptides. Four of the seven homologous fragments were somewhat collinear along the two genomes but the other three fragments were not. Thus, although the two viruses are related, many changes have occurred since the divergence of these two baculovirus subgroups.

---

**Figure 3** (*a*) Linearized physical map of PbGV (isolate ArGV1) DNA. The sites for BglI, PstI, and SmaI are all shown on the same line. The zero point is the 5' end of the granulin coding sequence, the location of which is marked. Distances of sites from the zero point are given in kbp. (*b*) Locations of genotypic differences between ArGV1 and ArGV isolates 2 to 6 and 9. Boxes identify regions, delineated by adjacent restriction sites, within which occurred insertions (+) or deletions (–) of specified size (in bp) relative to ArGV1. Additional (+) or lost (–) restriction sites are indicated as follows: B, BamHI; E, EcoRI; H, HindIII; K, KpnI; P, PstI. (Reproduced in modified form from Refs. 79 and 89.)

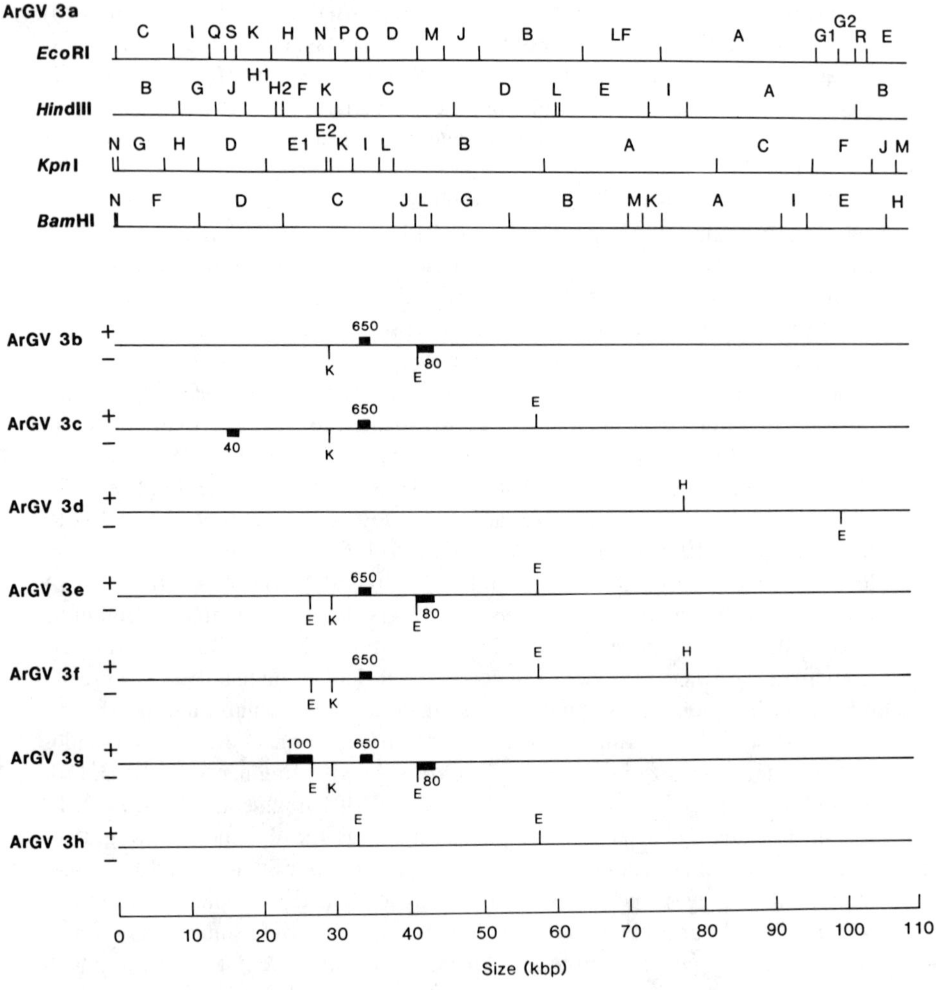


**Figure 4** Comparative physical map of eight genotypes cloned from ArGV obtained from a single field-collected diseased larva. The complete map for four enzymes is shown for the predominant genotype ArGV3a. A modified fragment lettering system is used (79) in which some fragments have a letter–number or two-letter combination so that the majority of fragments are lettered identically for all ArGV genotypes (cf. Figure 3*a*). The lower part of the figure shows locations (as defined in Figure 3) of differences between ArGV3a and each of the other seven genotypes. (Reproduced in modified form from Ref. 24.)

Smith and Summers (76) compared the DNA from 17 baculoviruses, including 4 GVs, by dot-blot hybridization. Under stringent conditions (37°C, 50% formamide), 1–3% homology was detected between DNAs from TnGV, SfGV, HaGV, and PiGV, and 0.1–4% homology was detected between these DNAs and 12 different NPV DNAs. Southern blots of AcMNPV EcoRI fragments probed with DNA from TnGV, HaGV, and PiGV showed that all had homology with AcMNPV EcoRI-I (containing the polyhedrin gene), L and M, and HaGV and PiGV had homology with AcMNPV EcoRI-P (containing the p10 gene).

## Granulin Gene Sequence Analysis

The only region of any GV genome to have been sequenced is the granulin gene; this has been done for PbGV (45), TnGV (44), and CpGV (46). In a recent review, Rohrmann (90) compared the structure of the PbGV and TnGV granulins with several polyhedrins that have been sequenced. The structure of CpGV granulin fits well with the general pattern of structural similarities and differences that were found for the two other granulins in comparison to polyhedrins. The granulins with 247 or 248 amino acids (Table 3) are all slightly larger than the polyhedrins with 245 or 246 amino acids, the difference being primarily due to an *N*-terminal insertion of 3 amino acids relative to most polyhedrins. The three granulins show much closer homologies to each other than to polyhedrins (Table 3), suggesting that in the evolution of baculoviruses, the GVs split from the NPVs well before divergence of the GVs occurred.

The promoter regions of the polyhedrin and granulin genes are of considerable interest because of the very high levels of expression of these proteins. The 5' flanking region of all of these genes contains a highly conserved sequence of 12 nucleotides followed by an AT-rich sequence up to the start of the coding region (90). For polyhedrin genes, the 12-mer has a consensus sequent AATAAG-TATTTT and is usually 41 nucleotides from the start codon. For the 3 granulin genes, there are an additional 2 conserved nucleotides adjacent to the 12-mer,

**Table 3** Comparison of Granulin Sequences

| | Size (amino acids) | Similarity of nucleotides (%)[a] | | |
|---|---|---|---|---|
| | | TnGV | PbGV | AcMNPV |
| CpGV | 248 | 78 (85) | 77 (83) | 61 (55) |
| TnGV | 248 | | 73 (77) | 58 (53) |
| PbGV | 247 | | | 58 (55) |

[a]Percentage for amino acids appears in parentheses.

```
PbGV                   { TAATTTT---AACAAAAT- }        15
                       { |||||||   ||||| ||  }
CpGV   TTTATAAGGAATTT  { TAATTTTAATAACAACATC } ATG    19
                       {   || | || ||||      }
TnGV                   { --ATATCAA-AACA----- }        11
```

**Figure 5** Comparison of sequences of the promoter regions of granulin genes from PbGV, CpGV, and TnGV. Gaps to align the sequences are indicated by dashes. The number of nucleotides between the conserved 14-mer, which probably contains the transcription initiation site, and the ATG translation start site is shown on the right.

giving a 14-mer, which is identical in all three genes and has the sequence TTATAAGGAATTT. The region following this sequence is, however, much shorter (11–19 nucleotides) and more variable than for polyhedrin genes (Figure 5), though with a similarly high A+T content of 82–93% and a complete absence of G.

## Transcription and Translation

Very little work has been done on GV gene expression largely because of the lack of suitable cell cultures. Even in those cases where replication of GVs in cell culture has been achieved, only a proportion of the cells are susceptible and it has not been possible to achieve synchronous infection of the whole of a cell population. One study, however, has been carried out on transcription of a GV using infected larvae as a source of mRNA. Dwyer and Granados (91) infected late fourth instar *P. rapae* larvae with PrGV and then extracted mRNA after 40 and 88 h. Since it is not possible to infect all susceptible tissues in an insect at the start of the experiment, transcripts and their proteins detected at different time points may not reflect the true temporal expression of viral genes. Nevertheless, mRNAs transcribed late in the replication cycle are likely to accumulate in the insect more slowly than early transcripts and therefore provide an indication of which are early and late genes. Moreover, if we wish to understand the overall course of infection at the molecular level, it is just as important to study transcription and translation in the host insect as it is to study them in isolated cells. Over 100 transcripts were detected and mapped by Northern blot analysis of poly(A)$^+$ RNA isolated from infected larvae, using labeled EcoRI restriction fragments as probes. Many differences in sizes and abundances of viral RNAs at the two time points were observed. There was no indication of splicing of any transcripts when S1 nuclease-treated DNA-RNA hybrids were analyzed by two-dimensional gel electrophoresis using neutral and then denaturing alkaline conditions. In some regions of the genome, the number of transcripts exceeded the coding capacity of the DNA. Since no splicing was detected, it seemed likely that overlapping sets of different size transcripts occur, as has been shown for AcMNPV (92–94). Evidence for this in the granulin gene region was obtained by two-dimensional electrophoresis of S1 nuclease-treated DNA–RNA hybrids followed by probing with a 0.6 kbp KpnI fragment contained totally within the granulin coding region. Three bands of 0.75, 1.0, and 2.0 kbp were detected, indicating that some of the sequences within the transcripts must overlap (91).

A wide range of PrGV translation products was detected using three methods to select the viral polypeptides (91). Immunoprecipitation of polypeptides produced by in vitro translation of total poly$(A)^+$ RNA yielded 15 viral structural proteins ranging in $M_r$ from 50,000 to 14,000 and including the granulin and major capsid protein. In vitro translation of poly$(A)^+$ RNA isolated by hybrid selection using total PrGV DNA yielded 18 polypeptides ranging from 74 to 15 kD, and when various individual restriction fragments were used for hybrid selection at least 35 polypeptides ranging from 68 to 15 kD were detected and mapped along the genome.

## PATHOGENESIS

### Infectivity

Studies on GV replication in the host insect have tended to concentrate on viruses infecting either tortricids or noctuids with fewer examples of viruses infecting larvae from other families. Even within these limited examples, considerable differences have been observed in the processes of infection and replication. For neonate larvae, the median lethal dose ($LD_{50}$) can vary from less than five capsules (60,83,95) to greater than $10^4$ capsules (84); the median lethal time ($LT_{50}$) can vary from less than four days (95) up to 25 days (97) or even 43 days in the case of a GV infecting an alternate host (98). In general, it appears that GVs can be divided into two categories on the basis of their pathogenicity. Either they are extremely infectious and kill rapidly, such as those infecting *C. pomonella*, *P. rapae*, and *P. xylostella*, where $LD_{50}$ values for neonates are no more than five capsules (and even late instars are usually only a few hundred capsules), or they require much larger doses to cause infection and death is often more prolonged. This latter category includes many of the GVs of noctuids such as *T. ni*, *H. armigera*, *L. oleracea*, and *A. segetum*. Allaway and Payne (84) noted that these categories correlated with the occurrence or nonoccurrence of an NPV for the host: insects in which the GV was highly infectious appeared to have no naturally occurring homologous NPV, whereas in insects susceptible to both a GV and an NPV, the $LD_{50}$ for the GV was usually much greater, at least in terms of number of OBs. If, however, $LD_{50}$ values were expressed in terms of numbers of nucleocapsids, the values obtained for GVs and NPVs infecting the same species were similar (99). It was speculated that since in these insects the number of nucleocapsids required to initiate infection was large, an NPV OB with perhaps several hundred nucleocapsids would provide a more successful strategy for survival of the virus than OBs containing a single nucleocapsid, as in the case for GVs. In addition, as noted later, the pathogenicity of different GVs correlated with the extent of infection of various organs in the larvae.

## Host Range

The host range of GVs appears to be restricted to a fairly small number of species within the same family as the original host (7). Not many GVs have been tested in more than a few species and even in those cases where cross-transmission appears to have occurred, the dose required has usually not been quantified and the progeny virus has rarely been identified. Indeed, in the few cases of apparent cross-transmission where the identity of the progeny virus has been established, it has often been found that the insects died from their homologous virus rather than the one used for infection; e.g., virus isolated from *C. pomonella* larvae following infection with either *C. leucotreta* GV (100) or *Epinotia aporema* GV (73) was found to be CpGV although CpGV did genuinely cross-infect *C. leucotreta* larvae (100). *Pieris brassicae* infected with very high doses of PrGV or AsGV died of either PbGV or a variety of genotypic variants which appeared to be recombinants between PrGV and PbGV (101). Contamination of the inoculum seemed very unlikely in the case of AsGV since this virus failed to cause any infection in *P. rapae* larvae, which would be much more susceptible to any contaminating PrGV or PbGV than would *P. brassicae*. These results suggest that GVs can exist in some inapparent form within an insect population and that inoculation with another virus can induce replication and lead to normal disease symptoms. Such results cast doubt on the interpretation of all host range studies in which the progeny virus has not been accurately identified.

Despite these reservations, it is of interest that at least three GVs appear to have a moderately broad host range and cause infection in a number of species from several genera (Table 4).

## Pathology and Replication In Vivo

Infection with GVs normally takes place by ingestion of capsules by larvae of the host insect, although transmission by parasites or transovarially may sometimes be important (see next section). Following ingestion, the OB protein dissolves in the alkaline gut juices, releasing virus particles which interact with cells of the midgut epithelium. The most likely route of entry appears to be fusion of the virus envelope with the microvillar membrane of columnar epithelial cells releasing the nucleocapsids into the cell cytoplasm (109). The replication of GVs is a complex process and descriptions of the stages observed by electron microscopy vary considerably. Observations on the infection of cells by TnGV (110)

**Table 4** Host Range of GVs from *Cydia pomonella*, *Heliothis armigera*, and *Scotogramma trifolii*

| Source of isolated virus | Additional hosts susceptible to virus | Reference |
|---|---|---|
| *C. pomonella* | *C. molesta* | 102 |
| | *C. nigricana* | 69 |
| | *Rhyacionia buoliana* | 103 |
| | *R. frustrana* | 104 |
| | *Grapholitha funebrana* | 103 |
| | *Lathronympha strigana* | 103 |
| | *Cryptophlebia leucotreta*[a] | 100 |
| *H. armigera* | *H. virescens* | 105 |
| | *H. zea*[a] | 106 |
| | *S exigua*[a] | 106 |
| | *S. frugiperda*[a] | 106 |
| | *T. ni*[a] | 106 |
| *S. trifolii* | *A. californica* | 107 |
| | *T. ni* | 107 |
| | *P. unipuncta* | 23 |
| | *S. exigua* | 108 |
| | *H. zea* | 108 |

[a]Confirmed by restriction enzyme analysis.

and EaGV (109) suggest that the nucleocapsids uncoat at the nuclear pore, releasing the DNA into the nucleus. However, intranuclear nucleocapsidlike structures were observed in GV-infected fat body cells of *S. frugiperda* (29) and midgut cells of *C. pomonella* (28), indicating that in some cases nucleocapsids may uncoat within the nucleoplasm. In general, the early stages of replication take place in the nucleus, but the nuclear membrane soon disintegrates so there is no clear distinction between nuclear and cytoplasmic components. However, the GVs from *A. argyrospila* (35) and *A. orana* (97) have been found to have two distinct modes of replication in fat body cells. One type is typical of most GVs in which replication begins in the nucleus and then, after disruption of the nuclear membrane, continues in the mixed nuclear and cytoplasmic components. A second type appears to replicate only in the cytoplasm without any apparent cytopathological changes in the nucleus.

Another interesting feature of GV infection in *A. argyrospila* was observed by Pinnock and Hess (111), who found replication in midgut epithelial cells in only 1 insect of 50 examined. Incomplete replication appeared to take place in the midgut of other larvae, resulting in empty capsidlike structures in the nuclei of

**Table 5** Tissue Specificity of GVs

| Host species | Tissues infected | Lethal time (days)[a] | Reference |
|---|---|---|---|
| *T. ni* | Midgut epithelium, fat body, prothoracic glands (hemocytes?) | 17 | 114–116 |
| *S. frugiperda* | Fat body | | 29,117 |
| *Euxoa auxiliaris* | Fat body, tracheal matrix, epidermis | 8–10 | 118 |
| *Plathypena scabra* | Fat body, tracheal matrix, epidermis | 3–9 | 119 |
| *A. orana* | Fat body | 25 | 97 |
| *H. brillians* | Midgut epithelium | 8–17 | 120 |
| *P. xylostella* | Fat body, epidermis, brain, Malpighian tubules | 5–6 | 36,121 |
| *A. argyrospila* | Fat body, tracheal matrix, hemocytes, epidermis | | 35 |
| *Carposina niponensis* | Fat body, tracheal matrix, epidermis, Malpighian tubules | | 122 |
| *H. cunea* | Fat body | 4–20 | 123,124 |
| *S. trifolii* | Fat body | | 107 |
| *C. pomonella* | Midgut epithelium, fat body, tracheal matrix, epidermis, Malpighian tubules | 3–6 | 28,125 |
| *C. fumiferana* | Fat body, tracheal matrix, epidermis, hemocytes | 12 | 126 |

[a]Accurate values (e.g., $LT_{50}$) are rarely quoted. These figures provide only an approximate guide to lethal times.

some cells. By contrast, Hess and Falcon (28) observed all stages of replication in the midgut epithelial cells of GV-infected *C. pomonella*. By 24 h postinfection, columnar and endocrine cells were infected and producing enormous amounts of budded virus. By 32 h postinfection, production of capsules was observed in the midgut cells and goblet cells were also infected.

Infected midgut epithelial cells release budded virus into the hemocoel, spreading the infection to other cells. The cytopathology of GV infections has been reviewed in detail by Granados (109), Tanada and Hess (112), Consigli et al. (113), and Hess and Falcon (28). In most infections, the fat body is the main site of virus multiplication and OB production, although with many GV infections several other tissues also become infected, especially the tracheal matrix and epidermis. Table 5 shows tissues that have been observed to support virus replication for a selection of GV infections. *Harrisina brillians* GV is the only

GV that has been observed to replicate principally in the midgut epithelium cells (120). This produces unusual external symptoms as the infected insects develop diarrhea, releasing a discharge containing OBs (108), which can lead to rapid spread of infection. *H. brillians* larvae infected as first instar took an average of 17.15 days to die, whereas larvae not infected until the third instar took an average of only 7.88 days to die. Normally, however, earlier instar larvae succumb more rapidly to GV infection than later instars. An indication of lethal times is given in Table 5. The way in which this has been estimated varies somewhat for different species but in general it appears that GVs with more limited tissue tropisms, especially when the tracheal matrix and epidermis do not become infected, take longer to cause mortality than GVs that infect a wider range of organs.

## In Vitro Replication

The first successful report of GV replication in a cell line was by Naser et al. (127), who demonstrated complete replication, including formation of OBs, for CpGV in a cell line originating from embryonic *C. pomonella* cells. However, only 2 of the 81 *C. pomonella* cell lines established in this way showed any cytopathic effect eight days after inoculation with hemolymph from infected *C. pomonella* larvae and a further 7 inoculated cell lines showed some reaction to one or both of two monoclonal antibodies raised against CpGV, although the extent of GV replication in these cell lines was not investigated further (128).

Replication of TnGV has now also been achieved in cell lines obtained from *T. ni* eggs by a method similar to that used for the *C. pomonella* cell lines (129). Using a peroxidase–antiperoxidase assay with antibody raised against virus particles purified from TnGV capsules, Granados et al. found that 15 of 31 cell lines supported replication when tested after 12 passages, but only 2 of these lines were still susceptible to TnGV infection by about passage 25. Only 5 of the *C. pomonella* cell lines had been tested for virus susceptibility by passage 12 or earlier, the majority of lines tested having been passaged at least 20 times. It is therefore possible that a much higher proportion would have been susceptible if tested at an earlier passage.

Granados et al. (129) noted two features common to all their susceptible cell lines:

1. They were all attached cells. No cells in suspension were susceptible. This confirmed the finding of Miltenburger et al. (128) that the cells which best supported replication were those passaged by scraping, i.e., the most firmly attached.
2. The cells were slow growing. This probably accounted for the loss of susceptibility at higher passage levels since faster growing nonsusceptible cells would become dominant.

Both the *C. pomonella* and *T. ni* cell lines were also tested for their ability to support replication of an NPV, TnSNPV in *T. ni* cells and *C. murinana* NPV in *C. pomonella* cells. Much higher proportions of both cells types supported NPV replication, confirming that GVs are more fastidious in their requirements than NPVs.

Attempts have also been made to obtain cell lines susceptible to PrGV replication. Dwyer et al. (130) screened over 150 primary sublines obtained from embryonic *P. rapae* for susceptibility to PrGV without observing any cytopathic effect in any of the cells. In some early passage sublines, a small proportion of cells (< 5%) gave a positive reaction when immunologically stained by the peroxidase–antiperoxidase procedure, but these permissive cells became less abundant as the cell lines were passaged and became undetectable within five subcultures.

Establishment of stable cell lines highly susceptible to GV replication is clearly still a difficult problem. Establishment of cloned cell lines from primary embryonic cell cultures may lead to more susceptible and homogeneous cell lines, the lack of which still hampers detailed biochemical and genetic studies of GV replication.

## ECOLOGY AND TRANSMISSION

### Epizootiology

Despite the large number of different baculoviruses that have now been isolated, observations on the natural incidence of baculovirus disease are limited to relatively few viruses. Indeed, there are no records at all of the natural occurrence of many of these viruses since they have been isolated from laboratory insect stocks. One GV that has been frequently recorded as the cause of epizootics is PrGV. This virus has provided natural control of *P. rapae* in New Zealand (131–133), Hawaii (134), Japan (135), Australia (136), and Canada (137) and has been reported in *P. rapae* and *P. brassicae* in several European countries and in India (138). It is now known that there is considerable variation between isolates of this virus from different areas and that whereas all isolates infect *P. rapae*, only a few isolates are capable of infecting *P. brassicae* (21).

A particularly extensive GV epizootic was observed by Federici in *S. trifolii* in southern California (107). Over 90% of larvae collected at 38 locations over an area of about 1000 $km^2$ were infected with the GV. It is of interest that the GV was able to infect both *T. ni* and *A. californica* in the laboratory since an earlier study by Hall (139) from the same laboratory had reported the occurrence of a GV in *A. californica* in several areas of southern California, including some of the sites where the StGV occurred. It is therefore possible that these two reports relate to the same virus causing epizootics in different insects. Whether one or both of these viruses is the same or similar to TnGV is unknown, which emphasizes the need for accurate identification of all new virus isolates. Another GV from *S. trifolii*, collected in central California, was shown to infect *P. unipuncta*, and the published restriction enzyme analysis of the progeny virus (23) suggested it was not closely related to the isolate of TnGV examined by Smith and Summers (75,76).

The studies by Federici and by Hall noted similar findings with respect to the occurrence of the virus at different larval densities. Unlike many virus epizootics which occur when the insects reach a high population density, these two GVs were found to occur in a high proportion of larvae even at very low host densities. Both studies also recorded the presence of NPVs in the larvae, although the incidence of NPV was much lower than the GV. In *A. californica* there was no instance of a double infection with the GV and the NPV in the same larva (139) but in *S. trifolii*, where relatively few NPV infections were observed, all were also infected with the GV (107).

Mixed GV and NPV epizootics were also observed in *P. unipuncta* by Tanada (49), who showed that late instar larvae were over 50 times more susceptible to the mixed viruses than the NPV alone, although the GV alone was also considerably more infectious than the NPV. In a study of mixed GV and NPV infections in larvae of *C. fumiferana*, Bird (126) found no evidence that an individual cell could become infected with both viruses, although the two viruses were frequently observed replicating in adjacent cells. Although double infection often caused higher mortality than single infection, there was no evidence of synergism between the viruses.

Another GV epizootic occurring in California infected *A. argyrospila* and was studied by Pinnock and Hess over a three year period (35). This study was unusual in that in addition to normal GV replication, aberrant replication processes were observed in some cells of all the larvae and the type of aberration was distinctive for each year. In the first year, large capsules containing two or more singly enveloped nucleocapsids were present; in the second year, enlarged rugose capsules were observed; in the third year, multiply enveloped nucleocapsids occurred containing two nucleocapsids per envelope.

Several other GVs were listed by Huger (25) as causing high mortality in the field, but most of these have received very little recent attention.

## Transmission and Persistence

There is virtually no direct evidence for assessing the relative importance of different methods of transmission of baculoviruses during the natural spread of these diseases. It seems highly likely that feeding on contaminated material is the major mechanism by which virus is transmitted at high host density, but at low host density and among solitary insects other mechanisms may be more important.

Levin et al. (140) showed in the laboratory that the parasitoid *Apanteles glomeratus* can transmit GV to a high proportion of *P. rapae* larvae in which it oviposits following oviposition in an infected larvae. They also showed that *A. glomeratus* which developed in GV-infected *P. rapae* transmitted the virus to 84% of larvae in which they subsequently oviposited as adults (141). Evidence for transmission in the field is circumstantial, but Wilson suggested that this might have been important in the spread of PrGV in field trials since field populations of *P. rapae* were heavily parasitized (142).

Laboratory tests on the transmission of a GV to larvae of *H. brillians* by *Apanteles harrisinae* and *Sturmia harrisinae* suggested that these parasites may transmit the virus (120), and this was also the implication from field observations where the extent of virus infection correlated with the degree of parasitism in the population.

Another mechanism for spread of virus for which there is some circumstantial evidence is transovum or transovarian transmission. Several authors have shown that adults infected or contaminated with GV can pass it on to their offspring, but it appears that this is most likely to occur by surface contamination of the eggs (143–146). Smith et al., however, reported the presence of capsules in cells of the gut epithelia of embryos of *H. brillians* but nevertheless concluded that there remained a reasonable doubt as to the identity of the particles seen (120).

## DIAGNOSIS

The later stages of granulosis disease are frequently characterized by the larvae becoming inactive and ceasing to feed (118,147). They may become swollen (36,117,118) or in other cases shrunken (120), but usually they become paler and may change color (36,107,117,118,120). Hunter and Hoffman (42) observed different color changes when two different GVs infected *P. interpunctella*—PiGV caused the larvae to become opaque white while CcGV caused them to turn pinkish yellow. In moribund and dead larvae the integument is generally very fragile and ruptures easily, releasing the liquefied body contents. The cadavers darken rapidly and soon disintegrate.

Many of these features are also characteristic of nuclear polyhedrosis disease, but diseases caused by the two baculovirus subgroups can be readily distinguished by microscopic examination of a diluted smear of the body contents in a wet mount. NPV OBs are most readily observed under phase contrast, whereas GV OBs are more readily seen with dark field illumination when they appear as highly refractile particles moving rapidly and randomly due to Brownian motion. Spherical vesicles may sometimes be seen containing a large number of vibrating capsules. The diagnosis of a granulosis disease may be confirmed by mixing a small volume of antisera to GV capsules with the body fluids and reexamining under the microscope (148). The GV capsules clump together and stop moving, due to aggregation by the antibody. Sufficient cross-reaction occurs between antisera to capsules of different GVs that the test is likely to be effective even when heterologous antiserum is used.

A histological staining method was developed for GVs by Huger but these days it is likely that for most purposes labeled antibody techniques would be more sensitive and specific, such as the use of peroxidase-labeled antibody, which has been used for the detection of GV in cell cultures, as discussed earlier (129,130). Labeled antibody has also been used in the ELISA technique for the detection and quantification of GVs (72). The method was very sensitive, detecting down to a few nanograms of capsules even in the presence of large amounts of larval material. The specificity of the technique varied according to the method and conditions used. The double antibody sandwich method could distinguish between different GVs even using antisera to whole capsules but the same antisera used at a higher concentration in the indirect ELISA was unable to discriminate between any of the GVs used or even between the GVs and an NPV.

Most of these techniques do not permit accurate identification of the GV, and analysis of the structural proteins or preferably the DNA is necessary to identify the virus. Although polypeptide profiles obtained by polyacrylamide gel electrophoresis of virus particles are distinct for different viruses, profiles obtained for the same virus by different people can vary so much that positive identification is very difficult. For example, published profiles for PbGV virus particles (22,58) or for PrGV nucleocapsids (22,63) obtained from different laboratories show little similarity. Factors such as method of purification, gel concentration and buffer system, and staining procedure can all affect the profile obtained, making it difficult to identify an isolate by comparison with published profiles.

Restriction enzyme analysis of the virus DNA provides much more consistent profiles and, as it is capable of identifying different isolates of the same virus as well as different viruses, it is much the best method of identification, as noted earlier. However, even with this technique care must be taken when comparing profiles obtained from gels of different agarose concentration or run with different buffers or at different voltage gradients, since all these factors affect the relative migration of different sized DNA fragments.

## CONCLUSION

In general, GVs have been less well studied than the closely related NPVs. This is partly because GVs have received less attention as microbial pesticides and also because it has been more difficult to obtain cell cultures that support GV replication. Now that at least one GV (CpGV) is again being developed for insecticidal use by a number of companies, and cell cultures have been obtained which support its replication, it is likely that it will attract increased interest for both applied and fundamental studies.

Despite the apparently close relationship between subgroup A and B baculoviruses, it seems likely that many differences will be found between them. Although the subdivision is based on the morphology of the occlusion body, the low levels of homology between GV and NPV DNAs and the production of virus in the cytoplasm rather than the nucleus of cells suggest that there are many important differences between the two groups. The next few years should see a considerable increase in the number of GV genes that are characterized. It will be of great interest to see the extent to which GV gene structure and function parallel those of NPVs.

## REFERENCES

1. Paillot, A. Sur une nouvelle maladie du noyau au grasserie des chenilles de *P. brassicae* et un nouveau groupe de micro-organismes parasites. *C. R. Acad. Sci.* 182, 180, 1926.
2. Paillot, A. Un nouveau type de maladie à ultravirus chez les insectes. *C. R. Acad. Sci.* 198, 204, 1934.
3. Steinhaus, E. A. A new disease of the variegated cutworm *Peridroma margaritosa* Haw. *Science* 106, 323, 1947.
4. Bergold, G. H. Über die kapselvirus-krankheit. *Z. Naturforsch.* 3b, 338, 1948.
5. Martignoni, M. E. and Iwai, P. J. *A Catalog of Viral Diseases of Insects, Mites, and Ticks*, 4th ed., Gen. Tech. Rep. PNW-195, U.S. Department of Agriculture, Forest Service, Pacific Northwest Research Station, Portland, OR, 1986.
6. Hull, R., Brown, F., and Payne, C. C. *Virology: Directory and Dictionary of Animal, Bacterial and Plant Viruses*, Macmillan, London, 1989.
7. Gröner, A. Specificity and safety of baculoviruses, in *The Biology of Baculoviruses*, vol. 1, Granados, R. R. and Federici, B. A., Eds., CRC Press, Boca Raton, FL, 1986, p. 177.
8. Payne, C. C. Insect viruses as control agents. *Parasitology* 84, 35, 1982.
9. Payne, C. C. Insect pathogenic viruses as pest control agents. *Fortschr. Zool.* 32, 183, 1986.
10. Reed, E. M. and Springett, B. P. Large-scale field testing of a granulosis virus for the control of the potato moth (*Phthorimaea operculella* [Zell.] [Lep., Gelechiidae]). *Bull. Entomol. Res.* 61, 223, 1971.
11. Zethner, O. Control of *Argotis segetum* (Lep.: Noctuidae) in root crops by granulosis virus. *Entomophaga* 25, 27, 1980.
12. Kinsinger, R. A. and McGaughey, W. H. Stability of *Bacillus thuringiensis* and a granulosis virus of *Plodia interpunctella* on stored wheat. *J. Econ. Entomol.* 69, 149, 1976.
13. Hunter, D. K., Collier, S. J., and Hoffman, D. F. The effect of a granulosis virus on *Plodia interpunctella* (Lepidoptera: Pyralidae) infestations occurring in raisins. *J. Stored Prod. Res.* 15, 65, 1979.
14. Cowan, D. K., Vail, P. V., Kok-Yokomi, M. L., and Schreiber, F. E. Formulation of a granulosis virus of *Plodia interpunctella* (Hübner) (Lepidoptera: Pyralidae):

Efficacy, persistence and influence on oviposition and larval survival. *J. Econ. Entomol.* 79, 1085, 1986.
15. Crook, N. E., Brown, J. D., and Foster, G. N. Isolation and characterization of a granulosis virus from the tomato moth, *Lacanobia oleracea*, and its potential as a control agent. *J. Invertebr. Pathol.* 40, 221, 1982.
16. Sato, T., Oho, N., and Kodomari, S. Utilization of granulosis viruses for controlling leafrollers in tea fields. *Jn. Agric. Res. Q.* 19, 271, 1986.
17. Schonherr, J. Field experiment on the biological control of *Choristoneura murinana* (Huebn.) by bergoldiavirus. *Entomophaga* 14, 251, 1969.
18. Liu, N. C., Liang, D. R., and Zhang, Q. L. The isolation, purification, identification, ultrastructure and application of a granulosis virus of the cabbage butterfly, *Pieris rapae*. *Wuhandaxue Xuebao* 2, 49, 1981.
19. Huber, J. Use of baculoviruses in pest management programs, in *The Biology of Baculoviruses*, vol. 2, Granados, R. R. and Federici, B. A., Eds., CRC Press, Boca Raton, FL, 1986, p. 181.
20. Cunningham, J. C. Baculoviruses: Their status compared to *Bacillus thuringiensis* as microbial insecticides. *Outlook Agric.* 17, 10, 1988.
21. Crook, N. E. Restriction enzyme analysis of granulosis viruses isolated from *Artogeia rapae* and *Pieris brassicae*. *J. Gen. Virol.* 67, 781, 1986.
22. Crook, N. E. A comparison of the granulosis viruses from *Pieris brassicae* and *Pieris rapae*. *Virology* 115, 173, 1981.
23. Harvey, J. and Tanada, Y. Characterization of the DNAs of five baculoviruses pathogenic for the armyworm, *Pseudaletia unipuncta*. *J. Invertebr. Pathol.* 46, 174, 1985.
24. Smith, I. R. L. and Crook, N. E. In vivo isolation of baculovirus genotypes. *Virology* 166, 240, 1988.
25. Huger, A. Granulosis of insects, in *Insect Pathology, An Advanced Treatise*, vol. 1, Steinhaus, E. A., Ed., Academic Press, New York, 1963, chap. 16.
26. Ackermann, H. -W. and Smirnoff, W. A. A morphological investigation of 23 baculoviruses. *J. Invertebr. Pathol.* 41, 269, 1983.
27. Summers, M. D. Baculoviruses (Baculoviridae), in *The Atlas of Insect and Plant Viruses*, Maramorosch, K., Ed., Academic Press, New York, 1977, p. 3.
28. Hess, R. T. and Falcon, L. A. Temporal events in the invasion of the codling moth, *Cydia pomonella*, by a granulosis virus: An electron microscope study. *J. Invertebr. Pathol.* 50, 85, 1987.
29. Walker, S., Kawanishi, C. Y., and Hamm, J. J. Cellular pathology of a granulosis virus infection. *J. Ultrastruct. Res.* 80, 163, 1982.
30. Longworth, J. F., Robertson, J. S., and Payne, C. C. The purification and properties of inclusion body protein of the granulosis virus of *Pieris brassicae*. *J. Invertebr. Pathol.* 19, 42, 1972.
31. Minion, F. C., Coons, L. B., and Broom, J. R. Characterization of the polyhedral envelope of the nuclear polyhedrosis virus of *Heliothis virescens*. *J. Invertebr. Pathol.* 34, 303, 1979.
32. Witt, M. A. and Manning, J. S. A phosphorylated 32-kDa protein and a subpopulation of polyhedrin are thiol linked to the carbohydrate layer surrounding a baculovirus occlusion body. *Virology* 163, 33, 1988.

33. Federici, B. A. Ultrastructure of baculoviruses, in *The Biology of Baculoviruses*, vol. 1, Granados, R. R. and Federici, B. A., Eds., CRC Press, Boca Raton, FL, 1986, p. 61.
34. Arnott, H. J. and Smith, K. M. Ultrastructure and formation of abnormal capsules in a granulosis virus of the moth *Plodia interpunctella* (Hbn.). *J. Ultrastruct. Res.* 22, 136, 1968.
35. Pinnock, D. E. and Hess, R. T. Morphological variations in the cytopathology associated with granulosis virus in the fruit-tree leaf roller, *Archips argyropila*. *J. Ultrastruct. Res.* 63, 252, 1978.
36. Asayama, T. Pathology and morphogenesis of a granulosis virus of the diamondback moth, in *Diamondback Moth Management, Proceedings of the First International Workshop Tainan, Taiwan, 11–15 March 1985*, 1986, p. 205.
37. Meynadier, G., Poitout, S., and Kuhl, G. Une granulose, nouvelle affection de *Mamestra oleracea* L. [Lepidoptera: Noctuidae]. *Entomophaga* 14, 421, 1969.
38. Stairs, G. R., Parrish, W. B., Briggs, J. D., and Allietta, M. Fine structure of a granulosis virus of the codling moth. *Virology* 30, 583, 1966.
39. Falcon, L. A. and Hess, R. T. Electron microscope observations of multiple occluded virions in the granulosis virus of the codling moth, *Cydia pomonella*. *J. Invertebr. Pathol.* 45, 356, 1985.
40. Smith, K. M. and Brown, R. M., Jr. A study of the long virus rods associated with insect granuloses. *Virology* 27, 512, 1965.
41. Whitlock, V. H. Long virus-like rods associated with a granulosis virus of the false codling moth *Cryptophlebia leucotreta* (Meyr.). *Phytophylactica* 12, 181, 1980.
42. Hunter, D. K. and Hoffmann, D. F. Cross infection of a granulosis virus of *Cadra cautella*, with observations on its ultrastructure in infected cells of *Plodia interpunctella*. *J. Invertebr. Pathol.* 20, 4, 1972.
43. Tweeten, K. A., Bulla, L. A., Jr., and Consigli, R. A. Applied and molecular aspects of insect granulosis viruses. *Microbiol. Rev.* 45, 379, 1981.
44. Akiyoshi, D., Chakerian, R., Rohrmann, G. F., Nesson, M. H., and Beaudreau, G. S. Cloning and sequencing of the granulin gene from *Trichoplusia ni* granulosis virus. *Virology* 141, 328, 1985.
45. Chakerian, R., Rohrmann, G. F., Nesson, M. H., Leisy, D. J., and Beaudreau, G. S. The nucleotide sequence of the *Pieris brassicae* granulosis virus granulin gene. *J. Gen. Virol.* 66, 1263, 1985.
46. Crook, N. E. and Barker, R. Unpublished results, 1987.
47. Summers, M. D. and Smith, G. E. *Trichoplusia ni* granulosis virus granulin: A phenol-soluble phosphorylated protein. *J. Virol.* 16, 1108, 1975.
48. Tweeten, K. A., Bulla, L. A., Jr., and Consigli, R. A. Structural polypeptides of the granulosis virus of *Plodia interpunctella*. *J. Virol.* 33, 877, 1980.
49. Tanada, Y. A synopsis of studies on the synergistic property of an insect baculovirus: A tribute to Edward A. Steinhaus. *J. Intertebr. Pathol.* 45, 125, 1985.
50. Derksen, A. C. G. and Granados, R. R. Alteration of a lepidopteran peritrophic membrane by baculoviruses and enhancement of viral infectivity. *Virology* 167, 242, 1988.
51. Yamamoto, T. and Tanada, Y. Phospholipid, an enhancing component in the

synergistic factor of a granulosis virus of the armyworm (*Pseudaletia unipuncta*). *J. Invertebr. Pathol.* 31, 48, 1978.
52. Tweeten, K. A., Bulla, L. A., Jr., and Consigli, R. A. Characterization of an alkaline protease associated with a granulosis virus of *Plodia interpunctella*. *J. Virol.* 26, 702, 1978.
53. Nagata, M. and Tanada, Y. Origin of an alkaline protease associated with the capsule of a granulosis virus of the armyworm, *Pseudaletia unipuncta* (Haworth). *Arch. Virol.* 76, 245, 1983.
54. Langridge, W. H. R. and Balter, K. Protease activity associated with the capsule protein of *Estigmene acrea* granulosis virus. *Virology* 114, 595, 1981.
55. Zummer, M. and Faulkner, P. Absence of protease in baculovirus polyhedral bodies propagated in vitro. *J. Invertebr. Pathol.* 33, 383, 1979.
56. Summers, M. D. and Smith, G. E. Baculovirus structural polypeptides. *Virology* 84, 390, 1978.
57. Smith, G. E. and Summers, M. D. Application of a novel radioimmunoassay to identify baculovirus structural proteins that share interspecies antigenic determinants. *J. Virol.* 39, 125, 1981.
58. Brown, D. A., Bud, H. M., and Kelly, D. C. Biophysicial properties of the structural components of a granulosis virus isolated from the cabbage white butterfly (*Pieris brassicae*). *Virology* 81, 317, 1977.
59. Yamamoto, T. and Tanada, Y. Comparative analysis of the enveloped virions of two granulosis viruses of the armyworm *Pseudaletia unipuncta*. *Virology* 94, 71, 1979.
60. Harvey, J. P. and Volkman, L. E. Biochemical and biological variation of *Cydia pomonella* (codling moth) granulosis virus. *Virology* 124, 21, 1983.
61. Boucias, D. G. and Nordin, G. L. Comparative analysis of the alkali-liberated components of the *Hyphantria cunea* and the *Diacrisia virginica* granulosis virus. *J. Invertebr. Pathol.* 36, 264, 1980.
62. Arif, B. M., Guangyu, Z., and Jamieson, P. A comparison of three granulosis viruses isolated from *Choristoneura* spp. *J. Invertebr. Pathol.* 48, 180, 1986.
63. Tweeten, K. A., Bulla, L. A., Jr., and Consigli, R. A. Characterization of an extremely basic protein derived from granulosis virus nucleocapsids. *J. Virol.* 33, 866, 1980.
64. Burley, S. K., Miller, A., Harrap, K. A., and Kelly, D. C. Structure of the baculovirus nucleocapsid. *Virology* 120, 433, 1982.
65. Wilson, M. E., and Consigli, R. A. Functions of a protein kinase activity associated with purified capsids of the granulosis virus infecting *Plodia interpunctella*. *Virology* 143, 526, 1985.
66. Wilson, M. E. and Consigli, R. A. Characterization of a protein kinase activity associated with purified capsids of the granulosis virus infecting *Plodia interpunctella*. *Virology* 143, 516, 1985.
67. Summers, M. D. and Hoops, P. Radioimmunoassay analysis of baculovirus granulins and polyhedrins. *Virology* 103, 89, 1980.
68. Chang, P. -M. and Tanada, Y. Serological study on the transmission of a granulosis

virus of the armyworm, *Pseudaletia unipuncta* (Lepidoptera: Noctuidae), to other lepidopterous species. *J. Invertebr. Pathol.* 31, 106, 1978.
69. Payne, C. C. The susceptibility of pea moth, *Cydia nigricana*, to infection by the granulosis virus of the codling moth, *Cydia pomonella*. *J. Invertebr. Pathol.* 38, 71, 1981.
70. Hunter, F. R. and Boraston, R. C. Application of the Laurell immunoelectrophoresis technique to the study of serological relationships between granulosis viruses. *J. Invertebr. Pathol.* 34, 248, 1979.
71. Yamamoto, T. and Tanada, Y. Protein components of two strains of granulosis virus of the armyworm *Pseudaletia unipuncta* (Lepidoptera: Noctuidae). *J. Invertebr. Pathol.* 32, 158, 1978.
72. Crook, N. E. and Payne, C. C. Comparison of three methods of ELISA for baculoviruses. *J. Gen. Virol.* 46, 26, 1980.
73. Ballard, J. and Ellis, D. J. Personal communication, 1988.
74. Harrap, K. A. and Payne, C. C. The structural properties and identification of insect viruses. *Adv. Virus Res.* 25, 273, 1979.
75. Smith, G. E. and Summers, M. D. Analysis of baculovirus genomes with restriction endonucleases. *Virology* 89, 519, 1978.
76. Smith, G. E. and Summers, M. D. DNA homology among subgroup A, B and C baculoviruses. *Virology* 123, 393, 1982.
77. Tweeten, K. A., Bulla, L. A., Jr., and Consigli, R. A. Restriction enzyme analysis of the genomes of *Plodia interpunctella* and *Pieris rapae* granulosis viruses. *Virology* 104, 514, 1980.
78. Dwyer, K. G. and Granados, R. R. A physical map of the *Pieris rapae* granulosis virus genome. *J. Gen. Virol.* 68, 1471, 1987.
79. Smith, I. R. L. and Crook, N. E. Physical maps of the genomes of four variants of *Artogeia rapae* granulosis virus. *J. Gen. Virol.* 69, 1741, 1988.
80. Burgess, S. EcoRI restriction endonuclease fragment patterns of eight lepidopteran baculoviruses. *J. Invertebr. Pathol.* 42, 401, 1983.
81. McIntosh, A. H. and Ignoffo, C. M. Restriction endonuclease patterns of three baculoviruses isolated from species of *Heliothis*. *J. Invertebr. Pathol.* 41, 27, 1983.
82. Cattano, S. P. and Langridge, W. H. R. Characterization of DNA from a granulosis virus of *Estigmene acrea*: (Lepidoptera). *Virology* 119, 199, 1982.
83. Crook, N. E., Spencer, R. A., Payne, C. C., and Leisy, D. J. Variation in *Cydia pomonella* granulosis virus isolates and physical maps of the DNA from three variants. *J. Gen. Virol.* 66, 2423, 1985.
84. Allaway, G. P. and Payne, C. C. Host range and virulence of five baculoviruses from lepidopterous hosts. *Ann. Appl. Biol.* 105, 29, 1984.
85. Qiu, B. -S., Pei, M. -Y., Ai, X. -L., Shi, Y. -H., and Wu, Z. -Y. The genomic libraries and physical map of *Agrotis segetum* granulosis virus. *Acta Microbiol. Sinica* 26, 210, 1986.
86. Fritsch, E. and Huber, J. The granulosis virus of the false codling moth, *Cryptophlebia leucotreta* (Meyr.), in *Fundamental and Applied Aspects of Invertebrate Pathology*, Samson, R. A., Vlak, J. M., and Peters, D., Eds., ICIP, Wageningen, The Netherlands, 1986, p. 112.

87. Upholt, W. B. Estimation of DNA sequence divergence from comparison of restriction endonuclease digests. *Nucleic Acids Res.* 4, 1257, 1977.
88. Fritsch, E. and Huber, J. Personal communication, 1986.
89. Smith, I. R. L. Genome Structure and Host Interactions of *Pieris* Granulosis Virus, Ph.D. diss., University of London, 1987.
90. Rohrmann, G. F. Polyhedrin structure. *J. Gen. Virol.* 67, 1499, 1986.
91. Dwyer, K. G. and Granados, R. R. Mapping *Pieris rapae* granulosis virus transcripts and their in vitro translation products. *J. Virol.* 62, 1535, 1988.
92. Lübbert, H. and Doerfler, W. Transcription of overlapping sets of RNAs from the genome of *Autographa californica* nuclear polyhedrosis virus: A novel method for mapping RNAs. *J. Virol.* 52, 255, 1984.
93. Friesen, P. D. and Miller, L. K. Temporal regulation of baculovirus DNA: Overlapping early and late transcripts. *J. Virol.* 54, 392, 1985.
94. Rankin, C., Ladin, B. F., and Weaver, R. F. Physical mapping of temporally regulated, overlapping transcripts in the region of the 10K protein gene in *Autographa californica* nuclear polyhedrosis virus. *J. Virol.* 57, 18, 1986.
95. Payne, C. C., Tatchell, G. M., and Williams, C. F. The comparative susceptibilities of *Pieris brassicae* and *P. rapae* to a granulosis virus from *P. brassicae*. *J. Invertebr. Pathol.* 38, 273, 1981.
96. Sheppard, R. F. and Stairs, G. R. Dosage-mortality and time-mortality studies of a granulosis virus in a laboratory strain of the codling moth, *Laspeyresia pomonella*. *J. Invertebr. Pathol.* 29, 216, 1977.
97. Schmid, A., Cazelles, O., and Benz, G. A granulosis virus of the fruit tortrix, *Adoxophyes orana* F.V.R. (Lep., Tortricidae). *Mitt. Schweiz. Entomol. Ges.* 56, 225, 1983.
98. Boucias, D. G. and Nordin, G. L. Susceptibility of *Hyphantria cunea* to the granulosis virus isolated from *Diacrisia virginica*. *J. Invertebr. Pathol.* 30, 377, 1977.
99. Allaway, G. P. Virus particle packaging in baculovirus and cytoplasmic polyhedrosis virus inclusion bodies. *J. Invertebr. Pathol.* 42, 357, 1983.
100. Fritsch, E. and Huber, J. Personal communication, 1988.
101. Smith, I. R. L. and Crook, N. E. Latent granulosis virus sequences in *Pieris brassicae*, in *Fundamental and Applied Aspects of Invertebrate Pathology*, Samson, R. A., Vlak, J. M., and Peters, D., Eds., ICIP, Wageningen, The Netherlands, 1986, p. 109.
102. Falcon, L. A., Kane, W. R., and Bethell, R. S. Preliminary evaluation of a granulosis virus for control of the codling moth, *J. Econ. Entomol.* 61, 1208, 1968.
103. Huber, J. The baculoviruses of *Cydia pomonella* and other tortricids, in *Invertebrate Pathology and Microbial Control, Proceedings Third International Colloquium on Invertebrate Pathology, 6–10 September 1982, Brighton, U.K.*, 1982, p. 119.
104. Stephen, F. M., Yearian, W. C., Young, S. Y., and Wallis, G. W. Preliminary evaluation of codling moth granulosis virus for suppression of Nantucket pine tip moth. *J. Georgia Entomol. Soc.* 17, 398, 1982.

105. Ignoffo, C. M., McIntosh, A. H., and Garcia, C. Susceptibility of larvae of *Heliothis zea*, *H. virescens*, and *H. armigera* [Lep.: Noctuidae] to three baculoviruses. *Entomophaga* 28, 1, 1983.
106. Hamm, J. J. Extension of the host range for a granulosis virus from *Heliothis armigera* from South Africa. *Environ. Entomol.* 11, 159, 1982.
107. Federici, B. A. Baculovirus epizootic in a larval population of the clover cutworm, *Scotogramma trifolii*, in southern California. *Environ. Entomol.* 7, 423, 1978.
108. Federici, B. A. Personal communication, 1986.
109. Granados, R. R. Infectivity and mode of action of baculoviruses. *Biotechnol. Bioeng.* 22, 1377, 1980.
110. Summers, M. D. Electron microscopic observations on granulosis virus entry, uncoating and replication processes during infection of the midgut cells of *Trichoplusia ni*. *J. Ultrastruct. Res.* 35, 606, 1971.
111. Pinnock, D. E. and Hess, R. T. Electron microscope observations on granulosis virus replication in the fruit tree leaf roller, *Archips argyrospila*: Infection of the midgut. *J. Invertebr. Pathol.* 30, 354, 1977.
112. Tanada, Y. and Hess, R. T. The cytopathology of baculovirus infections in insects, in *Insect Ultrastructure*, vol. 2, King, R. C. and Akai, H., Eds., Plenum Press, New York, 1984, p. 517.
113. Consigli, R. A., Russell, D. L., and Wilson, M. E. The biochemistry and molecular biology of the granulosis virus that infects *Plodia interpunctella*, in *The Molecular Biology of Baculoviruses*, Doerfler, W., and Böhm, P., Eds., Springer-Verlag, Berlin, 1986, p. 69.
114. Hamm, J. J. and Paschke, J. D. On the pathology of a granulosis of the cabbage looper, *Trichoplusia ni* (Hübner). *J. Insect Pathol.* 5, 187, 1963.
115. Tanada, Y. and Leutenegger, R. Multiplication of a granulosis virus in larval midgut cells of *Trichoplusia ni* and possible pathways of invasion into the haemocoel. *J. Ultrastruct. Res.* 30, 589, 1970.
116. Dougherty, E. M., Kelly, T. J., Rochford, R., Forney, J. A., and Adams, J. R. Effects of infection with a granulosis virus on larval growth, development and ecdysteroid production in the cabbage looper, *Trichoplusia ni*. *Physiol. Entomol.* 12, 23, 1987.
117. Hamm, J. J. Comparative histopathology of a granulosis and a nuclear polyhedrosis of *Spodoptera frugiperda*. *J. Invertebr. Pathol.* 10, 320, 1968.
118. Jackson, J. J. and Sutter, G. R. Pathology of a granulosis virus in the army cutworm, *Euxoa auxiliaris* (Lepidoptera: Noctuidae). *J. Kansas Entomol. Soc.* 58, 353, 1985.
119. Beegle, C. C. Histopathology and bioassay of *Plathypena scabra* granulosis virus. *J. Invertebr. Pathol.* 32, 325, 1978.
120. Smith, O. J., Huges, K. M., Dunn,. P. H., and Hall, I. M. A granulosis virus disease of the western grape leaf skeletonizer and its transmission. *Can. Entomol.* 88, 507, 1956.
121. Abdul-Kadir, H. Personal communication, 1988.
122. Sato, T., Oho, N., and Narita, H. A granulosis virus of the peach fruit moth,

*Carposina niponensis* Walsingham (Lepidoptera: Carposinidae). *Appl. Entomol. Zool.* 12, 365, 1977.
123. Watanabe, H. and Kobayashi, M. Histopathology of a granulosis in the larva of the fall webworm, *Hyphantria cunea*. *J. Invertebr. Pathol.* 16, 71, 1970.
124. Boucias, D. G. and Nordin, G. L. Interinstar susceptibility of the fall webworm, *Hyphantria cunea*, to its nucleopolyhedrosis and granulosis viruses. *J. Invertebr. Pathol.* 30, 68, 1977.
125. Tanada, Y. and Leutenegger, R. Histopathology of a granulosis-virus disease of the codling moth, *Carpocapsa pomonella*. *J. Invertebr. Pathol.* 10, 39, 1968.
126. Bird, F. T. Polyhedrosis and granulosis viruses causing single and double infections in the spruce budworm, *Choristoneura fumiferana* Clemens. *J. Insect Pathol.* 1, 406, 1959.
127. Naser, W. L., Miltenburger, H. G., Harvey, J. P., Huber, J., and Huger, A. M. In vitro replication of the *Cydia pomonella* (codling moth) granulosis virus. *FEMS Microbiol. Lett.* 24, 117, 1984.
128. Miltenburger, H. G., Naser, W. L., Harvey, J. P., Huber, J., and Huger, A. M. The cellular substrate: A very important requirement for baculovirus in vitro replication. *Z. Naturforsch. C.* 39, 993, 1984.
129. Granados, R. R., Derksen, G. C., and Dwyer, K. G. Replication of the *Trichoplusia ni* granulosis and nuclear polyhedrosis viruses in cell cultures. *Virology* 152, 475, 1986.
130. Dwyer, K. G., Webb, S. E., Shelton, A. M., and Granados, R. R. Establishment of cell lines from *Pieris rapae* embryos: Characterization and susceptibility to baculoviruses. *J. Invertebr. Pathol.* 52, 268, 1988.
131. Todd, D. H. Virus disease effectively controls white butterfly. *N.Z. J. Agric.* 100, 83, 1960.
132. Kelsey, J. M. Virus sprays for control of *Pieris rapae* L. *N.Z. J. Sci. Technol.* 38, 644, 1957.
133. Kelsey, J. M. Control of *Pieris rapae* by granulosis viruses. *N.Z. J. Agric. Res.* 1, 778, 1958.
134. Tanada, Y. Microbial control of imported cabbageworm. *Hawaii Farm Sci.* 4, 6, 1956.
135. Ito, Y., Miyashita, K., and Gotoh, A. Natural mortality of the common cabbage butterfly, *Pieris rapae* crucivora Boisduval, with considerations on the factors affecting it. *Jpn. J. Appl. Entomol. Zool.* 4, 1, 1960.
136. Hamilton, J. T. Seasonal abundance of *Pieris rapae* (L.), *Plutella xylostella* (L.) and their diseases and parasites. *Gen. Appl. Entomol.* 11, 59, 1979.
137. Harcourt, D. G. Major factors in survival of the immature stages of *Pieris rapae* (L.). *Can. Entomol.* 98, 653, 1966.
138. David, W. A. L. The granulosis virus of *Pieris brassicae* (L) and its relationship with its host. *Adv. Virus Res.* 22, 111, 1978.
139. Hall, I. M. The role of virus diseases in the control of the alfalfa looper. *J. Econ. Entomol.* 46, 1110, 1953.
140. Levin, D. B., Laing, J., and Jaques, R. Transmission of granulosis virus by

*Apanteles glomeratus* to is host *Pieris rapae*. *J. Invertebr. Pathol.* 34, 317, 1979.

141. Levin, D. B., Laing, J. E., Jaques, R. P., and Corrigan, J. E. Transmission of the granulosis virus of *Pieris rapae* (Lepidoptera: Pieridae) by the parasitoid *Apanteles glomeratus* (Hymenoptera: Braconidae). *Environ. Entomol.* 12, 166, 1983.
142. Wilson, F. The effectiveness of a granulosis virus applied to field populations of *Pieris rapae* (Lepidoptera). *Aust. J. Agric. Res.* 11, 485, 1960.
143. David, W. A. L. and Taylor, C. E. Transmission of a granulosis virus in the eggs of a virus-free stock of *Pieris brassicae*. *J. Invertebr. Pathol.* 27, 71, 1976.
144. Etzel, L. K. and Falcon, L. A. Studies of transovum and transstadial transmission of a granulosis virus of the codling moth. *J. Invertebr. Pathol.* 27, 13, 1976.
145. Melamed-Madjar, V. and Raccah, B. The transstadial and vertical transmission of a granulosis virus from the corn borer *Sesamia nonagrioides*. *J. Invertebr. Pathol.* 33, 259, 1979.
146. Tatchell, G. M. The transmission of a granulosis virus following the contamination of *Pieris brassicae* adults. *J. Invertebr. Pathol.* 37, 210, 1981.
147. Tatchell, G. M. The effects of a granulosis virus infection and temperature on the food consumption of *Pieris rapae* (Lep: Pieridae). *Entomophaga* 26, 291, 1981.
148. Payne, C. C. Personal communication, 1978.
149. Huger, A. Methods for staining capsular virus inclusion bodies typical of granuloses of insects. *J. Insect Pathol.* 3, 338, 1961.

# 3

# Molecular Biology of the Hz-1 and *Oryctes* Nonoccluded Baculoviruses

JOHN P. BURAND
*University of Massachusetts, Amherst, Massachusetts*

## INTRODUCTION

The nonoccluded baculoviruses are rod-shaped DNA-containing viruses that differ from other members of the family Baculoviridae in that they are not found embedded within a protein crystal occlusion body. Members of this prodigious and complex group of baculoviruses (subgroup C) (1) have been found associated with a number of arthropod hosts (2). The nonoccluded baculoviruses are important not only for the potential they hold in controlling insect pests but also for the interesting biological properties they possess. These properties include a host range for several economically important insects and infections that result in rapid cell lysis. Today, through the use of gene manipulation technology, the genetic information that these viruses carry can be exploited as a resource for specific biological properties that can be utilized to improve the efficacy of other baculoviruses which are currently used to control pest insects. To achieve these expectations, however, more information is needed about the molecular biology of these viruses. Furthermore, an increased understanding of the molecular biology of members of this group of insect viruses will certainly prove useful in their application as efficacious agents for the control of insect pests.

Except for Hz-1 and the *Oryctes* baculovirus very little is known about the molecular biology of the nonoccluded baculoviruses. This is due mainly to the lack of good cell culture systems for the in vitro replication of these viruses. This review concentrates on the molecular biology of the *Oryctes* and the Hz-1 nonoccluded baculovirus.

## *ORYCTES* BACULOVIRUS

### Physical and Biological Properties of the *Oryctes* Virus

The *Oryctes* baculovirus was first discovered in Malaysia by Huger (3). The virus originally recovered from a diseased rhinceros beetle (*Oryctes rhinoceros* L.) larvae has been used throughout the tropics to control the coconut palm rhinoceros beetle (4–6). Most recently the virus has been used to control this insect pest in coastal regions of Tanzania (7). The virus is an enveloped particle which has a taillike projection at one end (8). The genome of the virus is a supercoiled closed circular double-stranded DNA molecule of about 127 kb in length (9).

Once ingested by *O. rhinoceros* larvae, the virus invades midgut cells, the primary site of virus replication, then rapidly becomes systemic, infecting other tissues of the insect. First instar larvae die about nine days after ingesting the virus (10). In adult beetles apparently only midgut cells become infected. These infected cells are sloughed off, releasing the virus into the environment and often contaminating nesting sites. Adult beetles usually die within 60 days of ingesting the virus (11).

### Replication of *Oryctes* Virus in Cell Culture

The replication of *Oryctes* virus in a cell line derived from *Heteronychus arator* (HA) (Coleoptera: Scarabaeidae) has been studied extensively by Crawford and Sheehan (12). For these studies, a cloned isolate of virus strain PV505 was obtained by end point dilution. In these normally fibroblastic-like cells, the first visible sign of viral cytopathic effect (CPE) is the rounding of the cells. Complete cell lysis was observed 3–4 days postinoculation (p.i.).

Electron microscope (EM) studies (12) indicated that the virus enters cells by pinocytosis. The first sign of virus replication ultrastructurally was the appearance, by approximately 7 h p.i., of chromatin-free clear areas in the cell nucleus and the accumulation of envelope material in these cleared areas. At this time enveloped nucleocapsids were occasionally observed in the clear areas. However, by 12 h p.i., when virus replication was at its peak, a large number of enveloped particles were visible in the cleared areas (Figure 1C). Based on their electron microscopic observations Crawford and Sheehan (12) proposed that the *Oryctes* baculovirus assembly begins with the formation of an envelope and nucleocapsid shell, which is then filled with an electron-dense core containing the viral DNA (Figure 1A). The virus then enters the cytoplasm where between 12 and 36 h p.i. it buds from the cytoplasmic membrane acquiring a second unit membrane.

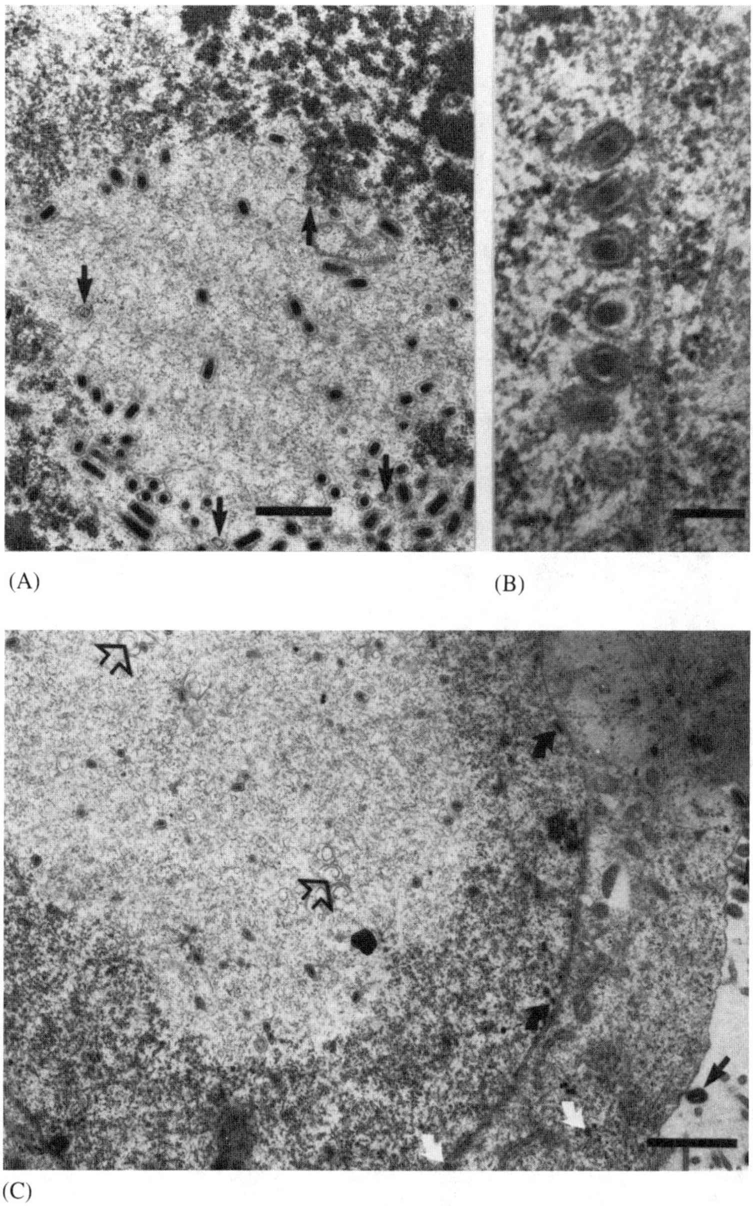


**Figure 1** Replication of *Oryctes* virus in HA cells. (A) Formation of virus particles in cleared area of cell nucleus. Arrows indicate empty nucleocapsids. Bar = 400 nm. (B) Virus particles along inner nuclear membrane at 16 h p.i. Bar = 150 nm. (C) Infected HA cells 16 h p.i. showing buildup of envelope material (open arrows), virus particles at nuclear membrane (black curved arrows), virus in cytoplasm (white arrows), and extracellular, budded virus (straight black arrows). Bar = 1 $\mu$m. (From Ref. 12, used with permission.)

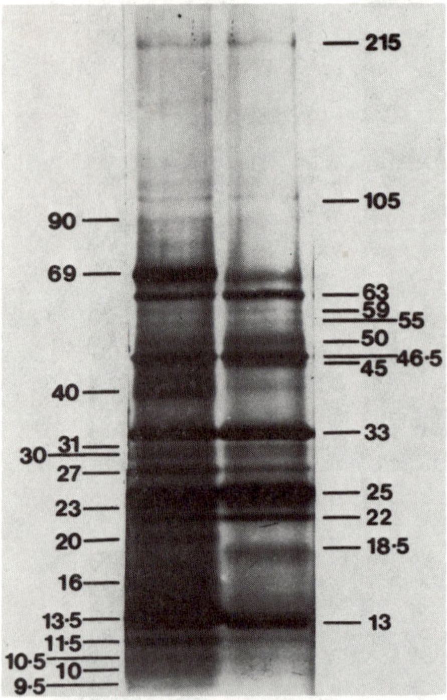


**Figure 2** *Oryctes* virus structural proteins of purified whole virus (A) and of nucleocapsids treated with 1% NP40 (B). Numbers are molecular weights × $10^{-3}$. (From Ref. 12, used with permission.)

## *Oryctes* Virus Structural and Intracellular Proteins

Crawford and Sheehan (12) identified 27 *Oryctes* virus structural proteins ranging in size from 9.5 to 215 kD (Figure 2). This is 15 more proteins than previously identified by Payne et al. (13). This difference could be due partly to the more sensitive staining method used by Crawford and Sheehan. They also found that treatment of the virus with Nonidet P40 (NP40) resulted in the complete or partial removal of 14 envelope proteins. The remaining 13 proteins are felt to be associated with the virus nucleocapsid.

*L*-[$^{35}$S]methionine pulse labeling of virus-infected HA cells at 2 h intervals for 36 h p.i. revealed the synthesis of only 8 (p46.5, p40, p27, p25, p22, p13, p11.5, and p10) of 27 virus structural proteins (12) (Figure 3). The inability to detect the synthesis of the remaining virus structural proteins or additional virus-specific intracellular proteins (VSIP) may be due in part to the large number of labeled cell proteins detected in these experiments. These background cell proteins are present because the *Oryctes* virus does not actively shut off HA cell protein synthesis during virus replication. Another possibility is that the remaining proteins are low or absent of methionine residues and were not highly labeled in these experiments.

In these time course studies, two proteins, p46.5 and p10, were first detected

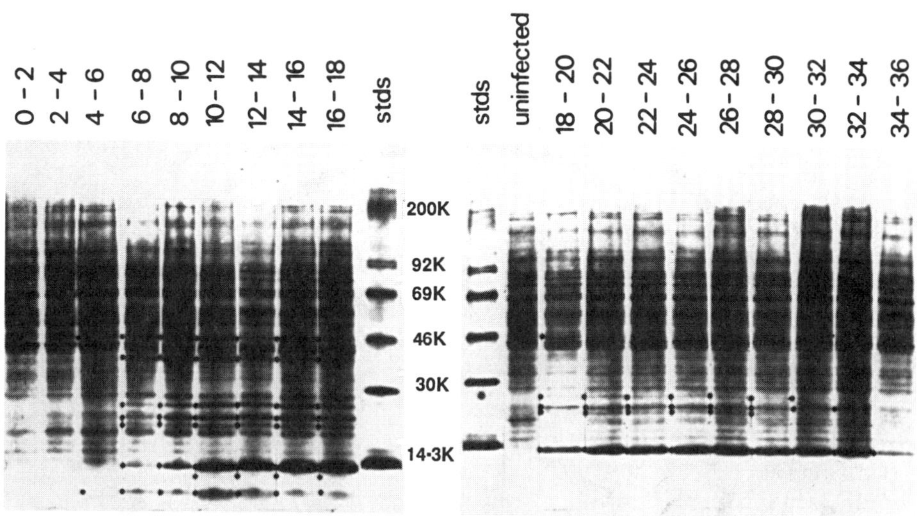


**Figure 3** Time course of *Oryctes* VSIP synthesis in HA cells pulse-labeled for 2 h (at times indicated above each lane) with [$^{35}$S]methionine. (From Ref. 12, used with permission.)

at 4 h p.i. All of the remaining virus proteins (p40, p27, p25, p22, and p13), except p11.5, were first detected at 6 h p.i. The p11.5 protein was first detected at 10 h p.i. No new viral proteins were detected later than 10 h p.i. It is interesting to note that Crawford and Sheehan (12) found that the synthesis of half of these proteins was not detectable by 20 h p.i. In ultrastructural studies, they observed the maximum release of enveloped virus particles budding from infected cells at 16 h p.i. This is in close agreement with their observation that all the virus proteins detected were synthesized by this time and that the synthesis of half of these proteins was no longer detected shortly after this 16 h time period. It will be important to determine what effects the continued synthesis of these remaining proteins have on the release of virus particles from infected cells.

## Physical Map of the *Oryctes* Virus Genome

A physical map of a cloned isolate of the *Oryctes* virus strain PV505 was reported by Crawford et al. (9). The circular genome of the virus was found to be approximately 130 kb in length. Digestion of purified viral DNA with the restriction enzymes BamHI, EcoRI, HindIII, and PstI along with Southern hybridization was used to generate this circular map (Figure 4). Hybridization experiments were also used to demonstrate that the *Oryctes* virus did not contain sequences homologous to the occlusion body gene common to the occluded baculoviruses. Crawford et al. (9) also demonstrated that there were six regions

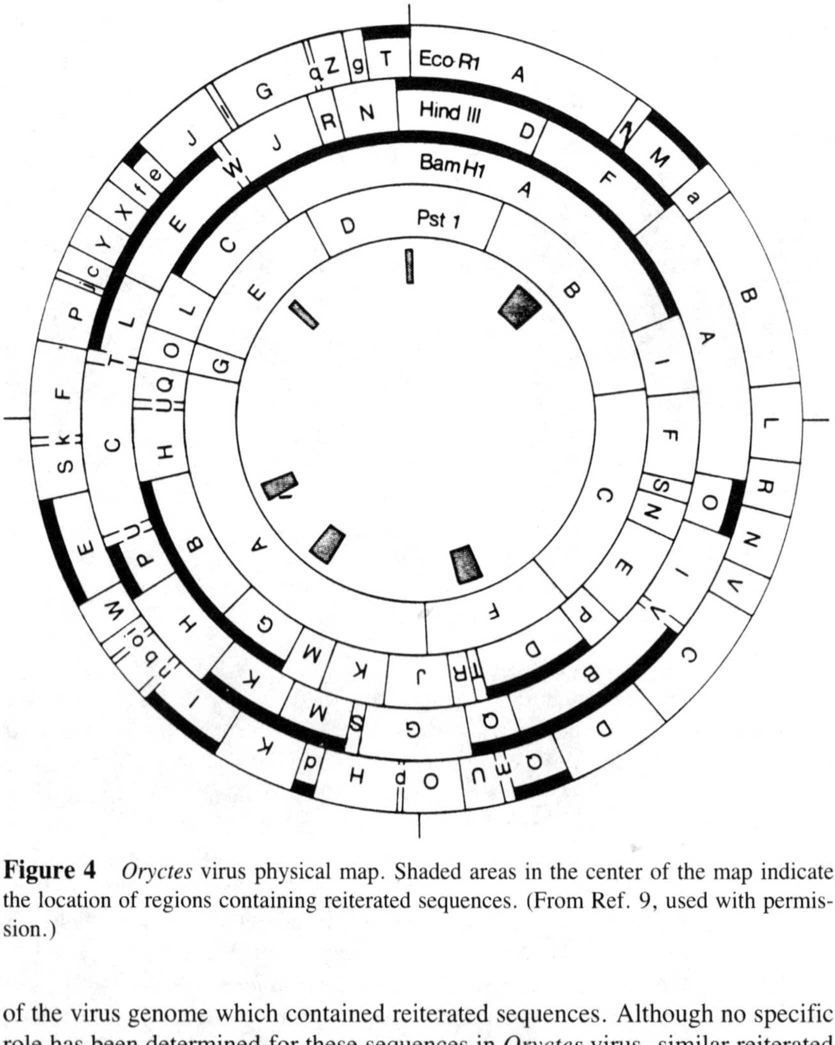


**Figure 4** *Oryctes* virus physical map. Shaded areas in the center of the map indicate the location of regions containing reiterated sequences. (From Ref. 9, used with permission.)

of the virus genome which contained reiterated sequences. Although no specific role has been determined for these sequences in *Oryctes* virus, similar reiterated sequences in the occluded baculovirus *Autographa californica* nuclear polyhedrosis virus (AcNPV) have been found to have enhancer activity (14).

## HZ-1 BACULOVIRUS

### Persistent Nature of the Hz-1 Virus

The nonoccluded baculovirus Hz-1 contains a supercoiled circular double-stranded DNA genome with an estimated size of 236 kbp in length (15, 16). This is almost twice the size of the *Oryctes* virus genome. Hz-1 was originally identified as a persistent agent in the IMC–Hz-1 tissue culture cell line (17). This cell line was established by Hink and Ignoffo (18) from adult ovarian tissue of *Heliothis zea*. It has been shown that unlike other baculoviruses, Hz-1 can infect a large number of cell lines derived from lepidopterans (17, 19, 20). Not only can the virus infect a number of different insect tissue culture cells but it can also persist in many of these cell lines (20, 21). TN-368 cells persistently infected with Hz-1 show very little evidence of replication of this virus and are permissive for AcNPV replication but resistant to superinfection by Hz-1 virus (20). This is a very common characteristic of other persistently infected cell cultures (22), particularly those persistent infections mediated by defective interfering (DI) particles (23, 24).

Burand and Wood (25) demonstrated that the establishment and maintenance of persistent Hz-1 infections depend on the presence and level of DI virus particles. The reader is referred to reviews by Burand et al. (26) and Wood and Burand (27) for a more complete description of Hz-1 virus persistence. The DI virus particles are much shorter and more heterogeneous than standard Hz-1 virus particles (28). Standard Hz-1 virus particles are $414 \pm 30 \times 80 \pm 3$ nm (Figure 5A) while DI particles are $370 \pm 76$ nm $\times$ $80 \pm 3$ nm (Figure 5B). The heterogeneity of DI particle preparations is also reflected in their DNA content. Restriction enzyme analysis of Hz-1 DI particle DNA has revealed a large number of deletions along the virus genome as evidenced by the occurrence of bands present in different stoichiometric amounts (27) (Figure 6).

## Replication of Hz-1 Virus in Cell Culture

Infection of the fibroblast-like cell line TN-368 results in a rounding of the cells by about 8 h p.i. and lysis of infected cells as early as 12 h p.i. (20, 21). EM studies by Granados (29) indicated that the virus enters cells by fusion of the virus envelope and the cytoplasmic membrane (although entry by viropexis could not be ruled out). Once in the cytoplasm the nucleocapsids migrate to the nuclear pore where uncoating takes place. Much like the *Oryctes* virus, the appearance of membrane material surrounding nucleocapsid shells in nuclei of infected cells is the first sign of Hz-1 virus replication. These shells are then filled with electron-dense material. Maturation of the virus proceeds with entry of the particle into the cytoplasm via openings which appear in the nuclear membrane and eventual release of virus particles by cell lysis.

A plaque-purified isolate of Hz-1 virus (designated B1) was prepared from virus recovered from IMC–Hz-1 cells in order to study the kinetics of virus replication in cell culture. An increase in the level of cell-associated virus (CAV) was first detected at 8 h p.i., followed by an increase in the level of extracellular virus (ECV) at about 12 h p.i. Both CAV and ECV levels increase logarithmically to about $4 \times 10^7$ plaque-forming units per milliliter by 20 h p.i. (16).

Viral DNA replication was followed using both DNA hybridization and [$^3$H]thymidine incorporation (25). From these results it was determined that Hz-1

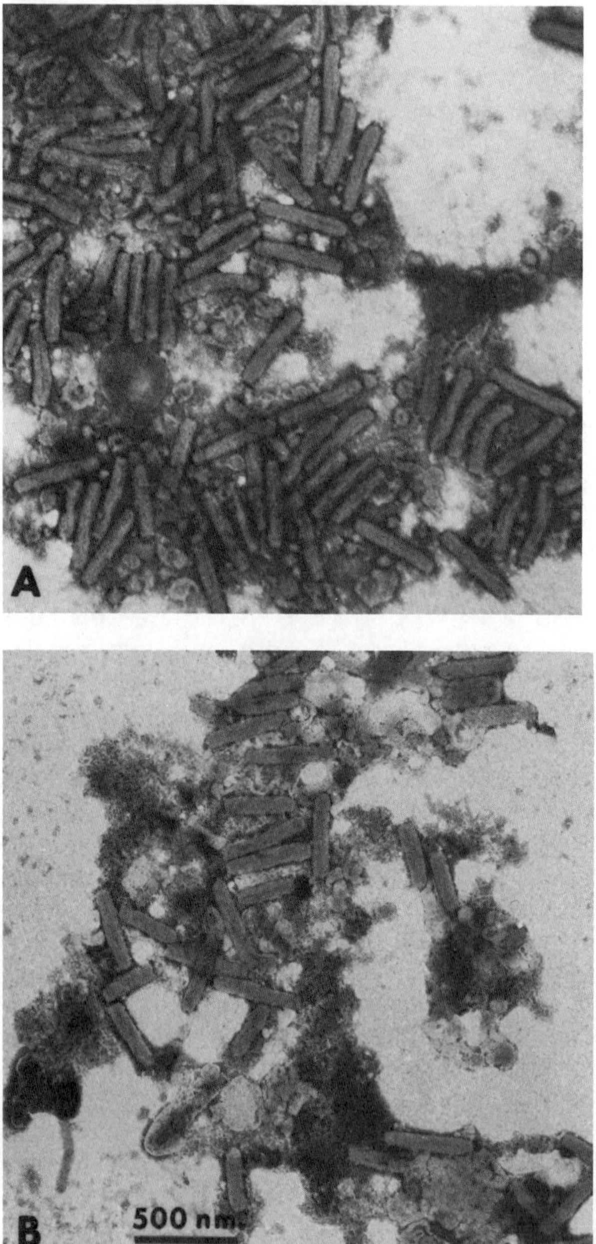


**Figure 5** Electron micrographs of standard (A) and defective (B) virus particles. (From Ref. 16, used with permission.)

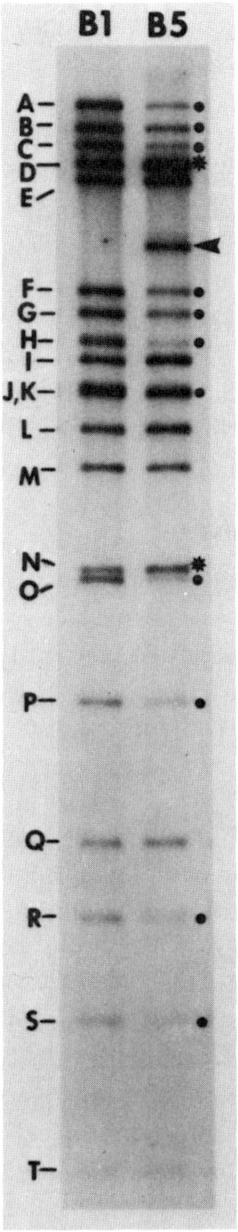


**Figure 6** EcoRI restriction enzyme fragment pattern of DNA from standard (B1) and defective (B5) Hz-1 virus. Letters at left identify individual DNA fragments. Small circles indicate bands present at lower concentrations in defective (B5) virus DNA. Arrow indicates additional DNA fragment in defective (B5) virus DNA. Asterisks indicate comigrating fragments. (From Ref. 25, used with permission.)

DNA replication begins at 4 h p.i. and is concomitant with the shutoff of host cell DNA replication.

## Hz-1 Virus Structural and Intracellular Proteins

Using [$^{35}$S]methionine labeling, Burand et al. (28) detected 28 Hz-1 virus structural proteins (Figure 7). This is almost the same number determined by

Crawford and Sheehan (12) for the *Oryctes* virus. Fourteen of these virus structural proteins were determined to be glycosylated based on labeling experiments using *N*-[$^3$H]acetylglucosamine and [$^3$H]mannose.

In time course experiments using [$^{35}$S]methionine pulse labeling, 38 VSIPs were detected during the first 12 h of virus replication (28). These proteins ranged in size from 14 to 200 kD (Table 1). Temporal regulation was detected with several of these proteins (mainly 72, 75, and 82 kD) based on their appearance and disappearance during the course of virus replication. Unlike the *Oryctes* virus, Hz-1 virus appears to actively turn off host cell protein synthesis. Pulse chase experiments revealed that two viral proteins, the 82 and the 109 kD proteins, undergo posttranslational modifications. Further studies have shown that the 109 kD protein, which is a highly glycosylated intracellular protein (28), is most likely deglycosylated to a 107 kD structural protein (21).

## Involvement of DI Particles in Hz-1 Virus Persistence

Several researchers have reported the induction of a persistent baculovirus in a variety of different insect populations (30–33). This type of virus persistence or latency appears to be common in insects (34). It has been suggested that Hz-1 virus serve as a model to study the persistence of other baculoviruses (26). This is strengthened by the fact that productive replication of Hz-1 virus can be induced in IMC–Hz-1 cells simply by challenging these cells with an occluded baculovirus (19). A variety of factors, such as the host immune response, play an important role in virus persistence in the insect. Since many of these factors are not present under in vitro conditions, care must be taken when extrapolating from the information gained by studying the persistence of Hz-1 virus in cell culture. Nonetheless, Hz-1 virus has proven to be a very useful tool in the study of baculovirus persistence.

It is clear from results reported by Burand and colleagues (25, 26) that DI particles play an important role in the establishment and maintenance of Hz-1 virus persistence. Unlike the virus isolated from IMC–Hz-1 cell line, the plaque-purified standard virus, B1, could not establish persistent infections in cell culture (26). Much like other animal viruses, DI particles evolved during serial passage of this Hz-1 isolate in culture. These DI particles were evident by the appearance of both shorter virus particles (16) (Figure 5) and smaller viral genomes (16, 25). Shorter viral genomes were evident in EcoRI digests of DNA purified from DI particles which had at least 11 submolar bands (Figure 6). These shortened genomes apparently arose through deletion of up to 92 kbp of DNA within the standard virus chromosome. This means that within the DI virus population there are particles that contain deletions of up to 40% of the standard virus genome. DI genomes were also found to be present in the nuclei of TN-368 cells persistently infected with Hz-1 virus (16).

Burand and Wood (25) examined the pattern of VSIP synthesis in cells infected with DI particle preparations (Figure 8). These cells were destined to

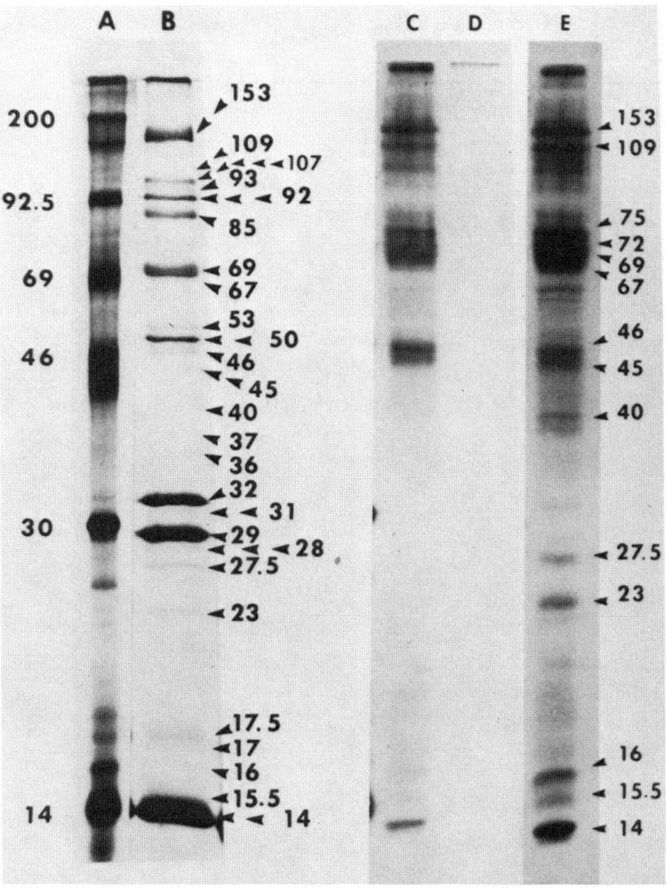


**Figure 7** Structural proteins of Hz-1 virus. Arrows indicate virus proteins and their apparent molecular weights ($10^{-3}$). Molecular weight markers (A), [$^{35}$S]methionine-labeled proteins (B), *N*-[$^{3}$H]acetylglucosamine- (D), and [$^{3}$H]mannose-labeled (E) glycoproteins. Lane D is Hz-1 virus labeled with *N*-[$^{3}$H]acetylglucosamine in the presence of the glycosylation inhibitor tunicamycin. (From Ref. 28, used with permission.)

become persistently infected with Hz-1 virus. They found that there were no differences in the number of virus proteins made in DI-infected cells compared to standard virus-infected cells. However, there was an alteration in the rate of synthesis of about half of the 38 VSIPs (Table 1). Since DI preparations contain both standard and defective virus genomes, the lower levels of individual proteins in DI-infected cells could be the result of a lower copy number of genes coding for these proteins. However, this does not explain the relative increase in

**Table 1** Intracellular Protein Synthesis During Standard and Defective Hz-1 Virus Replication

| Mol. wt($10^{-3}$) | Alteration in synthesis[a] | Structural protein[b] |
|---|---|---|
| 200 | N | |
| 153 | N | G |
| 139 | N | |
| 109 | – | G |
| 107 | + | + |
| 104 | N | |
| 93 | – | + |
| 92 | – | + |
| 90 | – | |
| 85 | – | + |
| 82 | N | |
| 87 | N | |
| 75 | + | G |
| 72 | N | G |
| 69 | – | G |
| 67 | N | G |
| 53 | N | + |
| 50 | – | + |
| 47 | N | |
| 46 | N | G |
| 45 | N | G |
| 41 | + | |
| 40 | – | G |
| 38 | N | |
| 37 | + | + |
| 36 | N | N |
| 32 | N | + |
| 31 | N | + |
| 30 | N | |
| 29 | N | + |
| 28 | N | + |
| 27.5 | – | G |
| 26.5 | N | |
| 23 | – | G |
| 17.5 | – | + |
| 17 | – | + |
| 16 | N | G |
| 15.5 | N | G |
| 14.2 | + | |
| 14 | – | G |

[a]+ and – indicates relative rates of synthesis during DI virus replication. N indicates no change detected. From Ref. 25, with permission.
[b]+ indicates a structural protein, G indicates a structural glycoproteins. From Ref. 28, with permission.

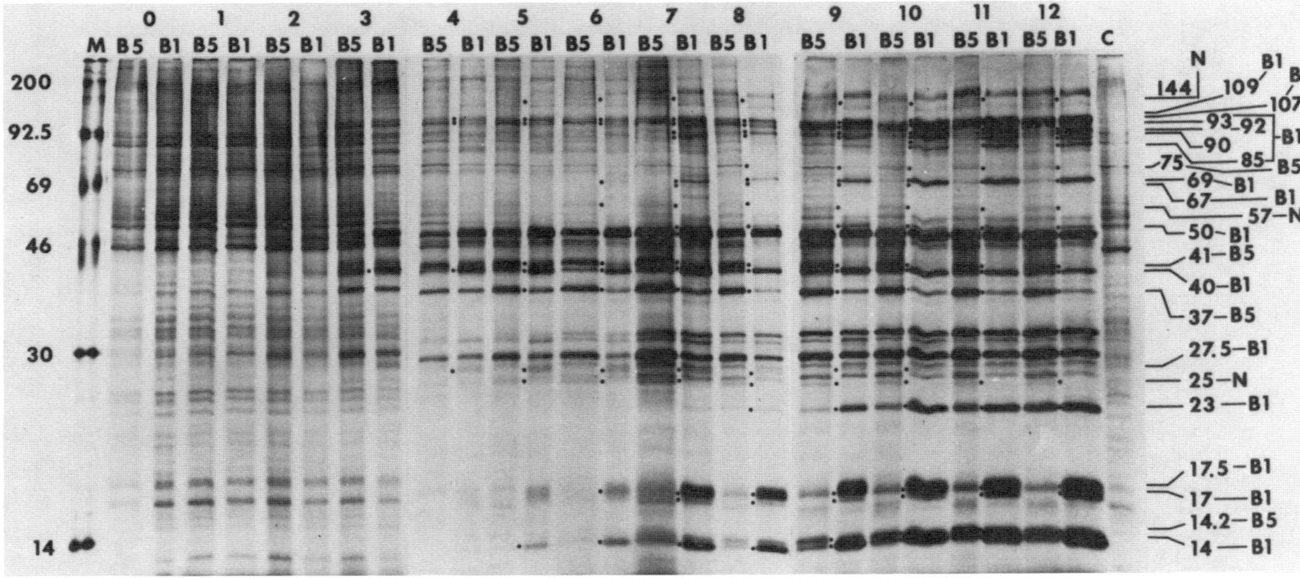


**Figure 8** Hz-1 VSIP in standard (B1) and defective (B5) infected cells pulse-labeled with [$^{35}$S]methionine for 1 h starting at times indicated. Numbers at right indicate the apparent molecular weights ($\times 10^{-3}$) of (N) "new" proteins and of proteins present in higher relative amounts in standard (B1) and defective (B5) virus-infected cells. Molecular weight markers (M) and mock infected cells (C). (From Ref. 25, used with permission.)

the rate of synthesis of proteins like 14.2, 37, and 41 kD, which are present at higher levels in DI-infected cells. These differences may reflect differences in the regulation of specific viral gene products responsible for the establishment of persistent infections by Hz-1 virus. It is also interesting to note that at least three "new" proteins were detected only in DI-infected cells (144, 57, and 25 kD). These could be viral proteins expressed only under conditions leading to persistence or they could be host cell proteins which are involved in the establishment of persistent infections. These questions can be answered only through further investigations into the molecular biology of Hz-1 virus persistence.

## CONCLUSION

The nonoccluded baculoviruses are a very interesting and important group of insect viruses. Only two members of this group, Hz-1 and the *Oryctes* virus, have been studied in any detail to date. Both of these viruses have demonstrated important practical applications. The *Oryctes* virus has been used successfully in the control of the rhinoceros beetle throughout the tropics, and Hz-1 currently serves as the model for the study of baculovirus persistent infections.

The potential of nonoccluded baculoviruses as control agents for insect pests and as a source of genetic information to improve other microbial pesticides has not been fully realized. The development of new cell culture systems coupled with improved isolation and screening methods will allow more research into the molecular biology of these viruses. It is hoped that future research with these viruses will provide information useful to baculovirology and virology as a whole.

## REFERENCES


1. Matthews, R. E. F. Classification and nomenclature of viruses. *Intervirology* 17, 52, 1982.
2. Crawford, A. M. and Granados, R. R. Non-occluded baculoviruses, *Proceedings Third International Colliquium on Invertebrate Pathology*, 6–10 September 1982, Brighton, U.K., 1982, p. 154.
3. Huger, A. M. A virus disease of the Indian rhinoceros beetle *Oryctes rhinoceros* (Linneaeus) caused by a new type of insect virus, Rhabdionvirus oryctes gen. n., sp. n. *J. Invertebr. Pathol.* 8, 38, 1966.
4. Marschall, K. J. Introduction of a new virus disease of the coconut rhinoceros beetle in Western Samoa. *Nature (London)* 225, 288, 1970.
5. Zelazny, B. Transmission of a baculovirus in populations of *Oryctes rhinoceros. J. Invertebr. Pathol.* 27, 221, 1976.
6. Bedford, G. O. Biology, ecology, and control of palm rhinoceros beetles. *Ann. Rev. Entomol.* 25, 309, 1980.

7. Purrini, K. Baculovirus oryctes release into *Oryctes monoceros* population in Tanzania, with special reference to the interaction of virus isolates used in our laboratory infection experiments. *J. Invertebr. Pathol.* 53, 285, 1989.
8. Payne, C. C. The isolation and characterization of a virus from *Oryctes rhinoceros*. *J. Gen. Virol.* 25, 105, 1974.
9. Crawford, A. M., Ashbridge, K., Sheehan, C., and Falkner, P. A physical map of the *Oryctes* baculovirus genome. *J. Gen. Virol.* 66, 2649, 1985.
10. Zelazny, B. Studies on Rhabdionvirus oryctes. I. Effect on larvae of *Oryctes rhinoceros* and inactivation of the virus. *J. Invertebr. Pathol.* 20, 235, 1972
11. Zelazny, B. Studies on Rhabdionvirus oryctes. II. Effect on adults of *Oryctes rhinoceros*. *J. Invertebr. Pathol.* 22, 122, 1973.
12. Crawford, A. M. and Sheehan, C. Replication of *Oryctes* baculovirus in cell culture: Viral morphogenesis, infectivity and protein synthesis. *J. Gen. Virol.* 66, 529, 1985.
13. Payne, C. C., Compson, D., and de Looze, S. M. Properties of the nucleocapsids of a virus isolated from *Oryctes rhinoceros*. *Virology* 77, 269, 1977.
14. Guarino, L. A. and Summers, M. D. Interspersed homologous DNA of *Autographa californica* nuclear polyhedrosis virus enhances delayed-early gene expression. *J. Virol.* 60, 215, 1986.
15. Huang, Y., Hedberg, J., and Kawanishi, C. Y. Characterization of the DNA of a nonoccluded baculovirus, Hz-1V. *J. Virol.* 43, 174, 1982.
16. Burand, J. P., Wood, H. A., and Summers, M. D. Defective particles form a persistent baculovirus infection in *Trichoplusia ni* tissue culture cells. *J. Gen. Virol.* 63, 391, 1983.
17. Granados, R. R., Nguyen, T., and Cato, B. An insect cell line persistently infected with a baculovirus-like particle. *Intervirology* 10, 309, 1978.
18. Hik, W. F. and Ignoffo, C. M. Establishment of a new cell line (IMC-Hz-1) from ovaries of the cotton bollworm moth, *Heliothis zea* (Boddie). *Exp. Cell Res.* 60, 307, 1970.
19. Kelly, D. C., Lescott, T., Ayres, M. D., Carey, D., Coutts, A., and Harrap, K. A. Induction of a nonoccluded baculovirus persistently infecting *Heliothis zea* cells by *Heliothis armigera* and *Trichoplusia ni* nuclear polyhedrosis viruses. *Virology* 112, 174, 1981.
20. Ralston A. L., Huang, Y., and Kawanishi, C. Y. Cell culture studies with the IMC-Hz-1 nonoccluded virus. *Virology* 115, 33, 1981
21. Burand, J. P. and Wood, H. A. Unpublished data, 1986.
22. Joklik, W. K. Mechanisms of establishment and maintenance of persistent infections, in *Microbiology 1977*, Schlessinger, D., Ed., American Society for Microbiology, Washington, D.C., 1977, p. 434.
23. Nayak, D. P. Defective interfering influenza viruses. *Ann. Rev. Microbiol.* 34, 619, 1980.
24. Huang, A. S. and Baltimore, D. Defective viral particles and viral disease processes. *Nature (London)* 228, 325, 1970.
25. Burand, J. P. and Wood, H. A. Intracellular protein synthesis during standard and defective Hz-1 virus replication. *J. Gen. Virol.* 67, 167, 1986.

26. Burand, J. P., Kawanishi, C. Y., and Huang, Y.-S. Persistent baculovirus infections, in *The Biology of Baculoviruses*, vol. 1, Granados, R. R. and Federici, B. A., Eds., CRC Press, Boca Raton, FL, 1986, p. 159.
27. Wood, H. A. and Burand, J. P. Persistent and productive infections with the Hz-1 baculovirus, in *The Molecular Biology of Baculoviruses. Current Topics in Microbiology and Immunology*, vol. 131, Doerfler, W. and Bohm, P., Eds., Springer-Verlag, Berlin, 1986, p. 119.
28. Burand, J. P., Stiles, B., and Wood, H. A. Structural and intracellular proteins of the nonoccluded baculovirus Hz-1. *J. Virol.* 46, 137, 1983.
29. Granados, R. R. Personal communication, 1985.
30. Longworth, J. F. and Cunningham, J. C. The activation of occult nuclear-polyhedrosis virus by foreign polyhedra. *J. Invertebr. Pathol.* 10, 361, 1986.
31. Jurkovicova, M. Activation of latent infections in larvae of *Adoxophyes orana* (Lepidoptera: Torticidae) and *Barathra brassicae* (Lepidoptera: Noctuidae) by foreign polyhedra. *J. Invertebr. Pathol.* 34, 213, 1979.
32. Maleki-Milani, H. Influence de Passages répétes du virus de la polyhedrose nucleaire de *Autographa californica* chez *Spodoptera litteralis* (Lep.: Noctuidae). *Entomophaga* 23, 217, 1978.
33. McKinley, D. J., Brown, D. A., Payne, C. C., and Harrap, K. A. Cross-infectivity and activation studies with four baculoviruses. *Entomophaga* 26, 79, 1981.
34. Evans, H. F. and Harrap, K. A. Persistence of insect viruses, in *Virus Persistence*, Mahy, W. J., Minson, A. C., and Darby, G. K., Eds., Cambridge University Press, London, 1982, p. 57.

# 4

# Protection Against Coccidiosis with Recombinant *Eimeria acervulina* Merozoite Antigen Expressed in Baculovirus

MARK C. JENKINS, EDWARD M. DOUGHERTY, AND
SUSAN K. BRAUN
*Agricultural Research Service, U.S. Department of Agriculture, Beltsville, Maryland*

## INTRODUCTION

Coccidiosis, an intestinal disease caused by protozoan parasites of the genus *Eimeria*, affects many avian and mammalian species. With more widespread and intensified rearing practices, economic losses inflicted by these parasites on various chicken industries worldwide are approaching $1.5 billion per year. These losses are due in part to unrealized weight gains and the cost of administering anticoccidial drugs. The rapid emergence of drug-resistant strains of coccidia (1–4) and the prohibitive cost of developing new anticoccidials have provided the impetus for developing a subunit vaccine for prevention of this parasitic disease.

The rationale behind vaccine development is that a single coccidiosis infection induces species-specific protection against challenge (5). An obstacle to generating an effective coccidiosis vaccine is that neither the targeted antigen(s) nor the immune mechanism(s) involved in a protective response is known. Cellular immune responses appear to play a major role in resistance (8, 9) involving either direct killing or the parasite by cytotoxic T lymphocytes or the release of lymphokines by helper T cells (10, 11), which either destroy the parasite or recruit other effector cells to do so. Although somewhat controversial, partial protection against disease can be obtained by immunizing susceptible chickens with nonviable extracts of infected tissue (6) or sporozoite antigen (7). Recent studies in our laboratory using x-irradiation of oocysts to study the minimal level of parasite development required to establish immunity suggest that sporozoite penetration, but not merozoite development, is the crucial event (unpublished observations). These findings are in conflict with studies showing that prevention of merogonous development through the use of stage-specific drugs disallows effective resistance to a challenge infection. It is therefore uncertain which developmental stage(s) of the parasite is targeted by the protective response. A general consensus is that an eventual vaccine preparation will have to be composed of antigens from several species and stages, both asexual (sporozoite and merozoite) and sexual (gametes), in order to confer complete protection against these protozoan parasites.

In recent years, many research groups using genetic engineering techniques have produced recombinant *Eimeria* antigens (12–16) that, in some instances, induce partial protection against a challenge infection with live coccidia (17, 18). In our laboratory we have identified and expressed a cDNA (14) from *E. acervulina* that encodes an immunodominant merozoite surface antigen (EAMZ250) which is composed of tandem-repeated amino acid sequences (19). The respective parasite protein appears to play some role in the immune response elicited by the host during a natural coccidial infection since the recombinant antigen is recognized by immune serum and induces the in vitro proliferation of immune T lymphocytes (14, 20). Furthermore, immunization of chickens with purified EAMZ250 or unpurified extracts of bacteria containing EAMZ250 induces significant humoral and cellular immune responses (unpublished observations). However, only moderate levels of protection against coccidial challenge are obtained by this immunization scheme. For this reason, eukaryotic expression of EAMZ250 in recombinant baculovirus-infected insect cells was investigated as a possible means of generating immunoreactive EAMZ250 antigen that would induce higher levels of protection against challenge infection.

## MATERIALS AND METHODS

### Cloning and Expression Studies

EAMZ250 insert DNA was isolated by cutting pEAMZ250 (14) plasmid DNA with EcoRI, electrophoresing the restriction digest fragments through agarose, electroelution in dialysis membrane, and passage over a NACS column (Bethesda Research Laboratories). EAMZ250 insert DNA was then subcloned into pUC19 in an orientation that would allow for in-frame excision relative to the upstream BamHI site of the plasmid. Recombinant plasmid containing the EAMZ250 gene in the correct orientation was prepared from transformant *Escherichia coli* JM101 cells and treated with both BamHI and PvuII, and insert EAMZ250 DNA was purified as described previously. Using T4 DNA ligase, purified EAMZ250 was inserted into the baculovirus transfer vector pAC311 (21, kindly provided by Dr. M. Summers), which had been previously digested with BamHI and treated wth bacterial alkaline phosphatase (Figure 1). The ligation mixtures were then treated with T4 DNA polymerase to fill in any nonligated PvuII or BamHI ends, phenol-chloroform extracted, ethanol pre-

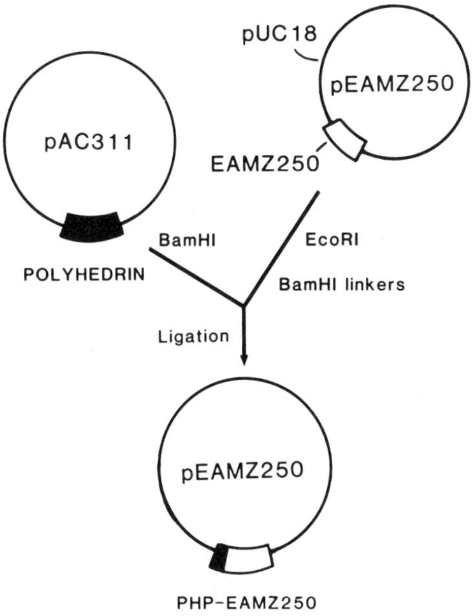


**Figure 1** Insertion of the gene for EAMZ250 merozoite antigen into the baculovirus transfer vector pAC311 by subcloning in frame with a short segment of the polyhedrin protein coding sequence. PHP, polyhedrin protein.

cipitated, and treated once again wth T4 DNA ligase to join blunt ends of the vector and insert. The ligation mixtures were used to transform *E. coli* DH5 using standard procedures (22). Transformants were analyzed by restriction enzyme digestion of miniplasmid preparations (23) for recombinant plasmid containing EAMZ250 in the correct orientation. The EAMZ250 gene sequence was inserted into the baculovirus genome by homologous recombination between PHP flanking sequences on the transfer vector that surrounded the parasite-specific insert DNA and the intact PHP sequence in the insect virus DNA. This insertion was accomplished by cotransfecting *Spodoptera frugiperda* insect cells with wild-type baculovirus and recombinant transfer vector pAC311 containing the EAMZ250 gene sequence using standard procedures (24).

Insect cells were cultured in 24 well plates (Falcon Plastics) for seven days in IPL-41 media containing 10% fetal bovine serum and analyzed for recombinant baculovirus by dot-blot hybridization. Prior to dot-blot analysis, the supernatant from each well was removed and saved as a source of recombinant virus. Viral DNA was prepared by alkaline treatment of the cultured insect cells using described procedures (24). An aliquot (10 $\mu$l) of neutralized supernatant from each well of the microtiter plate was pipetted onto Nytran paper in a gridlike array. The dot blots were baked at 80°C for 2 h in a drying oven and then treated with prehybridization buffer (23) for 4–6 h at 65°C. Insert EAMZ250 DNA was radiolabled with [$^{32}$P]dCTP by nick translation, denatured by heating in a boiling water bath for 3–5 min, and incubated with the dot blots in prehybridization fluid for 16 h at 65°C. The hybridized blots were washed three times in 0.2X standard saline citrate, 0.1% sodium dodecyl sulfate buffer at 50°C for 30 min per wash, and exposed to Kodak XAR film. Virus from microtiter wells showing hybridization to the EAMZ250 probe (Figure 2A) were then cloned by limiting dilution and analyzed for recombinant baculovirus DNA as described above (Figure 2B) in order to obtain a clonal population of high-titer recombinant virus.

Insect cells infected with recombinant or control baculovirus were analyzed for expression of immunoreactive EAMZ250 protein by sodium dodecylsulfate polyacrylamide gel electrophoresis (SDS-PAGE; 25) followed by immunoblotting onto nitrocellulose paper using standard procedures (26). Nitrocellulose paper impregnated with recombinant PHP-EAMZ250 and control PHP antigen extracts were immunostained with mouse anti-EAMZ250 serum, followed by biotinylated goat antimouse IgG (Vector Laboratories), avidin-peroxidase (Sigma Chemical Co.), and 4-chloro-1-napthol/$H_2O_2$ substrate.

## Vaccination Trials

Outbred seven-day-old sexsal chickens ($n$ = 15 per group) were administered either recombinant PHP-EAMZ250 or control PHP antigen by intramuscular injection of 1.0 $\mu$g antigen extract. Each group received an identical dose by intramuscular injection one week later and was then challenged in seven days with 5 × $10^5$ *E. acervulina* oocysts. Controls included chickens that were not immunized and either were challenged (UIC) or were not challenged (UIUC) with *Eimeria* parasites. Protection was assessed at day 6 by measuring weight gain during the course of the infection (from day of challenge to day of termination) and intestinal lesion scores using standard procedures (27).

## Enzyme-Linked Immunosorbent Assay

Serum was obtained by wing bleed prior to immunization (day 1), prior to parasite challenge (day 15), and on day of termination (day 21) using standard procedures (28). Serum samples from individual chickens in each group were pooled, diluted 1:100, and assayed for immunoglobulin specific for control PHP antigen and recombinant PHP-EAMZ250 antigen by enzyme-linked immunosorbent assay (ELISA). Individual wells of polystyrene microtiter plates (Immunlon II, Dynatech) were coated with 0.1 $\mu$g of PHP or PHP-EAMZ250

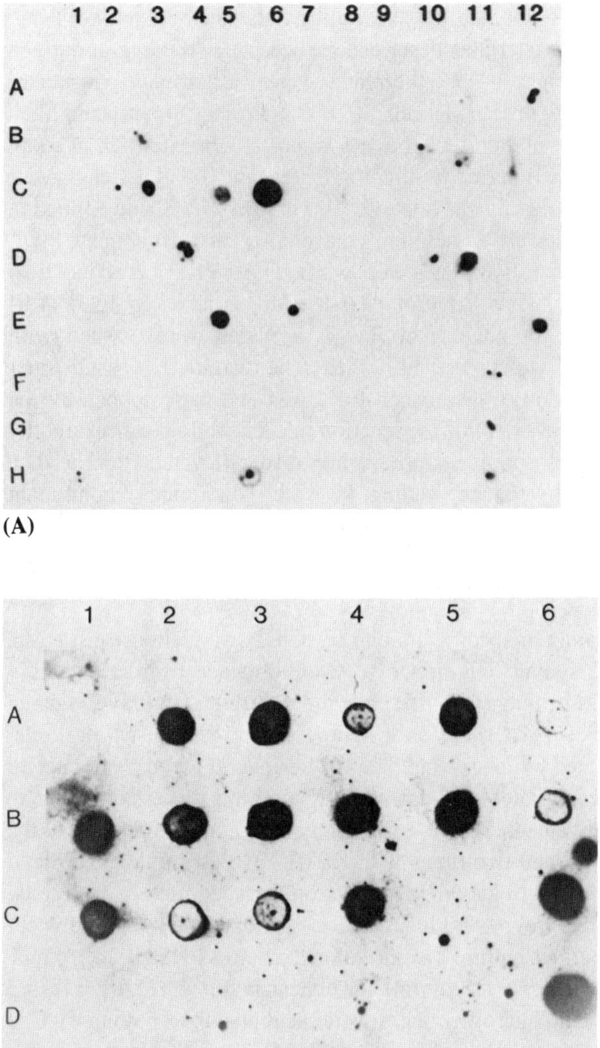


(A)

(B)

**Figure 2** Analysis of *Spodoptera frugiperda* insect cells for recombinant baculovirus containing gene sequences for *Eimeria acervulina* 250 kD merozoite protein (EAMZ250) by probing dot blots of infected cells with radiolabeled insert EAMZ250 DNA. Primary screen in microtiter wells (A) and secondary screen of clonal population grown in 24 well plates (B).

protein in 0.1 *M* sodium carbonate pH 9.5 buffer obtained by osmotic shock treatment of transformant bacteria as described earlier. After binding of antigen for 1 h at 37°C and overnight at 4°C, the plates were washed with phosphate-buffered saline containing 0.05% Tween 20 (PBS-TW20, Sigma) and then treated with PBS-TW20 containing 1% bovine serum albumin for 1 h at room temperature (RT) to reduce nonspecific binding of reagents in later steps. After the blocking step, immune serum was diluted 1:100 in PBS-TW20 and applied in duplicate for 1 h at RT to wells containing control PHP or recombinant PHP-EAMZ250 antigen. The microtiter plates were washed with PBS-TW20 and then individual wells were treated with peroxidase-labeled antichicken IgG, A,M (Miles Immunochemical) for an 1 h at RT. The plates were washed with PBS-TW20 and individual wells received *o*-phenylene diamine-$H_2O_2$ substrate. Color development was allowed to proceed for 5 min and then stopped by the addition of 1 *N* $H_2SO_4$. The extent of the reaction in each well was measured by absorbance at O.D. 492 nm on a microtiter plate reader (Dyantech). The data were expressed as the absorbance reading in wells containing recombinant PHP-EAMZ250 minus the reading in wells containing control PHP.

## T Cell Lymphoblastogenesis Assay

To determine whether antigen-specific cellular responses were elicited using the immunization protocol, splenic T lymphocytes were obtained from three chickens per group on day of challenge (day 15) and day of termination (day 21) using described procedures (29). To serve as a source of antigen-presenting cells, macrophages were isolated by incubating $5 \times 10^5$ peripheral blood mononuclear cells in the blastogenesis plates one day before initiating the assay. Prior to plating the nylon-wool nonadherent T cell-enriched splenic lymphocytes, the microtiter plates were washed two times with RPMI 1640 containing 10% fetal bovine serum (culture medium) leaving a population enriched for adherent cells. T lymphocytes from each group were incubated in triplicate at $5 \times 10^5$ cells per cell in the presence of either culture medium alone, control PHP (0.25 $\mu$g/ml), recombinant PHP-EAMZ250 (0.25 $\mu$g/ml), or concanavalin A (25 $\mu$g/ml) in a total volume of 0.2 ml. The microtiter plates were incubated for 5 days at 41°C in a 10% $CO_2$/90% air incubator. The cells were pulsed for 16 h prior to harvesting with 0.5 $\mu$Ci of thymidine ([$^3$H]TdR, Amersham) per well. T cell blastogenesis was measured by harvesting the lymphocytes on a PHD cell harvester and counting the insoluble cell fraction on a scintillation counter. Activation in the presence of various antigens as indicated by cell division and incorporation of [$^3$H]TdR was expressed as uptake in counts per minute corrected for the amount of thymidine incorporation by cells incubated with culture medium alone. The stimulation data were compared between groups using Student's *t* test of significance.

## RESULTS

### Expression of Immunoreactive PHP-EAMZ250

*Spodoptera frugiperda* cells infected with recombinant or control baculovirus were harvested and analyzed for production of PHP-EAMZ250 antigen by immunoblotting using mouse sera specific for the parasite portion of the fusion protein. A broad immunoreactive band about 35 kD in size was detected in recombinant virus-infected cells, but not in insect cells infected with control baculovirus (Figure 3, lanes 1 and 2, respectively). Silver staining of polyacrylamide gels containing recombinant PHP-EAMZ250 showed this 35 kD antigen to represent about 20% of the total cell protein (Figure 3, lane 3).

### Vaccination Trials with Recombinant PHP-EAMZ250

Attempts were made to immunize chickens against coccidiosis by intramuscular injection of recombinant PHP-EAMZ250 protein. Compared to PHP-immunized controls, chickens given recombinant EAMZ250 antigen showed significant ($P < 0.05$) protection against challenge with live coccidia as measured by weight gain (Figure 4A). This weight gain was similar to that observed in the un-

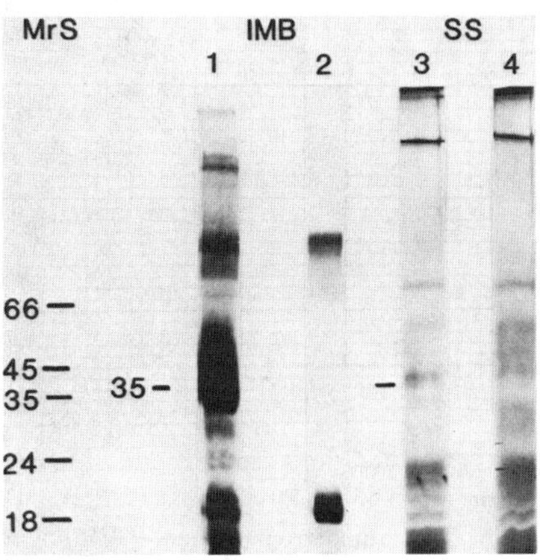


**Figure 3** Analysis of nonrecombinant polyhedrin protein (PHP) and recombinant PHP-EAMZ250 expression in baculovirus-infected *Spodoptera frugiperda* cells by SDS-PAGE and immunoblotting with mouse anti-EAMZ250 serum. Lanes 1 and 3, protein extract of recombinant baculovirus-infected insect cells; lanes 2 and 4, protein extract of wild-type baculovirus-infected insect cells; lanes 1 and 2, immunoblotting with immune mouse serum; lanes 3 and 4, silver staining. MrS, molecular weight standards in kilodaltons; IMB, immunoblot; SS, silver stain.

(A)

(B)

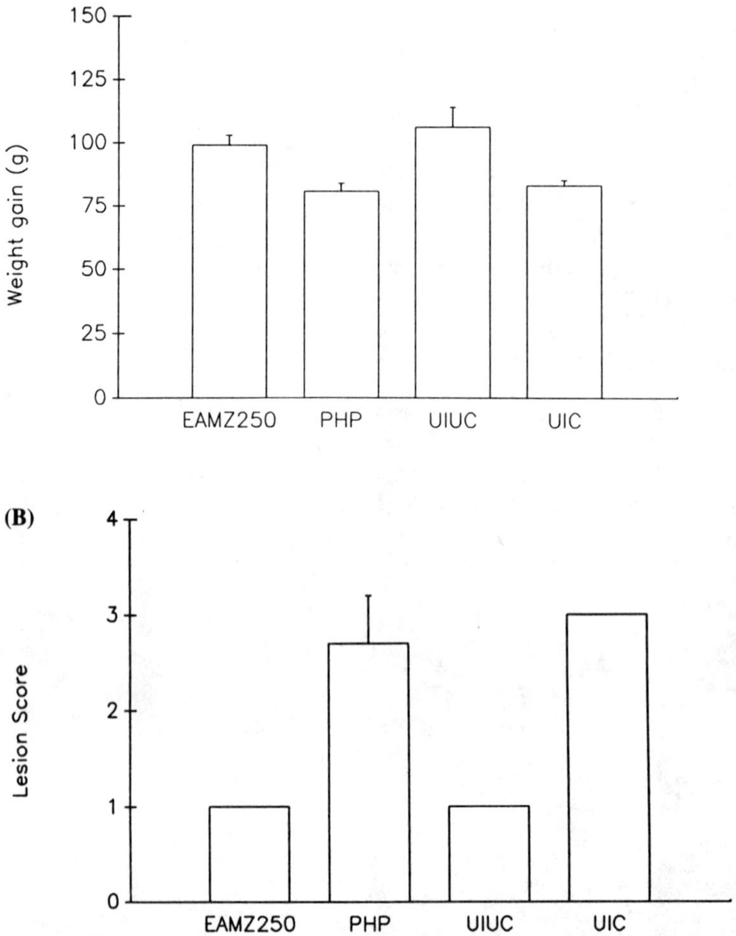


**Figure 4** Protective effects against weight loss (A) and lesions score (B) after *Eimeria acervulina* challenge with intramuscular injection of recombinant PHP-EAMZ250 antigen. EAMZ250 and PHP, recombinant PHP-EAMZ250 and nonrecombinant PHP antigen given by intramuscular injection; UIUC, unimmunized/unchallenged controls; UIC, unimmunized/challenged controls.

challenged control group (Figure 4A). Although a more subjective test, lesion scores in chickens immunized with recombinant PHP-EAMZ250 were reduced when compared to PHP-immunized controls (Figure 4B). Lesion scores were decreased from 2.7 ± 0.3 in the PHP-immunized control groups to 1.0 ± 0.0 in chickens immunized with recombinant PHP-EAMZ250.

## Humoral and Cellular Responses to PHP-EAMZ250

Prior to parasite challenge, chickens immunized with the recombinant merozoite antigen produced significant levels of immunoglobulin specific for PHP-EAMZ250 (Figure 5). Although the ELISA readings were corrected for binding in the presence of PHP extract alone, there appeared to be a measurable amount of nonspecific stimulation by the control protein preparation. Chickens immunized with control unfused PHP exhibited significant levels of anti-EAMZ250 immunoglobulin (Figure 5). Challenge with live coccidia served to boost the level of anti-EAMZ250 immunoglobulin in both recombinant and control antigen groups (Figure 5). This is not unexpected, since the EAMZ250 epitopes are present on the surface of merozoites that may have appeared during the injection.

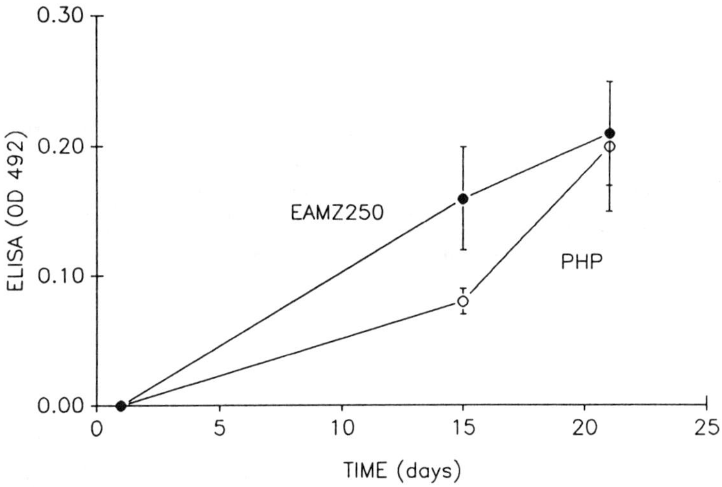


**Figure 5** Humoral response to recombinant PHP-EAMZ250 antigen in chickens immunized with recombinant PHP-EAMZ250 or nonrecombinant PHP antigen by intramuscular injection of baculovirus-infected *Spodoptera frugiperda* cell extract as measured by enzyme-linked immunosorbent assay. All data are expressed as absorbance readings in microtiter wells containing PHP-EAMZ250 minus the absorbance in wells containing PHP alone. Day 1, preimmunization; day 15, day of challenge; day 21, day of termination.

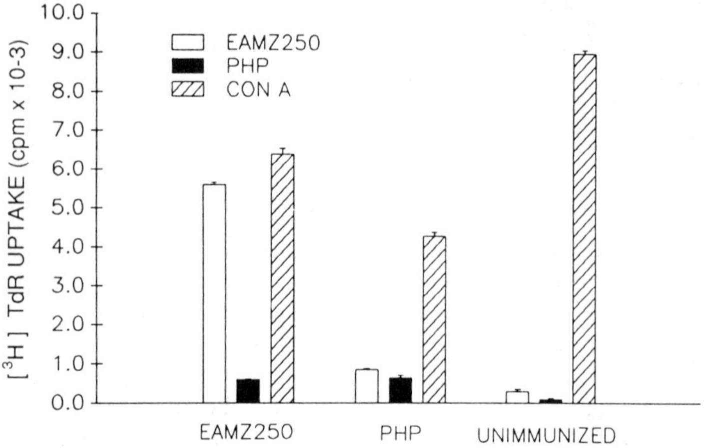


**Figure 6** In vitro stimulation of T lymphocytes obtained from chickens immunized with recombinant PHP-EAMZ250 or nonrecombinant PHP by intramuscular injection of baculovirus-infected *Spodoptera frugiperda* cell extract. Stimulating antigens, PHP-EAMZ250 and PHP (0.25 $\mu$g/well) or concanavalin A (25 $\mu$g/well). Activation expressed as [$^3$H]thymidine uptake in wells containing antigen or mitogen corrected for background stimulation in wells incubated with culture medium alone.

Immunization with recombinant merozoite antigen produced in baculovirus also elicited significant T cell stimulation. EAMZ250-immunized chickens showed EAMZ250-specific stimulation oF T lymphocytes in contrast to a much lower response observed in PHP-immunized chickens (Figure 6). It is unclear why PHP- or PHP-EAMZ250-immunized chickens did not exhibit any measurable level of in vitro activation by PHP antigen (Figure 6). It is possible that PHP is nonimmunogenic for T lymphocytes, in contrast to B cell responses, in this strain of chickens.

## DISCUSSION

This study demonstrated that eukaryotic expression of an *Eimeria* parasite antigen in baculovirus is a feasible approach to producing immunogenic proteins which can stimulate at least two effector arms of the immune system and also confer significant levels of protection against coccidial challenge. Previous attempts to vaccinate chickens by parenteral administration with EAMZ250 antigen produced in bacteria have yielded equivocal results (unpublished observations), leading us to hypothesize that the eukaryotic parasite protein was not being translated in a native configuration. Our goal in this study was to determine if production of merozoite antigen in a eukaryotic expression vector, such as baculovirus, would generate the appropriate epitopes on the parasite protein that are seen by the host immune system during a natural coccidial infection. Similar to immunization with purified EAMZ250 produced in bacteria, significant antigen-specific T cell and immunoglobulin responses were observed. It is possible that additional, more critical, epitopes were also generated since a significant level of protection was achieved.

Recent data from our laboratory suggest that protective immunization against this intestinal parasite depends on a number of factors and that the site of antigen delivery, in addition to the formation of the appropriate epitopes, plays an important role. Significant levels of protection have been achieved by immunization of chickens with live *Escherichia coli* transformants expressing recombinant EAMZ250 (18, 30). It is conceivable that recombinant antigens that are produced in bacteria and are released in vitro undergo a conformation change that destroys epitopes which would otherwise be recognized by the appropriate immune cells in the host. These same recombinant antigens might undergo different processing if they remain intracellular within the bacteria and are phagocytized by antigen-presenting cells and thus retain antigenic structure as found by eukaryotic expression. It will be critical to determine which specific immune cells are responding to the recombinant EAMZ250 antigen when produced in baculovirus since they may play an important role in immunity against disease.

## ACKNOWLEDGMENTS

We are grateful to Dr. M. Summers and colleagues for the generous supply of vectors provided for this study. We also acknowledge the input of Dr. M. Ruff and members of the Protozoan Diseases Laboratory during the vaccine trial phase of these studies.

## REFERENCES

1. McDougald, L. R. Anti-coccidial drug resistance in the southeastern United States: Polyether, ionophorous drugs. *Avian Dis.* 25, 600, 1981.
2. Jeffers, T. K. *Eimeria tenella*: Incidence, distribution, and anticoccidial drug resistance of isolates in major broiler-producing areas. *Avian Dis.* 18, 74, 1984.
3. Chapman, H. D. Drug resistance in avian coccidia (review). *Vet. Parasitol.* 15, 11, 1984.
4. Long, P. L. and Jeffers, T. K. Control of coccidiosis. *Parasitol. Today* 2, 236, 1986.
5. Rose, M. E. Immune responses of chickens to coccidia and coccidiosis, in *Avian Coccidiosis. Proceedings of the Thirteenth International Poultry Science Symposium*, Long, P. L., Boorman, K. N., and Freeman, B. M., Eds., British Poultry Science Ltd., Longman Group., Edinburgh,. p. 297.

6. McKenzie, M. E. and Long, P. L. Immunization of chickens against coccidiosis with extracts of *Eimeria*-infected tissues. *Poult. Sci.* 65, 892, 1986.
7. Murray, P. K., Bhogal, B. S., and Crane, M. S. *Eimeria tenella*: In vivo immunization studies with sporozoite antigen, in *Research in Avian Coccidiosis. Proceedings of the Georgia Coccidiosis Conference*, McDougald, L. R., Joyner, L. P., and Long, P. L., Eds., University of Georgia Press, Athens, p. 564.
8. Lillehoj, H. S. Effects of immunosuppression on avian coccidiosis: Cyclosporin A but not hormonal bursectomy abrogates host protective immunity. *Infect. Immun.* 55, 1616, 1987.
9. Rose, M. E., Wakelin, D., Joysey, H. S., and Hesketh, P. Immunity to coccidiosis: Adoptive transfer in NIH mice challenged with *Eimeria vermiformis*. *Parasite Immunol.* 19, 53, 1988.
10. Lillehoj, H. S., Kang, S. Y., Keller, L., and Sevoian, M. *Eimeria tenella* and *Eimeria acervulina*: Lymphokines secreted by an avian T cell lymphoma or sporozoite-stimulated immune T lymphocytes protects chickens against avian coccidiosis. *Exp. Parasitol.* 69, 54, 1989.
11. Rose, M. E., Wakelin, D., and Hesketh, P. Gamma interferon controls *Eimeria vermiformis* primary infection in BALB/c mice. *Infect. Immun.* 57, 1599, 1989.
12. Files, J. G., Leland, S. P., and Gabe, J. D. Identification and characterization of the gene for a major surface antigen of *Eimeria tenella*. *UCLA Symposium on Molecular and Cellular Biology*, vol. 42, *Molecular Strategies of Parasitic Invasion*. Allan R. Liss, New York, p. 713.
13. Brothers, V. M., Kuhn, I., Paul, L. S., Gabe, J. D., Andrews, W. H., Sias, S. M., McCanan, M. T., Dragon, E. A., and Files, J. G. Characterization of a surface antigen of *Eimeria tenella* sporozoites and synthesis from a cloned cDNA in *Escherichia coli*. *Mol. Biochem. Parasitol.* 28, 235, 1988.
14. Jenkins, M. C., Lillehoj, H. S., and Dame, J. B. *Eimeria acervulina*: DNA cloning and characterization of recombinant sporozoite and merozoite antigens. *Exp. Parasitol.* 66, 96, 1988.
15. Jenkins, M. C., Danforth, H. D., Lillehoj, H. S., and Fetterer, R. H. cDNA encoding an immunogenic region of a 22 kilodalton surface protein of *Eimeria acervulina* sporozoites. *Mol. Biochem. Parasitol.* 32, 153, 1989.
16. Profous-Juchelka, H., Liberator, P., and Turner, M. Identification and characterization of cDNA clones encoding antigens of *Eimeria tenella*. *Mol. Biochem. Parasitol.* 30, 233, 1988.
17. Miller, G. A., Bhogal, B. S., McCandliss, R., Strausberg, R. L., Jessee, E. J., Anderson, A. C., Fuchs, C. K., Nagle J., Likel, M. H., Strasser, J. M., and Strausberg, S. Characterization and vaccine potential of a novel recombinant coccidial antigen. *Infect. Immun.* 57, 2014, 1989.
18. Kim, K. S., Jenkins, M. C., and Lillehoj, H. S. Immunization of chickens with live *Escherichia coli* expressing *Eimeria acervulina* merozoite recombinant antigen induces partial protection against coccidiosis. *Infect. Immun.* 57, 2434, 1989.
19. Jenkins, M. C. A cDNA encoding a merozoite surface protein of the protozoan *Eimeria acervulina* contains tandem-repeated sequences. *Nucleic Acid Res.* 16, 9863, 1988.

20. Lillehoj, H. S., Jenkins, M. C., Bacon, L. D., Fetterer, R. H., and Briles, W. E. *Eimeria acervulina*: Evaluation of humoral and cellular responses to the recombinant coccidial antigens in B congenic chickens. *Exp. Parasitol.* 67, 148, 1988.
21. Muller, N., Vogel, M., Gottstein, B., Scholle, A., and Seebeck, T. Plasmid vector for overproduction and export of recombinant protein in *Escherichia coli*: Efficient one-step purification of a recombinant antigen from *Echinococcus multilocularis* (Cestoda). *Gene* 75, 329, 1989.
22. Hanahan, D. Studies on the transformation of *Escherichia coli* with plasmids. *J. Mol. Biol.* 166, 557, 1983.
23. Maniatis, T., Fritsch, E. F., and Sambrook, J. *Molecular Cloning: A Laboratory Manual*, Cold Spring Harbor Laboratory, Cold Spring Harbor, NY, 1982.
24. Summers M. D., and Smith, G. E. A manual of methods for baculovirus vectors and insect cell culture procedures, Texas Agricultural Experiment Station Bulletin No. 1555.
25. Laemmeli, U. K. Cleavage of structural proteins during assembly of the head of bacteriophage T4. *Nature* 227, 680, 1970.
26. Towbin, H., Staeheli, T., and Gordon, J. Electrophoretic transfer of proteins from polyacrylamide gels to nitrocellulose sheets: Procedure and some applications. *Proc. Natl. Acad. Sci. U.S.A.* 76, 4350, 1979.
27. Johnson, J. and Reid, W. M. Anticoccidial drugs: Lesion scoring techniques in battery and floor-pen experiments with chickens. *Exp. Parasitol.* 28, 30, 1970.
28. Shiigi, S. M. and Slomick, M. Preparation and testing of antisera, in *Selected Methods in Cellular Immunology*, Mishell, B. B. and Shiigi, S. M., Eds., W. H. Freeman and Co., San Francisco, p. 244.
29. Lillehoj, H. S. Immune response during coccidiosis in SC and FP chickens. I. In vitro assessment of T cell proliferation to stage-specific parasite antigens. *Vet. Immunol. Immunopathol.* 13, 321, 1986.
30. Jenkins, M. C., Castle, M. D., and Danforth, H. D. Protective immunization against the intestinal parasite *Eimeria acervulina* with recombinant coccidial antigen. *Poultry Sci.*, 199 (in press).

# 5

# The Polydnaviruses: Multipartite DNA Viruses from Parasitic Hymenoptera

PETER JOHN KRELL
*College of Biological Science, University of Guelph, Guelph, Ontario, Canada*

## INTRODUCTION

Polydnaviridae is the family name given for a most unusual and unorthodox group of viruses (8,50). Polydnaviruses have a somewhat complex genome organization and an intriguing life cycle. The family name of these insect DNA viruses is derived from their characteristic segmented (or polydisperse), often supercoiled, double-stranded DNA genome (50). Also, all polydnaviruses described to date have been found exclusively in the ovaries of different species of parasitic wasps (order Hymenoptera). The polydnaviruses can easily be classified on the basis of morphology into at least two subgroups (53). In one group (currently genus polydnavirus) the viruses have an ovoid shape with two membranes and a large lenticular nucleocapsid. In the second group the viruses can be somewhat polymorphic, consisting of a single membrane and cylindrical nucleocapsids of constant diameter but differing lengths. Polydnaviruses from the first group have thus far been found in parasite species belonging exclusively to the family Ichneumonidae while polydnaviruses from the latter group have been found in parasite species belonging only to the family Braconidae. The viruses from these two groups will therefore henceforth be referred to either as ichneumonid viruses or as braconid viruses. Although there is no consensus on the taxonomy of polydnaviruses yet, there is recognition that at least two genera (or families) of polydnaviruses exist and the differentiation, based on morphology, as described and correlating well with the family status of the parasite species from which they are derived, seems appropriate.

Since the ichneumonid and braconid polydnaviruses, although similar in some respects, are quite different from each other, they are described separately.

Several other viruses (or viruslike particles) have been found in association with polydnaviruses in certain species of parasitic wasps. Although these are presumably not polydnaviruses, they are described briefly, separately from the two polydnavirus groups.

## ICHNEUMONID POLYDNAVIRUSES

The most widely studied polydnaviruses are some which replicate in certain species of ichneumonid wasps. The morphology and genome organization of three ichneumonid polydnaviruses in particular are known. These include *Campoletis sonorensis* virus (CsV), isolated from *Campoletis sonorensis* female wasps (30,34); *Hyposoter exiguae* virus (HeV), derived from *Hyposoter exiguae* female wasps (33); and *Diadegma terebrans* virus (DtV), from *Diadegma terebrans* female wasps (27). Although other polydnaviruses from ichneumonid wasps have also been described, less is known about them and therefore I refer to them only briefly. For a complete list of ichneumonid species in which polydnaviruses or putative polydnaviruses have been described the reader is referred to the review by Krell (29).

The following description of ichneumonid polydnaviruses is based largely on CsV. However, since little is known about other ichneumonid polydnaviruses, CsV may not be entirely representative of this group. Thus it would be prudent at this time not to generalize about the ichneumonid polydnaviruses based on a review of CsV in particular.

### Virus Replication

The ichneumonid polydnaviruses replicate only in the ovaries of affected species of ichneumonid wasps (53). Virus replication is further restricted to a specialized region of the ovary between the ovarioles and the lateral oviducts called the calyx. In the ichneumonid wasp *Diadegma terebrans* practically all of the calyx epithelial cells are affected (27) (Figure 1A). These cells extend the full length of the calyx and completely encircle the ovary (i.e., calyx) lumen forming the major component of the calyx. Virus is secreted directly into the calyx lumen (Figure 1B) to form the major particulate component of the calyx fluid in which the parasite eggs are bathed.

The virus replication cycles of three ichneumonid polydnaviruses, CsV, HeV, and DtV, appear to be similar and a common morphogenetic pathway can be envisaged for them (27,33,34,36,37,51,54). The following proposed scheme for virus maturation is based on profiles of developing and "mature" nucleocapsids and virions in the nuclei and cytoplasm of infected calyx epithelial cells of *Campoletis sonorensis*, *Diadegma terebrans*, and *Hyposoter exiguae* (Figure 2).

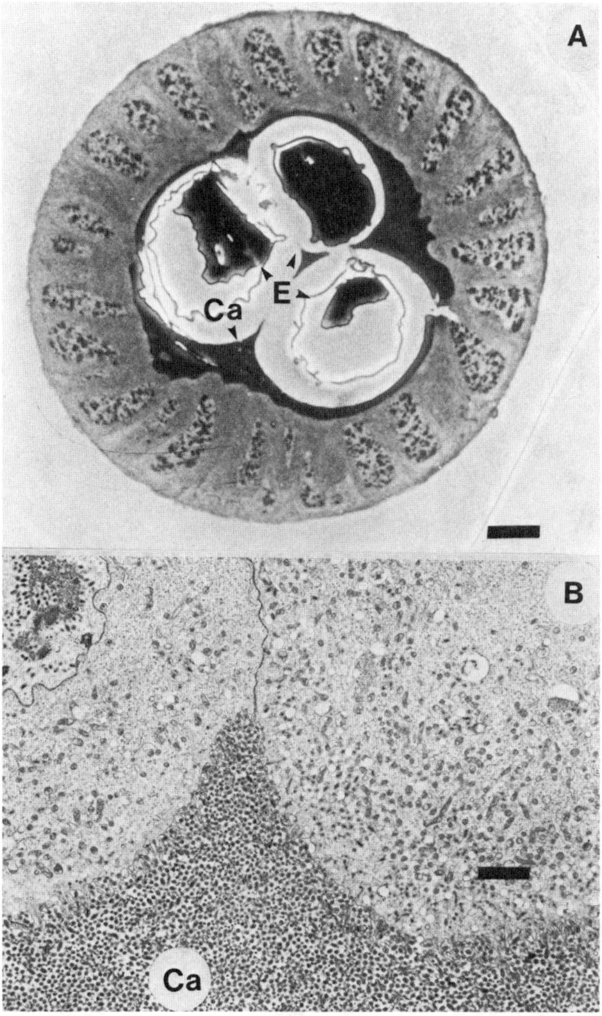


**Figure 1** Cross section through the middle of the ovary in the calyx region of the ichneumonid parasitoid *Diadegma terebrans*. Tissue was embedded in Epon 812 and processed for light microscopy (A) and electron microscopy (B). In (A) all the nuclei have a granular appearance due to replication of DtV in the calyx epithelial cells; three parasitoid eggs (E) are seen within the dark staining calyx fluid (Ca). In (B), which is an electron micrograph of the equivalent region seen in (A), a portion of the granular nucleus appears (upper left) with evidence of the replication of DtV nucleocapsids. Also evident in (B) are the abundant microvilli which abut directly into the calyx lumen (Ca) filled with DtV. (A) Bar = 50 $\mu$M; (B) bar = 500 nm. Figure is modified from Krell (27).

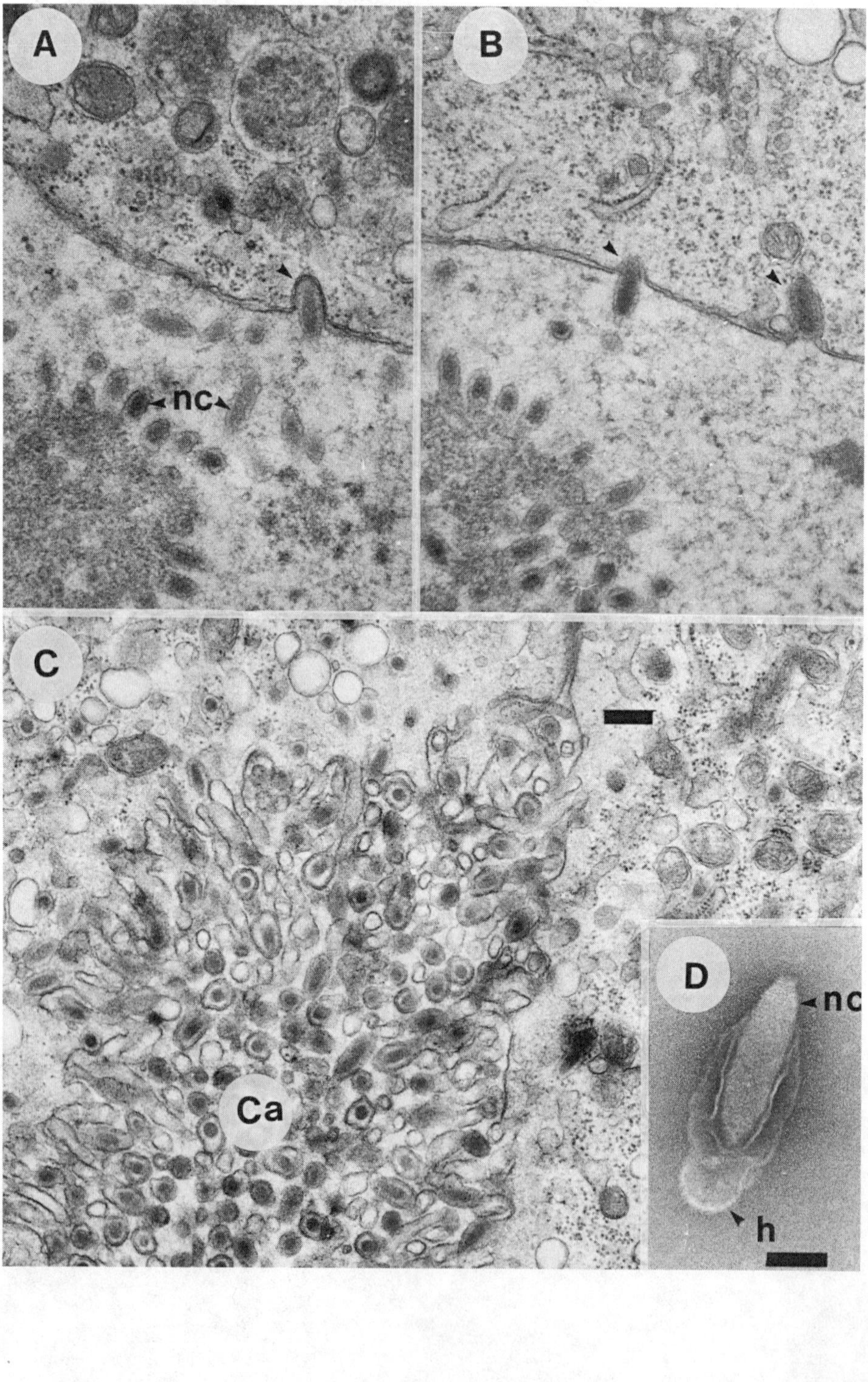
A
B
nc
C
Ca
D
nc
h

Polydnavirus replication can first be detected in nuclei only of calyx cells during an early phase of pupal development (specifically when only the eyes, head, and thorax were pigmented) (36). At this time the nuclear chromatin is altered and some nucleocapsids are detected. The nucleocapsids appear to develop by the condensation of material from the intranuclear fibrogranular stroma easily seen in infected calyx cells (Figure 2A,B). Membrane profiles appear to be assembled de novo and envelopment of the nucleocapsids appears to be coincident with condensation of the nucleocapsids. The mature intranuclear nucleocapsids thus consist of a largely cylindrical or ovicylindrical electron-dense core surrounded by a membrane derived de novo from within the nucleus (36,37,53,54). Late in development the enveloped nucleocapsids bud through both membranes of an extensive nuclear envelope (Figure 2A,B). The nuclei are greatly enlarged and the nuclear envelope of infected cells appears to be increased in surface area, giving a serrated appearance in some cross sections (29,33). Some of the nucleocapsids in the cytoplasm are transiently surrounded by nuclear envelope membranes but most consist only of the nucleocapsid and the surrounding de novo derived membrane assembled in the nucleus. Virus is released from the cell by budding through the membranes of the abundant microvilli that directly abut the calyx lumen (Figure 2C). The polydnavirus thus acquires a second envelope from the plasma membrane of the epithelial cells lining the lumen. There appears to exist a very efficient method of virus replication which ensures the production of copious amounts of virions and their directed secretion and accumulation in the calyx lumen without appearing to significantly alter the structural integrity of the calyx tissue (27). This correlates well with the proposed mutualistic relationship between the polydnavirus and the parasitoid (14, 53). I will elaborate on this mutualism later.

Although ichneumonid polydnaviruses are known to be coinjected with the egg into host lepidopteran larvae during oviposition, there has been no documented evidence of virus replication in any tissues of parasitized larvae. Replication thus seems to be restricted to the parasitoid species from which the polydnavirus was derived.

---

**Figure 2** Replication and morphology of DtV. (A,B) DtV nucleocapsids (nc) are assembled in a fibrogranular stroma within the nucleus and acquire a surrounding envelope before budding through the nuclear envelope (arrowheads). This stroma is equivalent to the intranuclear granules detected by light microscopy (Figure 1A). (C) Enveloped nucleocapsids bud through the plasma membrane in the abundant microvilli directly into the calyx lumen (Ca). (D) Partially disrupted DtV virion shows the shape and substructure of the DtV nucleocapsid (nc). The two envelopes of DtV and evidence of a hooklike structure (h) within the outer envelope are also readily apparent. (A,B,C) Bar = 200 nm; (D) bar = 100 nm. Figure is in part modified from Krell (27).

## Virus Structure and Composition

The ichneumonid polydnaviruses are large and have a complex symmetry. At least two morphological groups can be described, but good morphological data are available for only one of them. The morphology of the ichneumonid polydnaviruses CsV, DtV, HeV and *Hyposoter fugitivus* virus (HfV) has been described and all are structurally very similar (Figures 2D, 3) (27,29,36,37,51,53,54). The following description is a generalized one for these four polydnaviruses. Each ichneumonid polydnavirus consists of an elongated electron-dense nucleocapsid more than 300 nm long and approximately 80–90 nm in diameter (Figure 3D). Although their shape has variously been referred to as biconvex (54), lenticular (53), ovocylindrical (37), or quasicylindrical (8,51), none of these terms appears appropriate. Negative and positive stained preparations of disrupted virions adsorbed to Formvar-coated grids suggest that the ichneumonid polydnavirus nucleocapsids may be cylindrical and are capped at either end with different, roughly cone shaped, blunt-ended structures (Figure 3). The closest analogy in shape and size are the bacterial gas vacuoles (1). A regular surface substructure suggestive of a helical symmetry can also be seen on some negatively stained nucleocapsids (30) (Figure 2D).

The nucleocapsids are surrounded by two unit membranes. As described earlier, the inner membrane is derived de novo in the nucleus during nucleocapsid maturation and the outer is derived from the cell membrane during budding into the calyx lumen. In partially disrupted virions both membranes appear joined to each other at one end of the nucleocapsid (Figure 3B,C). In some negatively stained preparations a rigid, hooklike structure appears to run the length of the virus and may be trapped between the two envelopes (Figure 3A). Such a structure can also be seen to protrude at one end of the virions from DtV (27). In some positively stained partially disrupted (by trypsin) CsV a rigid 700 $\times$ 25 nm curved structure can be seen embedded in a membrane component at one end and capped by a blunt component at the other (Figure 3C,D). A regular substructure can readily be seen, suggesting that this structure is assembled from repeating subunits. This hooklike structure may be equivalent to the "tails" that were observed on ichneumonid polydnaviruses which were seen penetrating the basement membranes of parasitized larvae (54). These hooks may have a functional role in the transmission of the ichneumonid polydnaviruses to parasitized larvae by perhaps promoting membrane fusion as suggested for the tails of other ichneumonid polydnaviruses (53).

The number of virion polypeptides in the ichneumonid polydnaviruses is consistent with their structural complexity. The polypeptide composition of at least three ichneumonid polydnaviruses, HeV, CsV, and HfV, has been determined by polyacrylamide gel electrophoresis (26,33,34). Each of these viruses is composed of at least 20 different sized polypeptides. In all cases the

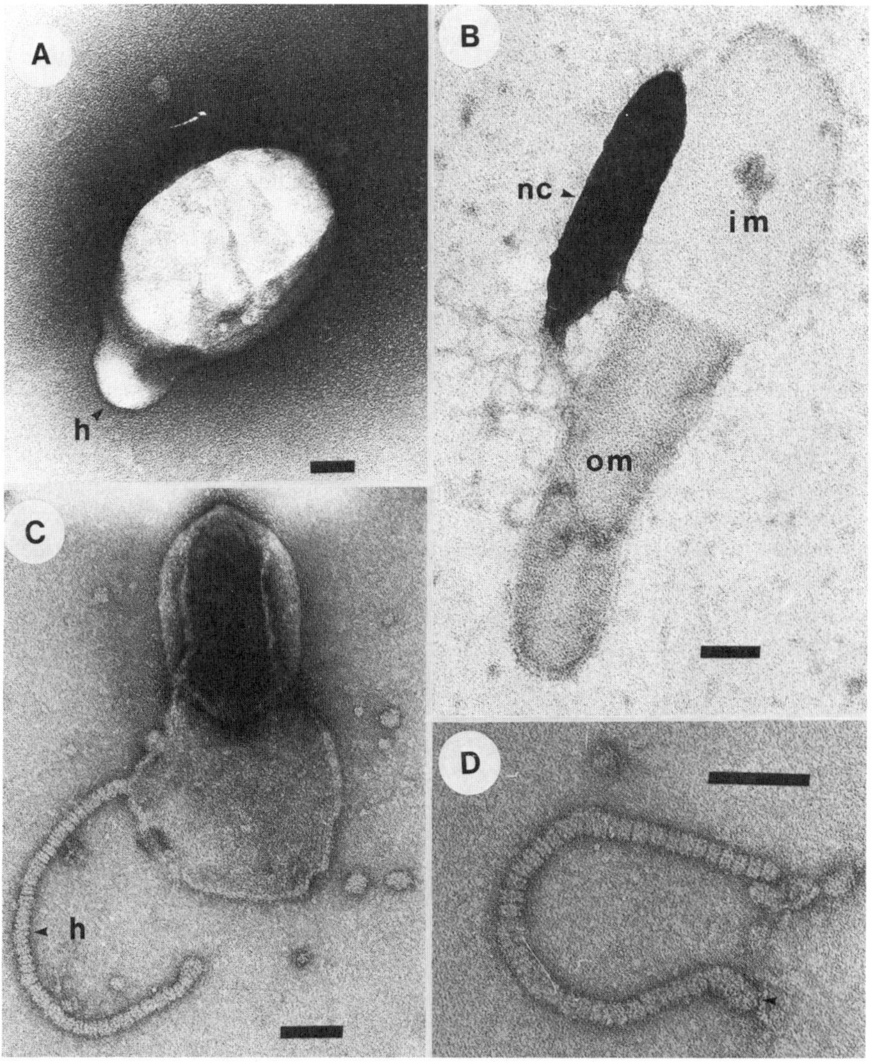


**Figure 3** Ultrastructure of the ichneumonid polydnavirus CsV. All are negatively stained samples following uranyl acetate staining. (A) Overall morphology of CsV. An internal structure appears to stretch the outer envelope at one end (bottom left) and along one edge (bottom edge). It is presumed that the hooklike structure (h) is trapped within the outer envelope. (B,C,D) CsV has been partially disrupted by limited trypsin digestion and reveals the electron-dense cylindrical nucleocapsid (NC) with conical ends attached to remnants of the outer (om) and inner (im) viral membranes. The hooklike appendage in (C) appears to be anchored to the inner envelope. The subunit structure and blunt end (arrow) of an isolated hook can be easily seen in (D). Bar = 100 nm in all micrographs.

aggregate molecular weight of polypeptides from one virus is just in excess of $10^6$. The specific location of any of the polypeptides in the virion is unknown. Presumably some of the major polypeptides (e.g., VP42 at molecular weight 42,000; 34) are derived from the nucleocapsids. A 25 kD glycoprotein identified in CsV may be a membrane component.

The minimum coding requirement to direct the synthesis of all these proteins has been calculated at approximately 30 kb. This calculation assumes that all proteins are unique and are coded on separate, nonoverlapping genes.

## Genome Organization

The family Polydnaviridae has been established largely due to the recognition that there exist DNA viruses with a bona fide polydisperse (or segmented) circular DNA genome (8,50). The viral DNA was sensitive to DNase digestion but insensitive to RNase digestion. Thus the genome of these viruses was shown to be DNA (26). The ichneumonid polydnaviruses were the first polydnaviruses in which the polydisperse nature of the genome was confirmed by Southern blot hybridization (33,34). For other parasitoid viruses the multipartite nature of the genome can be demonstrated by agarose gel electrophoresis of viral DNA. Usually there is such a high concentration of polydnaviruses in calyx fluid that agarose gel electrophoresis analysis can be done directly on calyx fluid without a prior need to further concentrate or purify the virus.

Although the multipartite nature of the polydnavirus genome is evident by agarose gel electrophoresis analysis (Figure 4), this method is inadequate to determine the conformation of the DNA. However, DNA from purified virions could be resolved into two bands by isopycnic centrifugation in cesium chloride and ethidium bromide (Figure 5B). This was suggestive of superhelical DNA, some of which would band at a lower density if nicked to a relaxed circular form. Final confirmation of the circular conformation of the virion DNA was by electron microscopy using the Kleinschmidt procedure (Figure 5A). HeV DNA from the lower band of a CsC1/EtBr gradient was in fact supercoiled. After relaxation using limiting treatment with DNase 1, relaxed circular forms were detected (33). The sizes of the circles as estimated by electron microscopy correlated well with size estimates based on migration rates in agarose gel electrophoresis relative to known plasmids (Figure 5C).

Because of the number of different sized circles of DNA in each virus and the potential of at least two conformational forms (superhelical and relaxed circular) for each, it is difficult from the AGE profile alone to determine the total number of circles per genome or the complete molecular weight range. Moreover, it is not always evident whether a given band consists of one or several different but comigrating bands. The genome of CsV, for example, was first thought to consist of 25 SH DNAs since "at least 25 SH DNA bands" could be identified

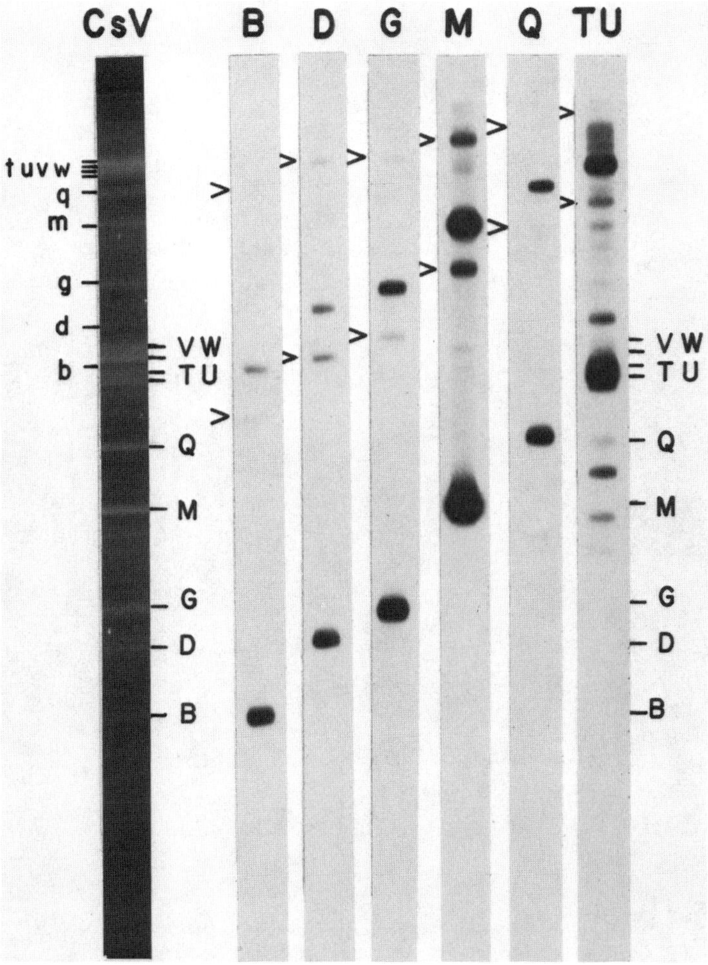


**Figure 4** Agarose gel electrophoresis of DNA from purified CsV (lane CsV) and Southern blot hybridization analysis of CsV using six different superhelical CsV DNAs as probes (lanes B to TU). In lane CsV only some of the major SH DNA bands (B, D, G, M, Q, T, U, V, and W) and the corresponding relaxed circular forms (b, d, g, m, t, u, v, and w) are indicated. In each of the Southern blots probe DNA isolated from SH bands B, D, G, M, Q, and TU, respectively, hybridized to the corresponding SH and relaxed circular DNAs as well as to two additional bands (marker >), which are presumed to be the SH and relaxed circular form of the dimer of the equivalent probe DNA. Some additional hybridization is noted for probe M and especially TU. In general, however, cross-hybridization for these probes was minimal. (Some of this work was done in collaboration with Dr. Max Summers, Texas A&M University.)

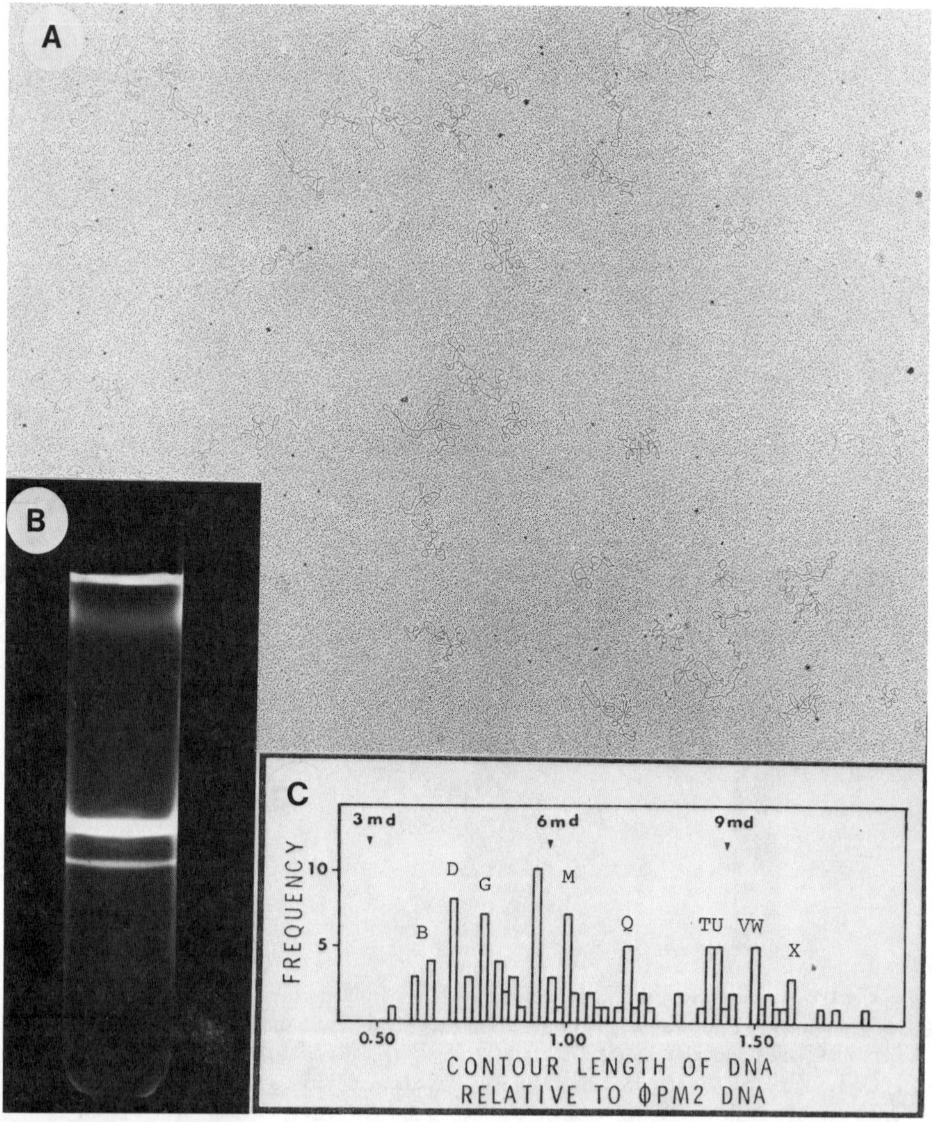


**Figure 5** Analysis of DNA from CsV. Kleinschmidt analysis of DNA from purified CsV is shown in (A) to have a circular conformation. In (B) DNA from purified CsV forms two bands after isopycnic centrifugation in CsCl–EtBr. The lower band contains SH CsV DNA; the upper band contains relaxed circular CsV DNA. In (C) the size of CsV DNA as determined by electron microscopy is shown in histogram form and correlates well with the size estimated by agarose gel electrophoresis. The letters above the bars (A to X) correlate to the SH DNA band designations in Figure 4 lane CsV.

(34) (Figure 4). Later Blissard et al. (4) identified three additional SH bands, $A^2$, $L^2$, and $O^2$, using higher amounts of CsV DNA in agarose gels. Another SH band, $C^2$, was detected by Southern blot hybridization (58). In addition, it is possible that band G consists of two comigrating DNAs ($G^1$ and $G^2$), which could be differentiated by SalI and Hind III digestion. It thus appears now that "at least 30 SH DNA bands" can be resolved for the CsV genome.

Another intriguing feature of the genomes of ichneumonid polydnaviruses is that not all of the different DNAs are in equimolar ratios. For CsV, for example, there appear to be at least three classes of DNAs based on relative abundance (Figure 6). In one, DNAs such as those in bands V, Q, M, G, D, and B are present in relatively high abundance; in another group of DNAs such as those in bands E, J, K, L, and TU they are present in intermediate abundance; in a third group are DNAs such as those in bands A, H, F, O, P, and S, which are present in low abundance. The significance of the nonequimolar relationship among the viral DNAs is unknown. From a simplistic approach this may be a convenient way of regulating gene dosage. Perhaps those genes coding for proteins required in high amounts are located on high-abundance circles, whereas genes from proteins needed in only small amounts are on low-abundance circles.

The genomes of two ichneumonid polydnaviruses, CsVand HeV, have been examined in some detail. As described earlier, the CsV genome consists of at least 30 different SH circles. These DNAs range in size from 6 to 28 kb. A crude estimate of the aggregate size of the CsV genome is 200–250 kb, based on AGE analysis of circular CsV DNA and of Csv DNA digested with restriction enzymes. This estimate may need to be increased as more circles are discovered. The average density of CsV DNA was calculated to be 1.71 g/cm$^3$, which corresponds to a G+C content of 50%.

The genome of HeV appears to be somewhat simpler than that of CsV (33). About 16 different SH DNA bands can be resolved for this ichneumonid polydnavirus. The sizes of the DNAs of HeV are smaller than those of CsV and fall in a narrow size range of 2–8 kb. The aggregate size calculated for the HeV genome is only about 75 kb, quite a bit less than that calculated for CsV. As with CsV, there appeared to be about three classes of SH DNAs of HeV based on relative staining intensities of different bands in agarose gels. The average density of HeV DNA was 1.689 g/cm$^3$, suggesting a G+C content of 29%.

It was first shown for HeV that at least some of the different sized circles were unique and that the smaller DNAs were not a subset of the larger ones. Although this Southern blot hybridization analysis did not include all the circles, the genome of HeV was shown to be multipartite for at least that proportion of circles which was analyzed. These data suggested also that the genome of HeV was functionally polydisperse with different genes segregated to different circles.

The genome of CsV has been studied much more extensively than that of any other polydnavirus. As described earlier, the CsV genome consists of at least 30 circles of DNA ranging in size from 6 kb to greater than 20 kb with an aggregate

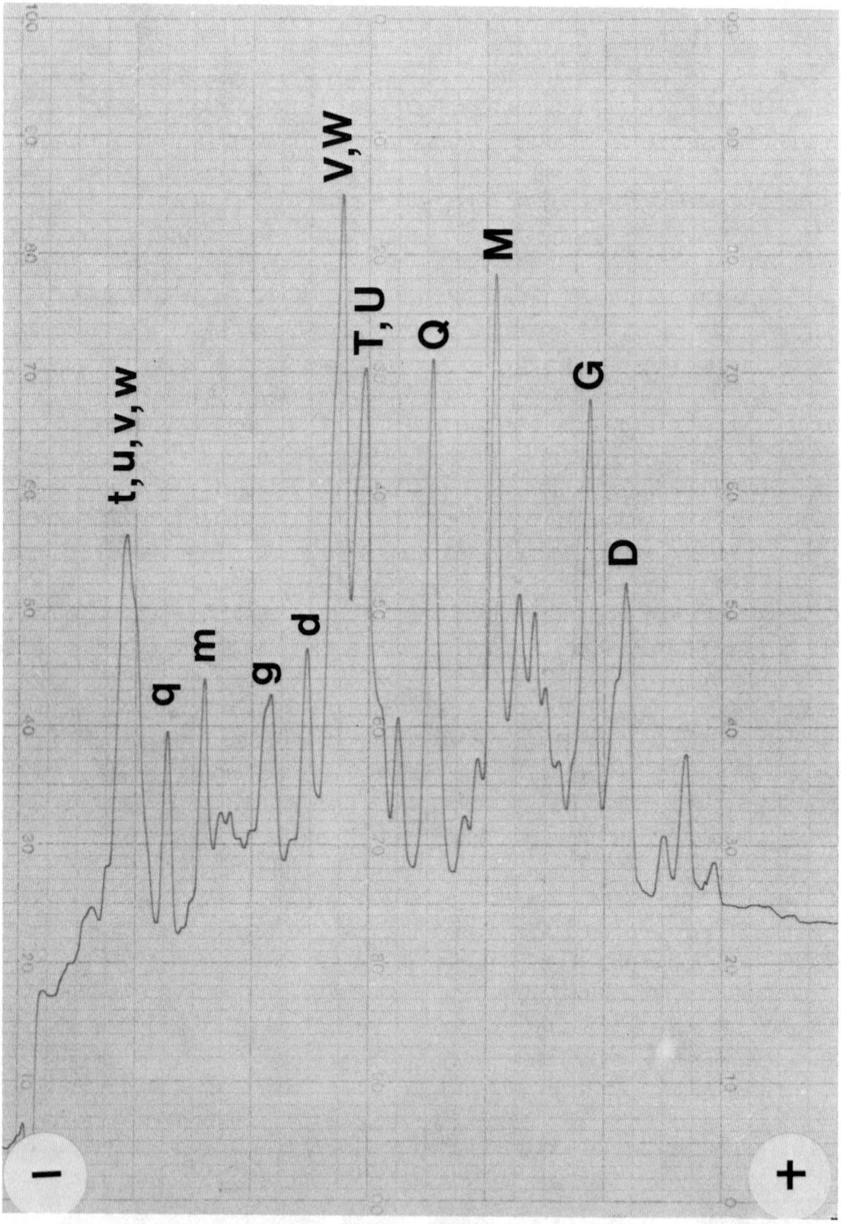


**Figure 6** Densitometer scan of a negative of a gel of CsV DNA similar to that shown in Figure 4, lane CsV. This scan shows the nonequimolar nature of the different CsV SH DNAs. Only the major bands are identified. Uppercase letters represent SH DNA bands; lowercase letters represent the corresponding relaxed circular DNA bands. (– is the cathode end of the gel; + is the anode end.)

size probably in excess of 250 kb. However, this estimate assumes that each circle is unique. In initial studies using stringent hybridization conditions little sequence homology could be detected among some of the CsV circular DNAs (34) (Figure 4). However, in a more extensive analysis using Southern cross-blot hybridization, Theilmann and Summers detected some degree of homology among most of the circles (58). Using cloned Sal1 or HindIII fragments of SH DNAs D, M, Q, T or U, and V or W, some cross-hybridization was detected for SH DNAs Q, T or U, and V or W, but not for D and M (Figure 7). Cloned D and M DNA hybridized to their corresponding SH and relaxed circular forms. The additional hybridization noted for these two cloned fragments was presumed to be due to the presence of dimer forms of the corresponding DNAs. Using cloned DNAs representing three cross-hybridizing DNAs (from bands $O^1$, H, and B), Theilmann and Summers (58) identified an imperfectly conserved repeat element of approximately 540 base pairs. Derivatives of the "consensus sequence" of this 540 bp repeat were found in several other circles as well. One copy was found in circle B, at least three in circle $O^1$, and at least five copies in circle H. Sequence analysis of these nine near repeats showed that despite some similarity no two were identical. In a detailed transcriptional analysis of circles B, M, H, and $O^1$, Blissard et al. (6) showed that all transcripts from circles H and $O^1$ and some from circles B and M contained sequences homologous to the imperfectly conserved 540 bp repeat. Moreover, the repeat mapped within the expected open reading frames and the 3' terminus of the repeat element coincides with the end of several ORFs. Summers and colleagues have suggested that all genes containing the repeat element represent a gene family and that the corresponding polypeptides represent a family of related proteins (6). Similar imperfect repeat elements may be distributed among other DNA circles and could account for the extensive cross-hybridization observed among the circles of DNA which comprise the CsV genome.

Perhaps significantly, the nine repeat elements shared an average sequence homology of up to only 60–70%. Even shorter regions were homologous to an extent no greater than 90%. The genesis or significance of these, or other, repeat sequences is unknown but is certainly open to speculation. Theilmann and Summers (58,59) suggest that they may be involved in genomic reorganization by mediating in intermolecular and intramolecular recombination. If this were the case, one would expect the genome of CsV to be quite peripatetic with genomic variants generated at high frequency. However, as judged by the apparently invariant nature of the AGE profile of CsV taken over a five year period, the genome seems quite stable. A converse view is that such "repeats" may have been generated as a result of very specific recombination events but not a prerequisite to them. Later Blissard et al. (6) suggested that the repeats were evidence of multiple recombination events. Since these divergent repeats appear to be part of an ORF, it is unlikely that recombination within the ORF would be favorable for the stability of the virus. Also, the repeat in one circle, B, does not

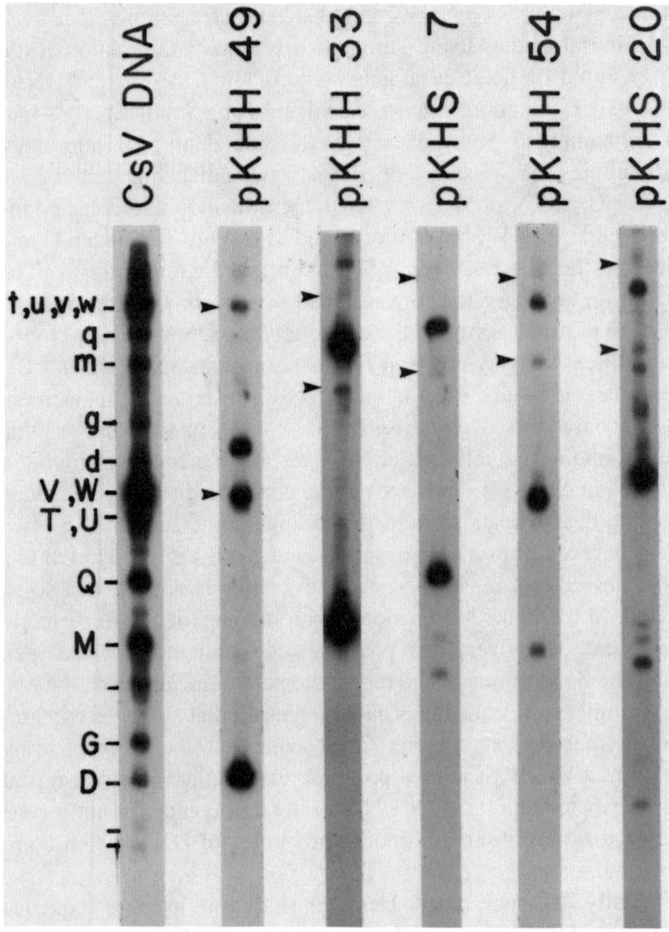


**Figure 7** Southern blot hybridization analysis of CsV DNA using CsV DNA as probe (lane CsV DNA) and five different cloned fragments of CsV DNA (lanes pKHH49 to pKHS20). CsV DNA was cloned into plasmid pKH47 (indicated by pKH in the acronym above the lanes) as either HindIII fragments (indicated by H) or SalI fragments (indicated by S). The plasmids pKHH49, pKHH33, pKHS7, pKHH54, and pKHS20 represent cloned fragments of SH CsV DNAs D, M, Q, T or U, and V or W, respectively. Hybridization was noted for the corresponding SH and relaxed circular forms of the corresponding intact circular DNA and their putative dimers (indicated by arrowheads) for all cloned fragments. Very little if any cross-hybridization was noted for cloned DNA pKHH49. Cross-hybridization with additional DNA bands was noted for probes representing bands M, Q, T or U, and V or W. This correlates with some of the cross-hybridization observed in Figure 4 using gel-isolated circular CsV DNAs as probes.

appear to be interrupted by the putative integration site (16,59). If the repeat was directly involved in integration, it would be expected to be part of the integration site and adjacent flanking sequences. An alternative view for the significance of the repeats is that these sequences may simply represent portions of duplicated and nonessential pseudogenes. There may be no direct selection against such a duplicated nonessential but expressed gene and thus the sequence could diverge to the extent that it bears only 60–70% homology to the original "essential gene." The existence of such selfish DNA sequences may easily be tolerated in a virus with such a large genome size. Whatever the significance or function, if any, of this characteristic of the CsV genome, it nevertheless remains intriguing.

It is quite likely that the replication of polydnaviruses will be found to be quite novel compared to other DNA viruses. The mechanism of viral DNA replication is unknown, and attempts to identify viral origins of DNA replication on viral circular DNAs have been futile (personal observations). This suggests that either no origins exist and that viral DNA circles do not replicate independently or that each circle does replicate independently as an individual replicon but either does not use a specific origin or replication or has an origin that is not detectable by the methods used (40). In the case of integrated viral DNA it is likely that at least some of the viral DNA circles are generated by an excision mechanism. However, whether these DNAs are amplified before excision (as part of chromosomal DNA) or after excision (as independent circular viral DNA replicons) is unknown.

Little is known about the packaging strategy for the genomes of polydnaviruses. The encapsidated (i.e., virion) DNA of HeV was shown to be in a superhelical conformation (26,33). This is presumably the case for other ichneumonid polydnaviruses since SH forms can be isolated readily from all species tested to date (26). The relaxed circular forms that are isolated from these polydnaviruses may have been generated by nicking of SH forms during the DNA extraction procedures. Since a single nick can convert an SH molecule to a relaxed circular form, even moderate shearing forces such as those encountered during manipulation of the viral DNA would suffice to relax some of the SH virion DNAs. The presence of a multipartite DNA genome presents a potential packaging problem for the ichenumonid viruses. The central question is, How are the different sized circles packaged in individual virions and in nonequimolar amounts? Among the possibilities are: (1) each nucleocapsid consists of at least one copy of each SH DNA circle that makes up the genome, (2) each nucleocapsid contains one or more copies of a single SH DNA, (3) each nucleocapsid contains a specific subset of certain SH circles but not all of them, or (4) packaging is entirely random except with respect to the total amount of DNA packaged. Under possibilities 2 and 3 the entire genome would be contained within a finite number of virions. Also, there would be certain subclasses of virions depending on the portion of the genome each encapsidated. However, no such subclasses could be resolved by either sucrose density gradient centrifugation or CsCl isopycnic centrifugation of CsV (34). In both gradients only a single band of purified CsV containing all the DNA circles was observed. Thus if subclasses do exist, they would be difficult to detect. A random packaging method is easiest to imagine and may be sufficient so that most virions would contain the entire genome.

The amount of DNA packaged by each virion is also not known. However, each nucleocapsid appears to be of equal size and therefore each virus would presumably encapsidate the same amount of DNA. A rough estimate of the potential amount of DNA encapsidated can be determined from the size of the nucleocapsid. For CsV the nucleocapsid has a volume of approximately $1.7 \times 10^6$ $nm^3$ (assuming the nucleocapsid is a cylinder 350 nm long with a diameter of 80 nm) (34). This is roughly four times the volume of a baculovirus nucleocapsid which encapsidates approximately 120 kb of DNA. Thus, at least for CsV, each virion could theoretically encapsidate in excess of 450 kb of DNA or at least twice the estimated aggregate size of the CsV genome. For HeV, which has a similar size, each virion could encapsidate more than five times the calculated aggregate size of the HeV genome. Thus random packaging of genomic DNA of HeV and perhaps CsV as well could be an effective way of ensuring that most virions have a good probability of each containing the entire genome. It was noted earlier that some circles were present in a higher proportion than others. A random packaging strategy would serve to maintain the nonequimolar nature of the different DNA circles if the relative amounts of each circle were maintained during the synthesis of viral DNA. It is clear that this aspect is still one of the outstanding questions concerning the polydnaviruses. However, resolution of the question of how the circles are segregated or combined in virions could easily open up new questions.

## Transmission

Polydnaviruses appear to be transmitted very efficiently. For any polydnavirus–parasite association examined, every female wasp of an affected species contains polydnavirus in her ovary. Thus transmission of polydnavirus to at least all female offspring is 100% efficient. Any model for transmission must satisfy this observation. Two major routes of polydnavirus transmission could be envisaged. One of these is per os transmission, which is also the main mechanism for horizontal transmission of many insect viruses. Polydnavirus could theoretically be transmitted when the developing parasitoid larva consumes polydnavirus. Since polydnavirus is coinjected into the hemocoel of parasitized larvae during oviposition it could theoretically be available for consumption by the parasitoid larva. However, several factors mitigate against per os transmission. As will be described in more detail later, most of the injected polydnavirus in the parasitized host larva is taken up by host cells and becomes uncoated in host cell nuclei. Therefore, there may be no extracellular polydnavirus available for consumption at the time the parasite egg hatches and the emergent parasitoid larva starts to feed. Moreover, per os transmission is notoriously inefficient, especially in the digestive secretion of the gut. Thus one would expect a least a few individual parasite larvae to escape infection. Also, if per os transmission was the mode of spreading the polydnavirus infection, one might expect male parasites to be infected as well. However, polydnavirus virions have not yet been detected in any male parasites examined by electron microscopy. One could argue that the infection itself is tissue-specific (i.e., the ovary) and thus male wasps could escape the infection. All in all, per os transmission is currently the least favored route, although that possibility still has to be considered.

An alternate route of virus transmission, which would be 100% efficient, is by vertical transmission through the germline of the specific parasite species. In this case the viral genome would be associated with the germline genetic material and replicate with it. This would ensure that each progeny egg would contain the viral DNA sequences required for virus replication. It was first reported by Krell and Lo (31) that at least some of the viral DNA was present in male *C. sonorensis* wasps and that the viral DNA in these wasps appeared to be integrated into the wasp genomic DNA (Figure 8A). In a more detailed study of viral DNA in different tissues of male and female *C. sonorensis* wasps, Fleming and Summers (16) concluded that, at least for superhelices B and Q, much of the viral DNA was integrated into cellular DNA. The conservation in sizes of B- and Q-specific restriction fragments of DNA from different tissues suggested that B and Q sequences are stably integrated at specific sites in both male and female chromosomal DNA.

In a similar but less extensive analysis, Stoltz et al. (48) found that for HfV viral DNA could be detected in nonovarian tissue and in male wasps. However, in contrast to the studies by Fleming and Summers (16), most of the viral DNA appeared to be in extrachromosomal circular form and not integrated. Therefore, different ichneumonid polydnaviruses may have evolved different strategies for transmission in the germline. Nevertheless, these studies suggest that in both situations the cellular and viral genomes are tightly associated and in either case the viral genome would be readily transmitted directly to the next generation.

Whichever mechanism holds, one related outstanding question remains: Why is virus replication (or synthesis) restricted to the calyx epithelial cells of the female ovary? This is especially germane if all cells have the complete viral genome. Norton and Vinson have demonstrated that virus replication can first be detected during the pupal stage (36). These results suggest that polydnavirus replication is tissue-specific and under temporal control. A simplistic explanation is that virus replication is a differentiation-specific event which is coordinated with the development of the female reproductive tract. Thus viral gene expres-

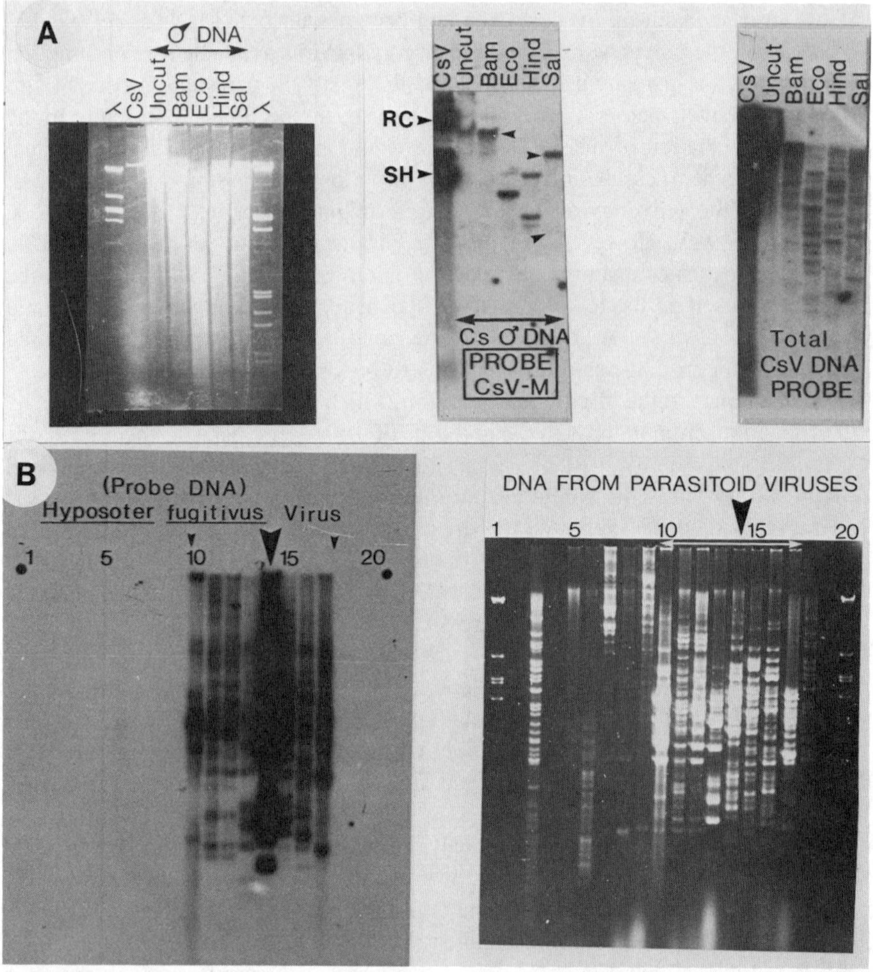


**Figure 8** (A) Evidence of integrated CsV DNA in *C. sonorensis* male DNA. Total nuclear DNA was isolated from male *C. sonorensis* and subjected to agarose gel electrophoresis (left panel) and Southern blot hybridization either intact (lanes uncut) or after digestion with BamHI (lanes Bam), EcoRI (lanes Eco), HindIII (lanes Hind), or SalI (lanes Sal). CsV DNA (lanes CsV) was also analyzed. Identical bidirectional blots were hybridized with either purified band M DNA (middle panel) or total CsV DNA (right panel). The hybridization analysis shows that CsV DNA is present in male wasps and that the majority of the homologous CsV DNA is in high molecular weight (lanes uncut) DNA, suggestive of integrated DNA. The hybridization of M DNA with enzyme-digested male DNA is suggestive specifically of integration of CsV band M DNA. The hybridization pattern is different from the expected for band M DNA in its circular form. The

sion, leading to synthesis and secretion of polydnavirus, might be correlated with the differentiation of the calyx tissue. If this hypothesis was correct, this would provide an ideal model for studying differentiation.

## Polydnavirus Expression

To date expression of polydnaviruses in the parasitoid itself has been investigated in only two studies (34,49). Also, in the absence of a permissive tissue culture system, a classical approach to understanding the replication cycle of polydnaviruses is not possible. Moreover, the difficulty of obtaining sufficient parasitoid tissue for molecular biology studies of virus replication in the ovary will hinder research in this important area. Consequently, much attention has been directed to the expression of polydnaviruses in the host of the parasites.

Early studies of the fate of ichneumonid polydnaviruses demonstrated that polydnavirus was injected with the parasite egg into the parasitized larvae during oviposition (53). It was also shown that at least for *C. sonorensis* the polydnavirus was essential for successful development of the parasite (14). In these experiments all eggs injected without virus into larvae were destroyed while at least some eggs coinjected with virus developed normally. It was shown by electron microscopy that several ichneumonid polydnaviruses, including HfV and CsV, can "infect" cells of the parasitized host. The nucleocapsids enter the nuclei, presumably through nuclear pores, and subsequently become activated. There has been no evidence for virus replication in any of these cells. Theilmann and Summers (57) noted that although viral DNA was not amplified in parasitized larvae it did not appear to be substantially degraded. Thus after uncoating in

---

position of the SH and relaxed circular form of CsV DNA in the same gel are indicated by SH and RC, respectively. (B) Relatedness of ichneumonid polydnaviruses. Viral DNA was isolated from the calyx fluid of several species of parasitic (ichneumonid) hymenoptera and analyzed by agarose gel electrophoresis (right panel). A blot of this gel was analyzed by Southern blot hybridization (left panel) using DNA from purified HfV from *Hyposoter fugitivus* wasps (position of HfV lane is indicated by the larger arrows). A diversity of AGE profiles of different ichneumonid polydnaviruses can readily be seen in the right panel. Hybridization with HfV DNA could be detected in lanes 10–17 containing viral DNA from *Diadegma terebrans*, *D. acronyctae*, *D. interruptum*, *Olesicampe spp.*, *H. fugitivus*, *H. annulipes*, *H. rivalis*, and *Hyposoter* spp., respectively. Hybridization was not detected to viral DNA in lanes 4–9 representing polydnavirus from *Campoplex* sp. 1, *Campoplex* sp. 2, *Glypta* spp., *Campoletis spp.*, *C. aprilis, and C. sonorensis*, respectively, or with DNA in lane 18 from an ascovirus of *Spodoptera*. (This study was done in collaboration with Dr. Don Stoltz, Dalhousie University.)

the nucleus the viral DNA appears to persist. Similarly, viral DNA from AfV persists in three different host species parasitized by *Hyposoter fugitivus* (48).

Much of the early work on the expression of CsV concentrated on transcriptional studies in the *H. virescens* host. Fleming et al. demonstrated that up to nine of the CsV SH DNAs were transcriptionally active in parasitized or CsV-injected *H. virescens* larvae and could be detected by cloned CsV DNA probes (17). Blissard et al. were able to resolve at least 10–12 size classes (0.6–3.2 kb) of CsV-specific RNA by Northern blot analysis (4,7). Some of these mapped to different CsV superhelical molecules, showing that CsV is transcriptionally multipartite. There also appeared to be some degree of temporal regulation of CsV transcription. Some RNAs were detected at early times post parasitism while other transcripts could be detected only at later times. This mimics closely the classical kind of temporal regulation of transcription observed for many DNA viruses.

These authors were also the first to show splicing of polydnavirus mRNA (5). At least two CsV transcripts 1.0 and 1.6 kb in size are derived by splicing from a larger transcript. Furthermore, sequence analysis of genomic DNA coding for the 1.6 kb transcript shows the predicted GT and AG consensus dinucleotides at the predicted splice junctions expected of spliced genes.

Examples of some degree of sequence relatedness among CsV mRNAs were revealed by using CsV cDNA as probes (7). In particular, one cloned cDNA hybridized to a 1.1 kb and a 1.4 kb transcript and a second cloned cDNA hybridized to two other transcripts, one of 1.0 kb and one of 1.6 kb. Of the latter two transcripts both the abundant 1.6 kb transcript and the less abundant 1.0 kb transcript mapped to separate regions of the same SH circle (circle W, 15.8 kb). Both of these transcripts were identified using a single cloned 5.3 kb HindIII fragment derived from at least one of the SH circles K or V. Not surprisingly, the sequence of the two transcripts, derived from the corresponding cDNAs, are fairly similar. In this case an element from one of the larger circles (W) was similar to another element on the same circle and also to element(s) present on other circles (e.g., V or K).

Theilmann and Summers (59) and Blissard et al. (6) identified several specific transcripts from parasitized larvae which hybridized to SH DNA B, H, M, $O^1$, and W. Some of these transcripts appeared to be under temporal control since the relative levels varied with time post parasitization.

In these same studies these authors demonstrated that part of the CsV genome was transcribed in the parasite as well. For example, five transcripts specific to SH-W DNA and six transcripts specific to SH-M DNA were detected in ovaries of *C. sonorensis*. Moreover, some of the genes were expressed differentially in *C. sonorensis* and in parasitized *H. virescens*. These authors reported that some genes were expressed only in the parasite, others were expressed only in the parasitized host, and still others were expressed in both backgrounds. Therefore, expression of CsV is not only temporally controlled but can also be under host-specific regulation. The host-specific regulation is similar to tissue-specific expression of eukaryotic genes. This differential expression in parasitized larvae could be important for successful parasitization by allowing expression of only those genes required for host regulation.

In an attempt to identify some of the gene products coded for by the CsV transcripts present in parasitized larvae, Blissard et al. (6) cloned the ORFs of the 1.6 kb transcript from WHv1 and the 1.0 kb transcript from WHv2 into baculovirus expression vectors. They determined that the 1.6 kb transcript codes for a nonglycosylated 23 kD polypeptide which can be secreted from cells infected with the recombinant virus. The 1.0 kb transcript, however, codes for a 24 kD precursor which is glycosylated to a 26 kD form and is also secreted from infected cells. Whether these two proteins are synthesized in parasitized *H. virescens* larvae and in what amounts were not determined in this study. However, the CsV-specific proteins produced in the baculovirus recombinants could be used to develop probes (e.g., monoclonal antibodies) to detect them in parasitized larvae.

The studies on CsV expression in host larvae demonstrate that CsV is transcriptionally active, that only certain genes are expressed, some of them under temporal control, and at least two of the expressed genes (WHv1 and WHv2) code for proteins that are secreted and could act in the parasitized larvae at sites remote from their initial synthesis. Proteins such as these may function in "host regulation" to allow successful development of the parasite in the parasitized host.

Several physiological effects on parasitized larvae have been attributed to the presence of ichneumonid polydnaviruses in their mutualistic relationship with the parasite (10–14,42–44,46,60,61). The virus clearly needs the parasite for its replication and transmission, but the parasite also needs the polydnavirus not only for its own individual survival to pupation, but also for the survival of the parasite species. In a landmark study Edson et al. (14) established that the successful development of *C. sonorensis* eggs depended on coinjection of CsV into the host larvae. In the absence of CsV, eggs of *C. sonorensis* failed to develop into larvae or emerge as adults in larvae of the usual host *H. virescens* injected with washed eggs. Eggs injected with CsV developed into larvae and some emerged as adult wasps. In the absence of CsV the parasite eggs were encapsulated while in the presence of CsV they were not. It was then recognized that, at least for CsV, there appears to be an obligatory mutualism between an ichneumonid polydnavirus and the parasite species in which it replicates and which ensures its transmission.

Vinson and his colleagues (14) suggested that CsV may be responsible for suppressing the host larva's immune response (e.g., encapsulation), allowing successful development of the parasite. This idea was supported by a study on differential hemocyte count and plasmatocyte activity in *H. virescens* parasitized by *C. sonorensis*. Davies et al. (11) demonstrated that as much as 75% of the circulating plasmatocytes (responsible for encapsulation) were removed and their spreading ability was decreased in parasitized larvae. Moreover, these characteristics could be duplicated by injection of purified CsV. These studies demonstrate that CsV has a direct and dramatic effect on the hosts' immune response to a developing parasite egg and larva.

Another gross physiological effect of parasitism by *C. sonorensis* is reduced weight gain and development of the parasitized host larva (61). This effect can be duplicated by the injection into normal *H. virescens* larvae of polydnavirus-containing calyx fluid. However, if the calyx fluid is UV-irradiated prior to injection, the host larvae develop normally. It was suggested that the UV irradiation inactivates the polydnavirus genome, thereby preventing its expression in host larvae. The exact mechanism of action of the polydnavirus on host growth, however, was not directly determined and is presumably quite complex.

Initially Dover et al. (12) demonstrated that developmental arrest in *H. virescens* larvae parasitized by *C. sonorensis* was due to the particulate component of the calyx fluid, which contained predominantly CsV. They also observed that parasitized larvae had depressed levels of ecdysteroid and that the arrested development could be reversed by injection of ecdysone. Only virus with access to the thorax (which contains the prothoracic gland) in ligated larvae stopped development. In a further analysis of the prothoracic gland of parasitized larvae, Dover et al. (13) demonstrated a degeneration of the prothoracic gland in larvae parasitized by *C. sonorensis* and in larvae injected by purified CsV alone. If CsV directly (by infection) or indirectly (by a secreted CsV-specific polypeptide) leads to the deterioration of the prothoracic gland, this would have a tremendous effect on the hormonal balance leading to decreased levels of ecdysteroid and arrested development. Such dramatic alterations in host physiology would be of obvious benefit to the developing parasite and could very well be manifested by expression of the CsV genome in parasitized larvae.

A third effect on parasitized larvae which has been attributed to expression of an ichneumonid polydnavirus is the inhibition of host phenol oxidase activity (44). Parasitism of either *Trichoplusia ni* or *H. virescens* larvae by either *H. exiguae* or *C. sonorensis* led to reduced phenol oxidase activity. For *H. exiguae* it was shown that purified HeV alone can inhibit phenol oxidase in *T. ni* larvae. Presumably the same effect might be shown for larvae injected with purified CsV.

There are therefore at least three manifestations of parasitism which appear to be related to the presence and presumably expression of an ichneumonid polydnavirus. As described earlier, polydnavirus DNA does persist in parasitized larvae and is expressed. It is likely that products of polydnavirus expression alter aspects of host physiology such as the immune response and host weight gain to allow successful development of the parasitoid. A major gap in our knowledge of polydnavirus interaction with the parasitized host is the molecular basis of the phenomenological effects on host larvae attributed to the polydnaviruses. Presumably one or more of the proteins specified by the CsV transcripts in parasitized larvae could modify host physiology. The two CsV-specific proteins described by Blissard et al. (6) and which could be secreted from the cells could move within the larva and directly affect remote tissues such as the prothoracic gland. Another possible candidate as an "effector" molecule is a 50–55 kD hemolymph glycoprotein that appears after parasitism of *H. virescens* larvae by *C. sonorensis* (10). This glycoprotein also appears in *Spodoptera frugiperda*, a habitual host species, after parasitism by *C. sonorensis*. This glycoprotein could be CsV coded since it is also synthesized after mechanical injection of calyx fluid or purified CsV into the larvae. If expression of CsV is inhibited by trioxsalen and UV light treatment (to crosslink the DNA), the glycoprotein is not synthesized. Similarly, the glycoprotein is not detected in hosts in which *C. sonorensis* does not develop. This latter observation may serve as a clue to the determinants of host larva specificity. A host that allows expression of a certain polydnavirus may serve as a habitual host species for the parasite which secretes that virus. Conversely, if a given polydnavirus cannot be expressed in a certain host species, then the egg and larva of the corresponding parasite species are destroyed by that host species. At a simplistic level this suggests that specificity of the parasite–host association could be dictated by the specificity of the ichneumonid polydnavirus for expression in a given host species.

## Relatedness of Ichneumonid Polydnaviruses

To date there is evidence that at least 21 different species of ichneumonid wasps contain polydnavirus (29). Doubtless more will be found. The relatedness of these viruses can be used as one tool in analyzing the question of the origin of the ichneumonid polydnaviruses. Their common morphology, similar complex genome organization, shared life cycle, and mutualistic relationship to the parasite give compelling evidence that the ichneumonid polydnaviruses evolved from a common prepolydnavirus. If there is a common ancestor, then it would be expected that some of the current polydnaviruses would be related to each other. Furthermore, there would be a constant pressure on the polydnaviruses to "co-evolve" with the parasite species in which they replicate. Thus as the parasite species evolves, so must the corresponding polydnavirus. The polydisperse nature of the viral genome is ideally suited for the fluidity required in such a co-evolutionary development since it allows reassortment of the different circles. Multiple copies of genetically important circles would allow the virus the flexibility to maintain alternate copies of similar genetic material, some of which may be more advantageous as the parasite evolves. The very similar sequences observed by Theilmann and Summers (58) on different circles of DNA may be due to just this kind of built-in genetic fluidity. A certain sequence may lose its advantage under certain circumstances, but as the parasite evolves, another related sequence, which may be more effective under the new conditions, would already be available.

This hypothesis of a "co-evolution" between polydnavirus and parasite can in part be tested by analyzing the relatedness of polydnaviruses from parasite species that are or are not related to each other. In the only study of this aspect it was shown by DNA hybridization that polydnavirus HfV from *Hyposter fugitivus* was closely related to polydnaviruses from three other *Hyposter* species and three *Diadegma* species but not to polydnaviruses from a *Campoplex* species, a *Glypta* species, and two *Campoletis* species. Moreover, the DNA banding patterns of polydnaviruses from the related species were similar to each other but different from that of the unrelated species (Krell and Stoltz unpublished; Figure 8B). A similar relatedness was inferred from an extensive Western blot analysis of some of these same polydnaviruses (9). In this analysis a polydnavirus from a given species is more similar to polydnaviruses from species in the same (e.g., *Hyposter*) or similar (e.g., *Diadegma*) genus than to polydnaviruses from unrelated species (e.g., *Campoletis*). It would be interesting to construct a "phylogentic" tree of related polydnaviruses and see how that compares with the phylogenetic tree of the parasite species from which they were derived.

## BRACONID POLYDNAVIRUSES

A second group of polydnaviruses consists of viruses that have so far been found in parasites belonging to the Braconidae family. These are collectively referred to here as braconid polydnaviruses. As for all members of the Polydnaviridae family, the braconid polydnaviruses have a multipartite genome and are found in the ovaries of affected species. They are easily distinguished from the ichneumonid polydnaviruses not only on the basis of the family status of the parasite species affected but also by their morphological differences.

### Morphology

The braconid polydnaviruses have a morphology very similar to that of the insect baculoviruses. It was for this reason that they were originally classified as subgroup C of the Baculoviridae (35). The braconid polydnaviruses consist of electron-dense rod-shaped nucleocapsids with a diameter of 35 nm and a length that varies as much as the variation in size of the DNA circles which comprise the genome (Figure 9). The range in the length of nucleocapsids depends on the species of braconid polydnavirus. For polydnavirus from the braconid wasp *Cardiochiles nigriceps*, for example, the nucleocapsids are fairly short and range

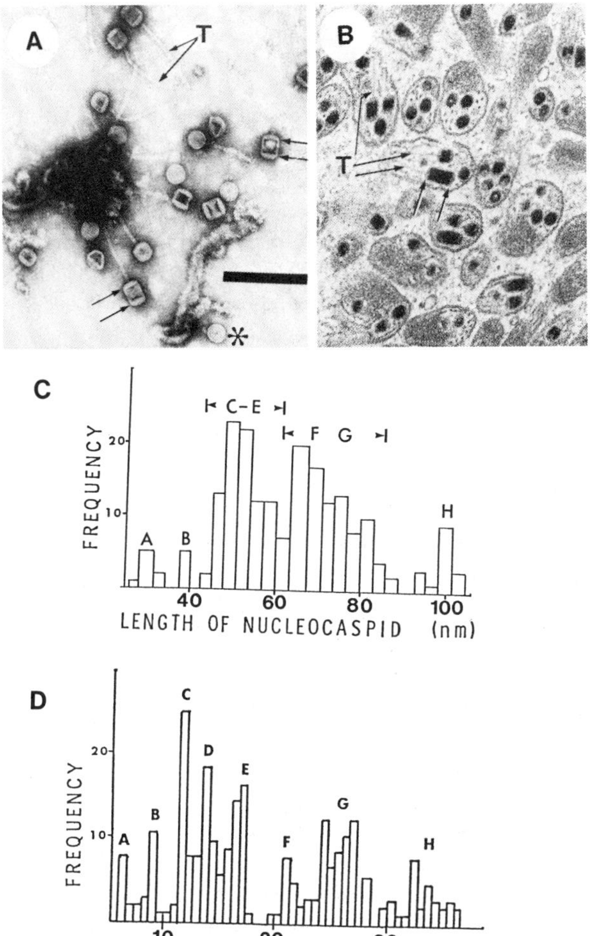


**Figure 9** Braconid polydnavirus. (A) Negative stain of liberated nucleocapsids of a braconid polydnavirus from *Cotesia* (= *Protapanteles*) *paleacritae* showing the end structures (arrows) of the cylindrical nucleocapsids and the tail (T) at one end of the nucleocapsid. The asterisk indicates an "end on" view of a nucleocapsid. (B) Thin section of *C. paleacritae* showing the nucleocapsid end structures (arrows) and tails (T). Several nucleocapsids of different lengths are seen within a common envelope. (C,D) Size distribution of nucleocapsid lengths (panel C) and viral DNA as measured by electron microscopy (panel D) of the braconid polydnavirus CmV from *Cotesia melanoscela*. The two histograms could be superimposed, suggesting that each nucleocapsid contains a single DNA species and that the length and frequency of that length correlate with the length of the encapsidated DNA and the frequency of that DNA in the population. Bar = 100 nm. (Figure is adapted from Ref. 52.)

in length between approximately 25 and 70 nm (55,62), whereas nucleocapsids of polydnavirus from *Protapanteles* (formerly *Apanteles*) *paleacritae* range in length between approximately 50 and 200 nm (52). The nucleocapsids are contained within a single limiting membrane which is derived de novo within infected nuclei. There may be only one nucleocapsid per envelope, as for *C. nigriceps* polydnavirus and polydnavirus from the braconid wasp *Chelonus texanus* and two *Apanteles* species, *A. crassicornis* and *A. fumiferaneae* (52,55). Braconid polydnavirus from the newly defined *Costesia* (formerly *Apanteles*) genus tend to have several nucleocapsids enveloped in each virion. This latter characteristic, that is, nucleocapsids being either singly enveloped or multiply enveloped, is also one which the braconid polydnaviruses share with baculoviruses.

## Genome

Although particles similar in morphology to the braconid polydnavirus have been identified in several species of braconid wasps, the genomes of only a few have been analyzed in sufficient detail to establish the multipartite nature of the genome. Of these the genome of *Cotesia melananoscela* virus (CmV) is perhaps the best understood, and even this is not as well defined as for some ichneumonid polydnaviruses. CmV is also of historical interest since it was the first virus identified as a polydnavirus (26,32). CmV DNA resolves into at least 16 different bands by agarose gel electrophoresis. Kleinschmidt analysis confirmed the circular nature of CmV DNA and showed that the circular DNAs ranged in size from 3 to 27 kb (32). The sequence heterogeneity of CmV DNA has not yet been extensively tested, for example, by Southern cross-blot hybridization. Nevertheless, in an unrelated study Stoltz et al. (48) showed that band *D* DNA of CmV was not homologous to band *A* DNA of CmV. A more systematic study would need to be done to determine the overall sequence complexity of the CmV genome and to determine if any of the CmV DNA circles are related to each other and if so, the extent of the relationship.

Stoltz et al. (48) report that unlike the ichneumonid polydnavirus CsV there is minimal sequence redundancy within individual DNA circles or between different CmV DNA circles. Nevertheless, many of the cloned CmV DNAs cross-reacted with very high molecular weight CmV DNA.

The DNA genomes of only a few braconid polydnaviruses, in addition to that of CmV, have been analyzed by agarose gel electrophoresis. These include polydnaviruses from the parasites *Cardiochiles nigriceps* (26), *Chelonus* near *curvimaculatus*, *Chelonus texanus* (21,22), and *Microplitis croceipes* (26). The DNA molecular size range for each of these braconid polydnaviruses was quite broad, generally from 5 kb up to greater than 25 kb.

Although some supercoiled DNA has been isolated from one braconid polydnavirus, CmV, this constituted only a minor fraction of the total viral DNA. Thus the nature of the encapsidated DNA is not known. It has been reported that for CmV, approximately 80% of the CmV genome is in the open circular conformation (18). However, the DNAs of the braconid polydnaviruses are quite large and although they may be supercoiled in situ (i.e., in the virus) during manipulation, they could easily be nicked to open circular forms.

The different size classes of CmV DNA are present in nonequimolar amounts with respect to each other. This is analagous to the situation that obtains for the ichneumonid polydnaviruses. For example, in CmV, DNA from bands 1–8 are present in much lower abundance than DNAs in bands 9–16 (26).

An interesting correlation exists between the variation in the length of CmV nucleocapsids and the sizes of circular CmV DNAs. A histogram of nucleocapsid lengths could easily be superimposed on a histogram of DNA lengths as determined by Kleinschmidt analysis (Figure 9C, D). For example, in this comparison, the frequency of nucleocapsids of 50 nm in length correlates with that of DNA with a size of approximately 12 kb (equivalent to the major band 9 by AGE and peak C by electron microscopy), the frequency of nucleocapsids 65 nm in length correlates with the frequency of DNA molecules with a size of approximately 18 kb (bands 13 and 14 and peak F), and nucleocapsids with a length of 40 nm correlate with DNA 9 kb in size (band 6 and peak B). Thus in braconid polydnaviruses each DNA could be separately encapsidated in individual nucleocapsids whose length is determined by the size of DNA encapsidated. This is an attractive hypothesis on the DNA packaging strategy of braconid polydnavirus that has yet to be tested.

## Proteins

The polypeptide profiles of braconid polydnaviruses, like that of the ichneumonid polydnaviruses, is relatively complex. At least 18 different size polypeptides have been resolved for CmV, 5 of which are from purified nucleocapsids (26,32). Two major nucleocapsid polypeptides of 38 and 32 kD have also been identified. Some of the other major polypeptides may be from the envelope. The aggregate molecular weight of the virion polypeptides is in excess of $10^6$ and would require a minimum coding capacity in excess of 30 kb assuming all proteins are unique and overlapping DNA sequences are not used. The polypeptide composition of some other braconid polydnaviruses has also been determined by polyacrylamide gel electrophoresis, but there has been no analysis of virion proteins beyond this level (26).

The braconid polydnaviruses, like the ichneumonid polydnaviruses, also replicate in calyx cells (52,53,55). The nucleocapsids are assembled within affected nuclei. These nucleocapsids become enclosed within an envelope which appears also to be assembled de novo within affected nuclei. These "mature virions" do not appear to bud through either the nuclear envelope or cell membrane, as is the case for ichneumonid polydnaviruses. However, many virions are seen in the calyx lumen along with remnants of cellular debris. This suggests that affected cells lining the lumen may become lysed, releasing their contents, including virus, directly into the lumen.

Braconid polydnaviruses are injected with the parasite egg during oviposition into the hemocoel of a host larva (52). The virus membrane may associate with cellular membranes in the parasitized host, thereby releasing the nucleocapsids. The nucleocapsids then enter the cells of the host and become internalized into the cytoplasm. Subsequently, the nucleocapsids orient themselves perpendicular to nuclear pores with the tail projecting through the pores. The nucleocapsids appear to be uncoated at the nuclear pore. It has been suggested that the viral DNA penetrates into the nucleus through a conduit in the nucleocapsid tail (26). Both the capsid and tail remain at the nuclear pore. There is some evidence that other secretions from the female ovary, namely from the venom gland, are required for uncoating and subsequent persistence of the viral DNA (47). Although the viral DNA persists, there is no evidence of viral replication.

## Host Regulation and Braconid Polydnaviruses

Associated with parasitism by endoparasitic hymenoptera are several physiological and biochemical changes, collectively referred to as host regulation (3), which ultimately benefit parasitoid development. Some aspects of host regulation include alterations in host growth rates, food utilization, hemolymph composition, and response to microbial and other forms of foreign challenge. The role of ichneumonid polydnaviruses in host regulation has already been alluded to. In this section I introduce some changes noted in parasitized larvae and demonstrate how the braconid polydnaviruses might be involved.

Some recent studies of larvae parasitized by braconid wasps have shown the appearance of new hemolymph proteins. Ferkovich et al. (15), for example, have shown that parasitism of *Spodoptera frugiperda* larvae by *Cotesia marginiventris* resulted in the appearance of two unique hemolymph proteins. Beckage and Templeton (3) have similarly shown that new hemolymph proteins appear in *Manduca sexta* larvae after parasitization by *Cotesia congregata*. Other changes in protein profiles revealed by polyacrylamide gel electrophoresis have also been repeated in these and other host–parasite associations. Jones et al. (21) noted that the changes in hemolymph proteins of *Trichoplusia ni* larvae following parasitization by the braconid wasp *Chelonus* spp. were similar to that present in last instar larvae. They further suggest that such changes result from a decline in juvenile hormone titers and perhaps other alterations in neurohormonal signals resulting after parasitization. Beckage and Templeton (3), however, noted that the level of juvenile hormone following parasitization of *M. sexta* by *C. con-*

*gregata* was sufficiently high to prevent larval metamorphosis. She also noted a decrease in the level of ecdysteroid in the same larvae. Thus the physiological integration that occurs in parasitized larvae could be attributed directly to the control of relative titers of insect hormones such as juvenile hormone and ecdysteroid in these larvae.

One aspect of host physiology that relates directly to parasitoid survival in the host is the host's response to foreign bodies (25). In particular, the encapsulation response of certain host hemocytes to large foreign bodies cannot be directed against the parasite egg or larva, or the parasite would be killed at an early developmental stage. For a parasite species ovipositing in a habitual host species the normal encapsulation response of the host against the injected egg or subsequent parasitoid larva is somehow abrogated. However, for the host to survive long enough for the parasite to develop and pupate, other aspects of the host's defense reaction should not be compromised by parasitization. Ross and Dunn (39), for example, noted that while *Manduca sexta* larvae parasitized by the braconid wasp *Cotesia congregata* were more sensitive to challenge with injected *Pseudomonas aeruginosa* compared to nonparasitized larvae, some antibacterial activity could still be detected in parasitized larvae. Parasitization in this case is accompanied by some (but not total) reduction in the immune defenses of the host larvae. A successful parasite then is one that can effectively suppress the immune response of a host against the parasitoid egg or larvae without extensively affecting the response to other foreign challenge. That there must be some specificity to this suppression is apparent when it is recognized that the parasite–host association itself is specific. Thus a parasite can survive in a habitual host species but not in a nonhabitual host species. In the nonhabitual host species the parasitoid egg or larva is destroyed, presumably because the system which prevents encapsulation in the habitual host is not operable in the nonhabitual one. It is indeed tempting to speculate that the polydnaviruses play some role in host regulation and are the agents that define the specificity of these interactions. Polydnaviruses are ideally suited for this role since "infection" of host cells by polydnaviruses may be specific to only certain hosts. The braconid polydnavirus may be involved in host regulation. Clearly the female parasite injects something that exerts control over host physiology following parasitization. Kitano and Nakatsuji (25) initially postulated that the oviposited eggs of *Apanteles glomeratus* produce an agent that directly inhibits encapsulation in parasitized host larvae of *Pieris rapae crucivora*. Later, however, Kitano recognized that components of both the calyx fluid (polydnavirus?) and venom were required to prevent encapsulation (23,24). Similarly, Guzo and Stoltz noted that polydnavirus from *C. melananoscela* was required for immunosuppression of encapsulation in *Orgyia leucostigma* larvae (19). Moreover, they established that venom also was required and suggested that some components(s) of venom may aid in the uncoating of the viral genome in cells of the parasitized host (47). If the

CmV is inactivated by psoralen treatment, the encapsulation response is maintained, suggesting that viral transcription is required for immunosuppression.

In a study similar to that described earlier for CsV and *H. virescens*, Tanaka et al. (56) analyzed the hormonal status of larvae of *Pseudaletia separata* parasitized by the braconid wasp *Apanteles kariyai*. Parasitized larvae were arrested in development and had lower titers of prothoracicotropic hormone (PTTH) and ecdysteroid than nonparasitized larvae. These effects could be duplicated by calyx fluid from *P. separata* only if it was injected with venom fluids into unparasitized larvae. These authors concluded that components of both the calyx and venom fluids inhibit the secretion or synthesis of PTTH and slow development by keeping the ecdysteroid levels low. This effect of a braconid polydnavirus (in association with venom fluid) on hormone levels in parasitized larvae appears very similar to that seen for the ichneumonid polydnaviruses described earlier. Thus both "genera" of polydnaviruses may use a common strategy to effect host regulation by manipulating the hormonal balance.

Although the available data on factors affecting host regulation are by no means complete, a general picture of how polydnaviruses are involved in this process is slowly emerging. The following is a rough framework of a working hypothesis that lends itself to further experimentation. In the simplest scenario, shortly after oviposition of a parasitized egg in the hemocoel of a habitual host, the corresponding polydnavirus "infects" a number of host cells. This same polydnavirus may be incapable of infecting analogous cells in a nonhabitual host due to the host specificity dictated by the virus. The viral genome is uncoated in some or all of the "infected" cells and this may require additional components from the venom gland, calyx fluid, or other accessory glands. Following uncoating, virus-specific transcription and translation ensue to provide effector molecules either directly (e.g., as proteins) or indirectly (e.g., as secondary metabolites) were synthesis at least in part is directed by viral-specific enzymes. These viral-synthesized or induced effector molecules affect host metabolism perhaps by an intermediary route involving regulation of endocrine levels or by acting directly on targeted tissues such as the prothoracic gland, which could interfere with ecdysone synthesis and release.

This is a very superificial scenario and this hypothesis would require a lot of fine tuning to account for some of the different changes observed in host regulation and also to account for differences noted in a variety of parasite–host associations. The main tenet in all of these is that the parasitoid polydnaviruses have a central role to play and have the genetic resilience to adapt to such a multifaceted interaction.

## UNUSUAL VIRUSES OF PARASITIC HYMENOPTERA

The parasitic hymenoptera species appear to serve as an excellent reservoir not only for polydnaviruses but for several other unusual viruses or, better, "viruslike particles." Although these may be unrelated to the poplydnaviruses described in this chapter, they are often found associated with them. In some cases these unusual parasitoid viruses may be partly responsible for some of the physiological changes observed in parasitized habitual host larvae. For these reasons it is worthwhile to at least introduce these viruses and put them into context with the more commonly encountered polydnaviruses.

Viruslike particles have been described in the braconid wasp *Phanerotoma flavitestacea* (38) and two ichneumonid wasps, *Bathyplectes anurus* and *B. curculionis* (20). However, these have been described in insufficient detail to determine whether they are polydnaviruses.

A virus from the ichneumonid parasitoid *Mesoleius tenthredinis* was first described by Stoltz et al. (51) as "similar to baculoviruses" in morphology. In a more extensive study of the morphology and replication of this virus, Stoltz provided convincing evidence based on the morphogenesis and genome of this virus to conclude that this virus is in fact a baculovirus infecting a species of parasitic hymenoptera (41).

A rather unusual, large, quasicylindrical agent, CmV2, was found to coexist, but as a relatively minor component, with the polydnavirus CmV in some strains of the braconid wasp *C. melanoscela* (45). This virus was initially detected by virtue of its ability to replicate in parasitized *Orgyia leucostigma* larvae. A more detailed microscopic and molecular analysis of CmV2 showed that it is not a polydnavirus but replicates in calyx epithelial cells (49). It also replicates in cells of several host species including *Lymantria dispar*, *Stilpnotia salicis*, *Estigmene acrea*, and *Hyphantria cunea* as well as cell lines derived from *L. dispar* and *E. acrea*. The latter makes CmV2 amenable for future studies in the replication of this virus. During replication of CmV2 sufficient viral DNA is replicated to visualize viral DNA-specific restriction fragments above the background of digested cell DNA. From this analysis, and in keeping with the large size of CmV2, it was concluded that the genome was approximately 125 kb in size.

Another group of morphologically related viruslike particles has been observed in the calyx of several species of both braconid and ichneumonid wasps. Long, rod-shaped particles have been observed in calyx cells of the braconid wasps *C. congregata*, *Microplitis croceipes*, and *Cotesia hyphantriae* (52,53). In these species the particles appear to replicate in calyx cells, as does the corresponding braconid polydnavirus, and are also seen as a component of the calyx fluid in those parasites.

Similar large viruslike particles have been observed in the calyx of the ichneumonid wasp *D. terebrans* (28). In this species the large (up to 2 $\mu$m) particles replicate in ovary cells near the junction between the ovarioles and the calyx while the polydnavirus DtV replicates in different cells within the calyx. These large particles replicate in the nuclei where they form 20 nm diameter "nucleocapsids," which become enveloped in the cytoplasm (Figure 10). As in the case for CmV2, these long particles form only a minor component of the

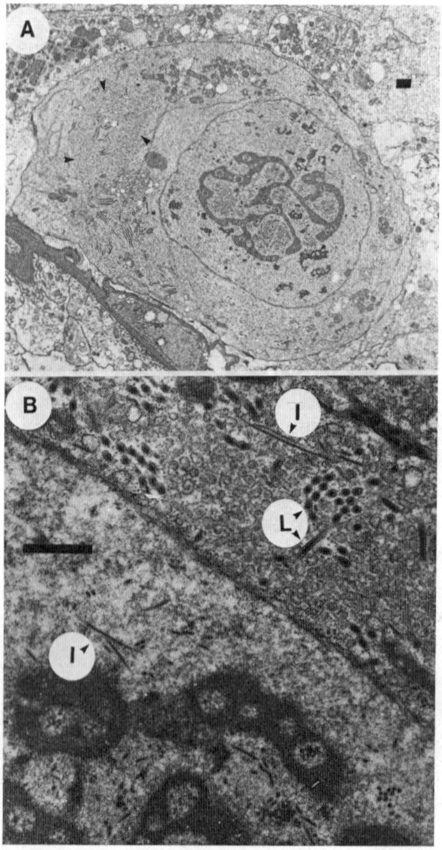


**Figure 10** Sections of ovary cells of the ichneumonid parasitoid *Diadegma terebrans* showing in (A) an overview of a cell infected with an unusual long viruslike particle (DtLV). A virogenic stroma can be seen in the nucleus. DtLV can be seen in the cytoplasm in areas replete with membrane-bound vesicles (arrowheads). Some long naked nucleocapsids (!) can be seen in both the nucleus and cytoplasm of an infected cell in (B). Different profiles of enveloped DtLV can also be seen in the cytoplasm (L). Bar = 50 nm.

calyx fluid. There is as yet no evidence for replication of these long particles in parasitized *Ostrinia nubilalis*. Recently N. Beckage observed similar particles in the ovary of the braconid wasp *Cotesia congregata* (personal communication).

Although I have described only a few "unusual viruses" of parasitic hymenoptera, doubtless more exist. Presumably some will be morphologically related to those described here. Additional morphological types of viruslike particles will probably be found if more parasite species are examined. Whether any of these are related to any of the polydnaviruses is not known. That they occupy the same niche (i.e., the calyx of parasitic wasps and the hemocoel of parasitized larvae) suggests that some may have a similar transmission strategy. In addition, some of these viruses may cooperate with the resident polydnavirus in parasitization and the preparation of an environment suitable for parasitoid development in the otherwise hostile confines of the host larva hemocoel. Whether they do is an academic question but the availability of at least one virus, CmV2, which can replicate both in host larvae and in vitro in insect cell lines, makes this question open to evaluation by an experimental approach.

## ACKNOWLEDGMENTS

Since their initial description as polydnaviruses, members of this group have been studied largely in the laboratories of Dr. Don Stoltz, Dalhousie University, Halifax, Nova Scotia, and Dr. Brad Vinson and Dr. Max Summers, both from Texas A&M University, College Station, Texas, and more recently in my own laboratory at the University of Guelph. I would like to acknowledge these fellow researchers and their co-workers for their continued enthusiasm and collaboration. Much of my own work was sponsored by grants from the Natural Sciences and Engineering Research Council of Canada and Agriculture Canada. I acknowledge both agencies for their continued support. I also thank Lesley Burke and Sandra Good for typing the manuscript.

## REFERENCES

1. Armstrong, R. E. and Walsby, A. E. The gas vesicle: A rigid membrane enclosing a hollow space, in *Organization of Prokaryotic Cell Membranes*, vol. 2, Ghosh, B. K., Ed., CRC Press, Boca Raton, FL, 1981, p. 95.
2. Beckage, N. E. Endocrine interactions between endoparasitic insects and their hosts. *Ann. Rev. Entomol.* 30, 371, 1985.
3. Beckage, N. E. and Templeton, T. J. Physiological effects of parasitism by *Apanteles congregatus* in terminal-stage tobacco hornworm larvae. *J. Insect Physiol.* 32, 299, 1986.
4. Blissard, G. W., Fleming, J. G. W., Vinson, S. B., and Summers, M. D. *Campoletis sonorensis* virus: Expression in *Heliothis virescens* and identification of expressed sequences. *J. Insect Physiol.* 32, 351, 1986.
5. Blissard, G. W., Smith, O. P., and Summers, M. D. Two related viral genes are located on a single superhelical DNA segment of the multipartite *Campoletis sonorensis* virus genome. *Virology* 160, 120, 1987.
6. Blissard, G. W., Theilmann, D. A., and Summers, M. D. Segment W of *Campoletis sonorensis* virus: Expression, gene products, and organization. *Virology* 169, 78, 1989.

7. Blissard, G. W., Vinson, S. B., and Summers, M. D. Identification, mapping, and *in vitro* translation of *Campoletis sonorensis* virus mRNAs from parasitized *Heliothis virescens* larvae. *J. Virol.* 57, 318, 1986.
8. Brown, F. The classification and nomenclature of viruses: Summary of results of meetings of the International Committee on Taxonomy of Viruses in Sendai. *Intervirology* 25, 141, 1986.
9. Cook, D. and Stoltz, D. B. Comparative serology of viruses isolated from ichneumonid parasitoids. *Virology* 130, 215, 1983.
10. Cook, D. I., Stoltz, D. B., and Vinson, S. B. Induction of new haemolymph glycoprotein in larvae of permissive hosts parasitized by *Campoletis sonorensis*. *Insect. Biochem.* 14, 45, 1984.
11. Davies, D. H., Strand, M. R., and Vinson, S. B. Changes in differential haemocyte count and *in vitro* behavior of plasmatocytes from host *Heliothis virescens* caused by *Campoletis sonorensis* polydnavirus. *J. Insect Physiol.* 33, 143, 1987.
12. Dover, B. A., Davies, D. H., Strand, M. R., Gray, R. S., Keeley, L. L., and Vinson, S. B. Ecdysteroid titer reduction and developmental arrest of last instar *Heliothis virescens* larvae by calyx fluid from the parasitoid *Campoletis sonorensis*. *J. Insect Physiol.* 33, 333, 1987.
13. Dover, B. A., Davies, D. H., and Vinson, S. B. Degeneration of last instar *Heliothis virescens* prothoracic glands by *Campoletis sonorensis* polydnavirus. *J. Invertebr. Pathol.* 51, 80, 1988.
14. Edson, K. M., Vinson, S. B., Stoltz, D. B., and Summers, M. D. Virus in a parasitoid wasp: Suppression of the cellular immune response in the parasitoid's host. *Science* 211, 582, 1981.
15. Ferkovich, S. M., Greany, P. D., and Dillard, C. Changes in haemolymph proteins of the fall armyworm, *Spodoptera frugiperda* (J. E. Smith), associated with parasitism by the braconid parasitoid *Cotesia marginiventris* (Cresson). *J. Insect Physiol.* 29, 933, 1983.
16. Fleming, J. G. W. and Summers, M. D. *Campoletis sonorensis* endoparasitic wasps contain forms of *C. sonorensis* virus DNA suggestive of integrated and extrachromosomal polydnavirus DNAs. *J. Virol.* 57, 552, 1986.
17. Fleming, J. G. W., Blissard, G. W., Summers, M. D., and Vinson, S. B. Expression of *Campoletis sonorensis* virus in the parasitized host, *Heliothis virescens*. *J. Virol.* 48, 74, 1983.
18. Guzo, D. and Stoltz, D. B. Obligatory multiparasitism in the tussock moth, *Orgyia leucostigma*. *Parasitology* 90, 1, 1985.
19. Guzo, D. and Stoltz, D. B. Observations on cellular immunity and parasitism in the tussock moth. *J. Insect Physiol.* 33, 19, 1987.
20. Hess, R. T., Poinar, G. O., Jr., Etzel, L., and Merritt, C. C. Calyx particle morphology of *Bathyplectes anurus* and *B. curculionis* (Hymenoptera: Ichneumonidae). *Acta Zool.* (*Stockholm*) 61, 111, 1980.
21. Jones, D., Jones, G., Rudnicka, M., and Click, A. Precocious expression of the final larval instar developmental pattern in larvae of *Trichoplusia ni* pseudoparasitized by *Chelonus* spp. *Comp. Biochem. Physiol.* 83B, 339, 1986.
22. Jones, D., Sreekrishna, S., Iwaya, M., Yang, J. N., and Eberely, M. Comparison of

viral ultrastructure and DNA banding patterns from the reproductive tracts of eastern and western hemisphere *Chelonus* spp. (Braconidae: Hymenoptera). *J. Invertebr. Pathol.* 47, 105, 1986.

23. Kitano, H. Effect of the venom of the gregarious parasitoid, *Apanteles glomeratus* on its hemocytic encapsulation by the host, *Pieris*. J. *Invert. Pathol.* 40, 61, 1982.
24. Kitano, H. The role of Apanteles glomeratus venom in the defensive response of its host, *Pieris rapae crucivora*. *J. Insect Physiol.* 32, 369, 1986.
25. Kitano, H., and Nakatsuji, N. Resistance of *Apanteles* eggs to the haemocytic encapsulation by their habitual host *Pieris*. *J. Insect Physiol.* 24, 261, 1978.
26. Krell, P. J. Viruses in Parasitoid Hymenoptera: A Preliminary Characterization, Ph.D. diss., Dalhousie University, Halifax, Nova Scotia, 1979.
27. Krell, P. J. Polydnavirus replication and tissue organization of infected cells in the parasitic wasp *Diadegma terebrans*. *Can. J. Microbiol.* 33, 176, 1987.
28. Krell, P. J. Replication of long virus-like particles in the reproductive tract of the ichneumonid wasp *Diadegma terebrans*. *J. Gen. Virol.* 68, 1477, 1987.
29. Krell, P. J. Polydnaviridae, in *Atlas of Invertebrate Viruses*, Adams, J. R., Ed., CRC Press, Boca Raton, FL, 1991.
30. Krell, P. J. and Beveridge, T. J. The structure of bacteria and molecular biology of viruses, in *International Review of Cytology, Supplement 17*, Bourne, G. H., Ed., Academic Press, New York, 1987, p. 15.
31. Krell, P. J. and Lo, R. Y. *Sixth International Congress of Virology, T-38 Insect DNA Viruses*, 1984. Sendai, Japan, Sept. 1-7
32. Krell, P. J. and Stoltz, D. B. Unusual baculovirus of the parasitoid wasp *Apanteles melanoscelus*: Isolation and preliminary characterization. *J. Virol.* 29, 1118, 1979.
33. Krell, P. J. and Stoltz, D. B. Virus-like particles in the ovary of an ichneumonid wasp: Purification and preliminary characterization. *Virology* 101, 408, 1980.
34. Krell, P. J., Summers, M. D., and Vinson, S. B. Virus with a multipartite superhelical DNA genome from the ichneumonid parasitoid *Campoletis sonorensis*. *J. Virol.* 43, 859, 1982.
35. Mathews, R. E. F., Ed. *Classification and Nomenclature of Viruses, 3rd Report of the International Committee on Taxonomy of Viruses*, S. Karger, A. G. White Plains, NY, 1979, p. 170.
36. Norton, W. N. and Vinson, S. B. Correlating the initiation of virus replication with a specific pupal developmental phase of an ichneumonid parasitoid. *Cell Tiss. Res.* 231, 387, 1983.
37. Norton, W. N., Vinson, S. B., and Stoltz, D. B. Nuclear secretory particles associated with the calyx cells of the ichneumonid parasitoid *Campoletis sonorensis* (Cameron). *Cell Tiss. Res.* 162, 195, 1975.
38. Poinar, G. O., Jr., Hess, R., and Caltagirone, L. E. Virus-like particles in the calyx of *Phanerotoma flavitestacea* (Hymenoptera: Braconidae) and their transfer into host tissues. *Acta Zool.* (*Stockholm*) 57, 161, 1976.
39. Ross, D. R. and Dunn, P. E. Effects of parasitism by *Cotesia congregata* on the anti bacterial responses of *Manduca sexta* larvae, in *Fundamental and Applied Aspects of Invertebrate Pathology*, Samson, R. A., Vlak, J. M., and Peters, D., Eds., ICIP, Wageningen, The Netherlands, p. 77.

40. Stinchcomb, D. T., Thomas, M., Kelly, J., Selker, E., and Davis R. W. Eukaryotic DNA segments capable of autonomous replication in yeast. *Proc. Natl. Acad. Sci. U.S.A.*, 77, 4559, 1980.
41. Stoltz, D. B. A putative baculovirus in the ichneumonid parasitoid, *Mesoleius tenthredinis*, *Can. J. Microbiol.* 27, 116, 1981.
42. Stoltz, D. B. Interactions between parasitoid-derived products and host insects: An overview. *J. Insect Physiol.* 32, 347, 1986.
43. Stoltz, D. B. Viruses of parasitic hymenoptera, in *Fundamental and Applied Aspects of Invertebrate Pathology*, Samson, R. A., Vlak, J. M., and Peters, D., Eds., ICIP, Wageningen, The Netherlands, 1986, p. 81.
44. Stoltz, D. B. and Cook, D. I. Inhibition of host phenoloxidase activity by parasitoid hymenoptera. *Experientia* 39, 1021, 1983.
45. Stoltz, D. B. and Faulkner, G. Apparent replication of an unusual viruslike particle in both a parasitoid wasp and its host. *Can. J. Microbiol.* 24, 1509, 1978.
46. Stoltz, D. B. and Guzo, D. Apparent haemocytic transformations associated with parasitoid-induced inhibition of immunity in *Malacosoma disstria* larvae. *J. Insect Physiol.* 32, 377, 1986.
47. Stoltz, D. B., Guzo, D., Belland, E. R., Lucarotti, C. J., and MacKinnon, E. A. Venom promotes uncoating *in vitro* and persistence *in vivo* of DNA from a braconid polydnavirus. *J. Gen. Virol.* 69, 903, 1988.
48. Stoltz, D. B., Guzo, D., and Cook, D. Studies on polydnavirus transmission. *Virology* 155, 120, 1986.
49. Stoltz, D. B., Krell, P., Cook, D., MacKinnon, E. A., and Lucarotti, C. J. An unusual virus from the parasitic wasp *Cotesia melanoscela*. *Virology* 162, 311, 1988.
50. Stoltz, D. B., Krell, P., Summers, M. D., and Vinson, S. B. Polydnaviridae: A proposed family of insect viruses with segmented, double-stranded, circular DNA genomes. *Intervirology* 21, 1, 1984.
51. Stoltz, D. B., Krell, P. J., and Vinson, S. B. Polydisperse viral DNA's in ichneumonid ovaries: A survey. *Can. J. Microbiol.* 27, 123, 1981.
52. Stoltz, D. B. and Vinson, S. B. Baculovirus-like particles in the reproductive tracts of female parasitoid wasps. II. The genus *Apanteles*. *Can. J. Microbiol.* 23, 28, 1977.
53. Stoltz, D. B. and Vinson, S. B. Viruses and parasitism in insects. *Adv. Virus Res.* 24, 125, 1979.
54. Stoltz, D. B. and Vinson, S. B. Penetration into caterpillar cells of virus-like particles injected during oviposition by parasitoid ichneumonid wasps. *Can. J. Microbiol.* 25, 207, 1979.
55. Stoltz, D. B., Vinson, S. B., and MacKinnon, E. A. Baculovirus-like particles in the reproductive tracts of female parasitoid wasps. *Can. J. Microbiol.* 22, 1013, 1976.
56. Tanaka, T., Agui, N., and Hiruma, K. The parasitoid *Apanteles kariyai* inhibits pupation of its host, *Pseudaletia separata*, via disruption of prothoracicotropic hormone release. *Gen. Comp. Endocrinol.* 67, 364, 1987.
57. Theilmann, D. A. and Summers, M. D. Molecular analysis of *Campoletis sonorensis* virus DNA in the lepidopteran host *Heliothis virescens*. *J. Gen. Virol.* 67, 1961, 1986.

58. Theilmann, D. A. and Summers, M. D. Physical analysis of the *Campoletis sonorensis* virus multipartite genome and identification of a family of tandemly repeated elements. *J. Virol.* 61, 2589, 1987.
59. Theilmann, D. A. and Summers, M. D. Identification and comparison of *Campoletis sonorensis* virus transcripts expressed from four genomic segments in the insect hosts *Heliothis virescens* and *Campoletis sonorensis. Virology* 167, 329, 1988.
60. Vinson, S. B., Davies, D. H., and Dover, B. A. Effects of the polydnaviridae in select parasitoid host insect systems, in *Fundamental and Applied Aspects of Invertebrate Pathology*, Samson, R. A., Vlak, J. M., and Peters, D., Eds., ICIP, Wageningen, The Netherlands, 1986, p. 74.
61. Vinson, S. B., Edson, K. M., and Stoltz, D. B. Effect of a virus associated with the reproductive system of the parasitoid wasp, *Campoletis sonorensis,* on host weight gain. *J. Invertebr. Pathol.* 34, 133, 1979.
62. Vinson, S. B. and Scott, J. R. Particles containing DNA associated with the oocyte of an insect parasitoid. *J. Invertebr. Pathol.* 25, 375, 1975.

# 6

# The Entomopoxviruses

BASIL M. ARIF
*Forest Pest Management Institute, Forestry Canada, Sault Ste-Marie, Ontario, Canada*
EDOUARD KURSTAK
*University of Montreal, Montreal, Quebec, Canada*

## INTRODUCTION

Insect poxviruses, known as entomopoxviruses (EPVs) have been isolated from at least 31 species of insects (1–3). EPVs are a complex group of viruses that have biochemical and morphological resemblance to poxviruses of vertebrates with the exception that EPVs have the distinguishing feature of being occluded in a paracrystalline proteinic matrix known as spheroid and they replicate at 28°C (4). The International Committee on Taxonomy of Viruses has included the genus entomopoxvirus in the family Poxviridae (5). The majority of EPV isolates come from insects belonging to the orders Lepidoptera and Coleoptera. They have been also known to infect members of Diptera and Orthoptera. We have summarized the host range and the morphological features of EPVs in Table 1.

In comparison to other viruses of insects and to poxviruses of vertebrates, EPVs have remained in relative obscurity; little is known about their biochemical and biophysical nature, composition, the details of their replication cycle, and cross-infection and serological properties. One reason for the lack of general interest is that they have not been considered pest-controlling agents because of their morphological similarities to poxviruses of vertebrates.

This chapter gives an overview of the structural characteristics, biochemical properties, and replication of EPVs.

## STRUCTURAL FEATURES

The virions are oval or brick-shaped, ranging in size from 150 to 470 nm long and 165 to 300 nm wide. Negatively stained virions exhibit an outer membrane

**Table 1** Morphological Properties of Certain Entomopoxviruses

| Host | Virus size (nm) | Inclusion body ($\mu$m) | Virus shape | Core shape, vertical section | Spindles | Reference |
|---|---|---|---|---|---|---|
| Lepidoptera | | | | | | |
| *Amsacta moorei* (Arctiidae) | 350 × 250 | 1–4 | Oval | Rectangular | Few | 6, 7 |
| *Oreopsyche angustella* (Psychidae) | 360 × 260 | 3–10 × 2–7 | Oval | Rectangular | Present | 8 |
| *Operophtera brumata* (Geometridae) | 400 × 350 | 3–15 | Oval | Rectangular | Present | 9, 59 |
| *Euxoa auxiliaris* (Noctuidae) | 260 × 165 | 4–5 | Oval | Rectangular | Absent | 11 |
| *Choristoneura biennis* (Tortricidae) | 400 × 300 | 2–3 | Oval | Rectangular | Present | 12 |
| *Choristoneura conflictana* (Tortricidae) | 273–235 | 7 × 4 | Oval | Rectangular | Present | 13 |
| *Choristoneura diversana* (Tortricidae) | 280 × 220 | 7 × 5 | Oval | Rectangular | Present | 14 |
| Coleoptera | | | | | | |
| *Melolontha melolontha* (Scarabaeidae) | 450 × 250 | 10–24 | Oval | Unilaterally concave | Present | 15–17 |
| *Othnonius batesi* (Scarabaeidae) | 470 × 265 | 5–10 | Oval | Unilaterally concave | Present | 4, 18 |
| *Demodena boranensis* (Scarabaeidae) | 420 × 230 | 8–11 | Oval | Unilaterally concave | Present | 19 |
| *Geotrupes silvaticus* (Scarabaeidae) | 366–416 × 255–286 | 3.5–11 | Oval | Unilaterally concave | Present | 20 |

| | | | | | | |
|---|---|---|---|---|---|---|
| *Dermolepida alborhirtum* (Scarabaeidae) | 420–450 × 220–240 | 3–5 | Oval | Unilaterally concave | Absent | 18, 21 |
| *Aphodius tasmaniae* (Scarabaeidae) | 380–430 × 250–300 | 5–12 | Oval | Unilaterally concave | Present | 21 |
| *Anomala cuprea* (Scarabaeidae) | 440 × 250 | 5 × 8 | Oval | Unilaterally concave | Present | 22 |
| *Phyllopertha horticola* (Rutelidae) | 400 × 240 | 6–25 | Oval | Unilaterally concave | Present | 23 |
| *Figulus sublaevis* (Lucanidae) | 330 × 290 | 1–5 | Oval | Unilaterally concave | Present | 19 |
| Diptera | | | | | | |
| *Chironomus luridus* (Chironomidae) | 330 × 230 × 110 | 4–7 | Cuboidal | Dumbbell | Absent | 24, 25 |
| *Chironomus attenuatus* (Chironomidae) | 330 × 250 × 130 | Up to 6 | Cuboidal | Dumbbell | Absent | 26 |
| *Camptochironomus tentans* (Chironomidae) | 220–250 × 270–300 | 2–8 | Cushion shape | Dumbbell | Absent | 9 |
| *Goeldichironomus holoprasinus* (Chironomidae) | 346 × 300 × 160 | 3 × 5 | Cushion shape | Dumbbell | Absent | 27 |
| *Aedes aegypti* (Culicidae) | 320 × 230 | | | Dumbbell | | 28 |
| Orthoptera | | | | | | |
| *Melanoplus sanguinipes* (Acrididae) | 320 × 250 | 2–11 | | Rectangular or dumbbell | Absent | 29 |

*Source:* Modified from Ref. 1.

that has been repeatedly folded to give a mulberrylike surface (Figure 1A) probably similar to the M form of poxviruses of vertebrates. The size of the spherical folds was estimated to be approximately 22 nm for *Melolontha melolontha* EPV (MmEPV) and 40 nm for *Amsacta moorei* EPV (AmEPV; 7, 30). The viral core is an electron-dense multilayer structure containing a double-stranded linear DNA. Depending on the genus, one or two lateral bodies are outside the core structure. Kurstak and Garzon (1) reported that three subgroups can be distinguished by their morphology and the size of the virion:

1. Lepidopteran and orthopteran EPVs measure approximately 250 × 350 nm, are generally smaller than those from Coleoptera, and possess a symmetrical cylindrical core and two discrete lateral bodies (Figure 1B).
2. Coleopteran EPVs are relatively large, measuring approximately 250 × 450 nm, and contain a single lateral body that presses against the core structure to give it a concave appearance (Figure 2).
3. Dipteran EPVs are the smallest, 230 × 320 nm; they are brick-shaped with a biconcave core and two lateral bodies (Figure 3A,B).

Figure 4 illustrates the morphology of vaccinia virus and EPVs from three genera.

Electron microscopic studies have also shown that EPVs of Lepidoptera and Coleoptera synthesize paracrystalline proteinic bodies called spindles. The spindles are devoid of virus particles. There are two types of spindles: macro- and microspindles. The latter are occluded in the spheroid much like virions (31) (Figure 5). Immunological studies have indicated that the spindle protein was antigenically different from the spheroidal protein spheroidin (32, 33). The role of the spindles, if any, in the viral replication cycle is not known. Microspindles do not appear to be as susceptible to dissolution by sodium carbonate as spheroidin and consequently band in sucrose density gradients with virions that have been slightly damaged by alkali (34) (Figure 6).

## THE VIRION

### Structural Polypeptides

The complexity of EPVs is evident from electron microscopic studies and from analysis of the virions by polyacrylamide gel electrophoresis (PAGE). EPVs of the *Choristoneura* spp. were shown by coomassie blue staining to be composed of at least 40 structural polypeptides ranging from 12 to 250 kD (35). The virion probably consists of a larger number of polypeptides, which would be visible if a more sensitive stain, such as silver stain, was used. The authors demonstrated that 12 polypeptides constituted approximately 95% of the total viral proteins (VPs). Purified viral cores contained one major protein, VP59 (59 kD), and to a

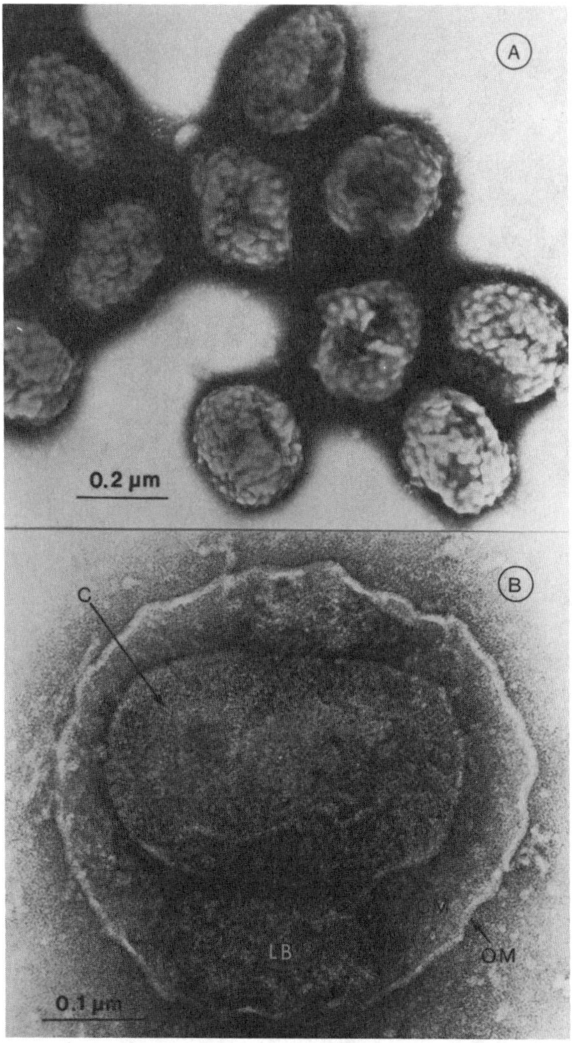


**Figure 1** EPV of *Choristoneura biennis*. (A) Virions liberated from spheroids with alkali and purified on density gradients. (B) The internal structure of the virion showing the lateral bodies (LB), rectangular core (C), the core membrane (CM), and the outer membrane (OM).

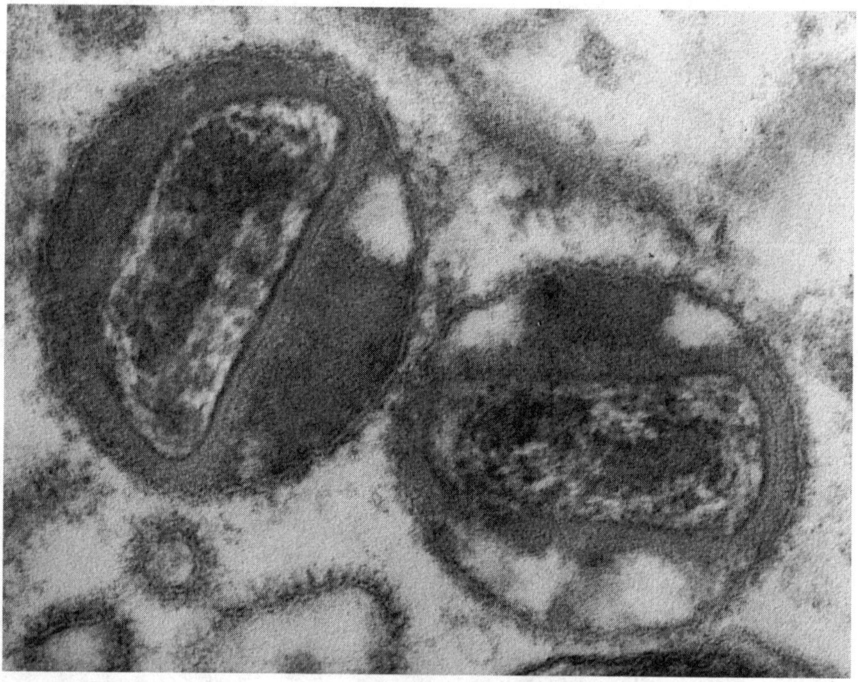

**Figure 2** Ultrastructure of the poxvirus (EPV) of *Melolontha melolontha* Coleoptera showing concave core and lateral body (× 162,000).

lesser extent VP102 and VP89. VP59 is probably an internal protein associated with the DNA. It would be interesting to identify the DNA binding proteins and illustrate their role in the transcription of the EPV DNA.

Langridge and Roberts (36) reported that virions of AmEPV, *Euxoa auxiliaris* (EaEPV), and *Melanoplus sanguinipes* (MsEPV) contained 36, 24, and 39 polypeptides, respectively. They also compared the EPV patterns to that of vaccinia polypeptides and reported no similarity between the two groups. The EPV of the lesser cornstalk borer, *Elasmopalpus lignosellus* (ElEPV), was composed of 32 structural polypeptides ranging in size from 13 to 145 kD. There also was no similarity to the polypeptide pattern of the baculovirus AcMNPV (37).

In a study where MsEPV was passaged in the homologous host and in *Locusta migratoria* and the polypeptides were analyzed by PAGE, three different bands were observed to be characteristic for each progeny virus (38). The proteins were either present in different amounts or unique. The authors speculated that the

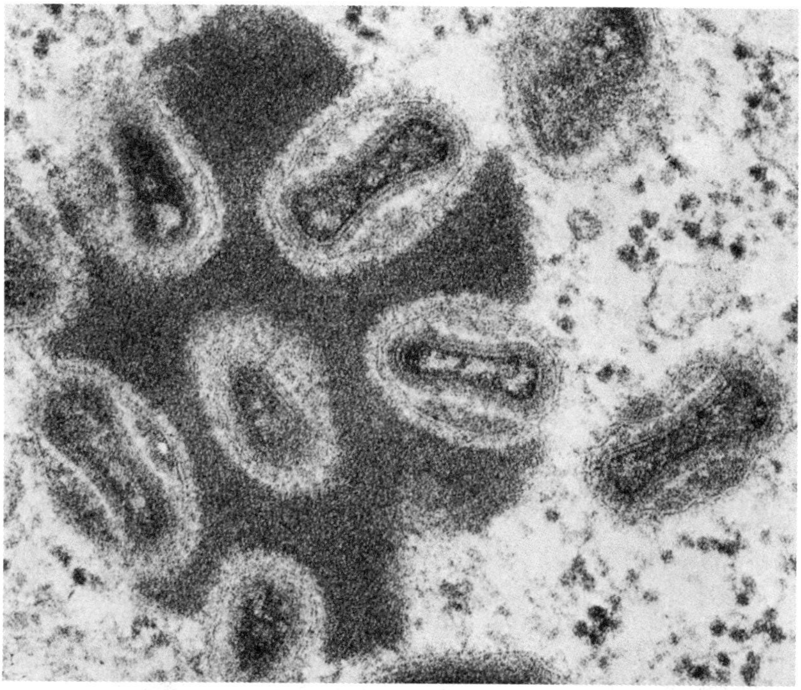

**Figure 3** The mature cuboidal viral particles of *Chironomus attenatus* Diptera poxvirus (EPV) (× 98,000). Courtesy of Dr. M. D. Summers.

unique polypeptides might have been the result of selective transcription of the viral DNA regulated by conditions within the host.

## Spheroidin

In comparison to the polyhedrin of baculoviruses, the estimated size of spheroidin is 102–110 kD, which is approximately three times larger (36, 37,39). Cysteine and methionine constituted a relatively high percentage of the total amino acids of three spheroidins, which may explain the need for a reducing agent and sodium carbonate to dissolve the spheroids completely (32, 36).

## The Viral Genome

The genome of EPVs is a double-stranded and linear DNA which constitutes approximately 5% of the virus particle (7, 24, 34, 40). The size of the genome measured from centrifugation through sucrose density gradients and contour length measurements and from electrophoresis of restricted DNA ranged from 135 to 200 × $10^6$ daltons (3, 34, 37, 41). The genomic size in subgenus B,

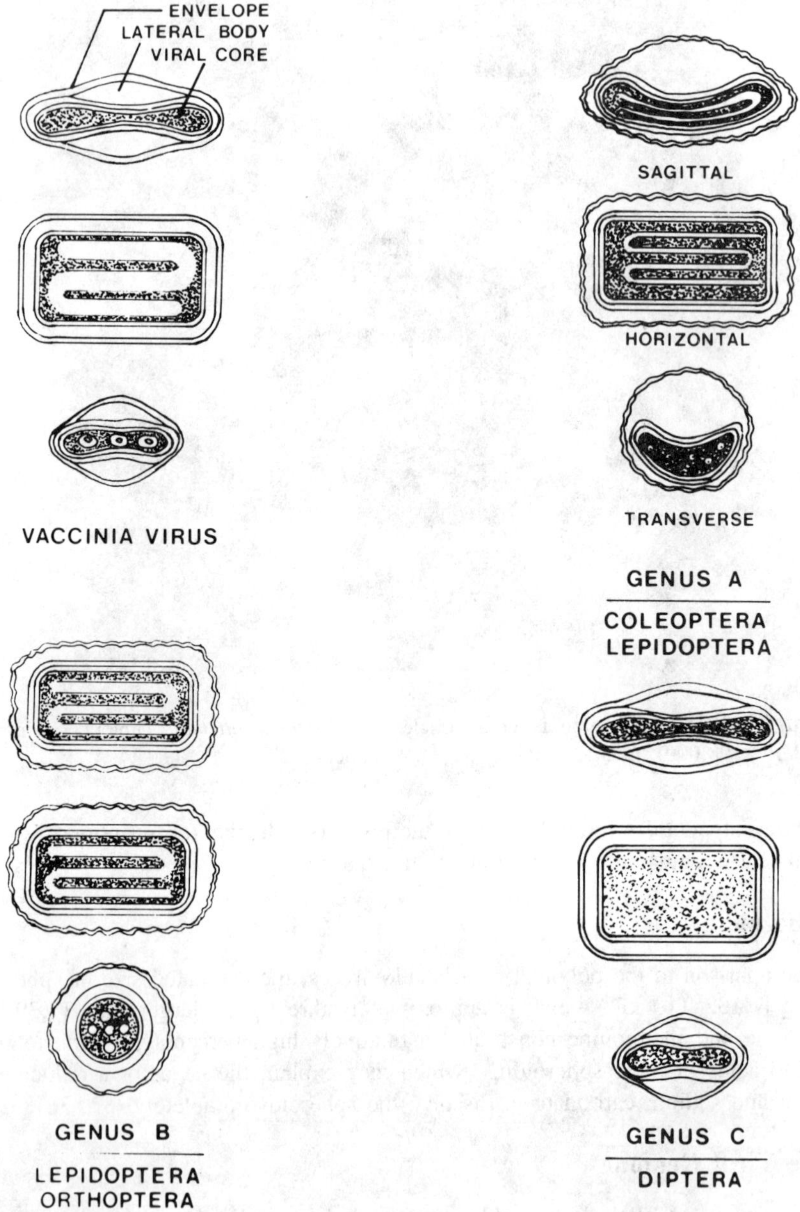


**Figure 4** Diagrammatic representation of EPVs from three genera and of vaccinia virus. (Reproduced from Ref. 2).

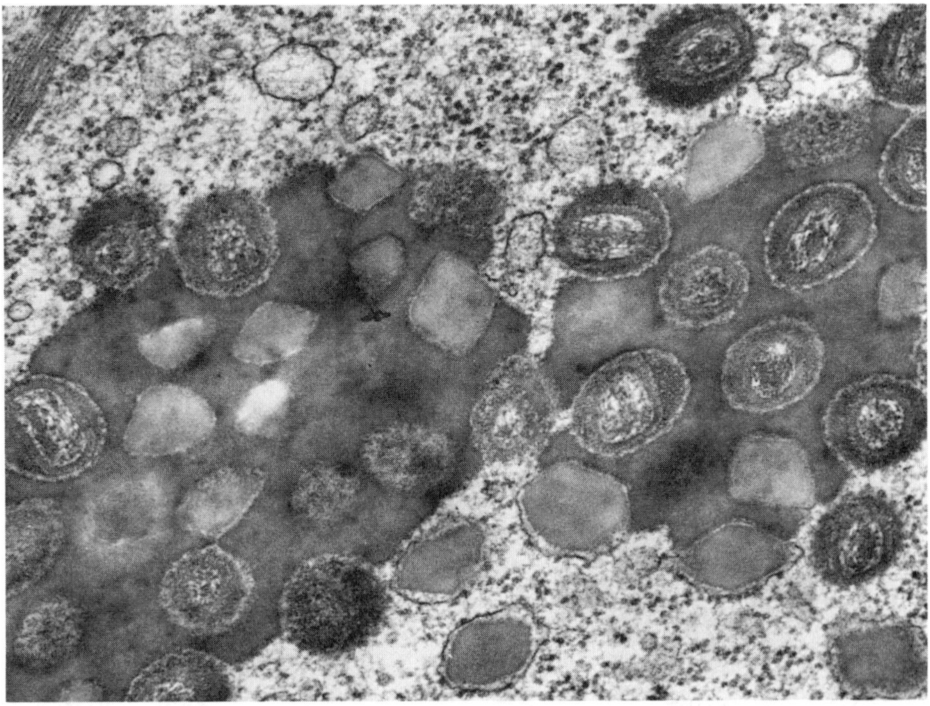

**Figure 5** Stages of morphogenesis of *Choristoneura fumiferana* Lepidoptera poxvirus (EPV). Random distribution of virions and spindles inside the spheroid (× 47,000).

Lepidoptera, is smaller than those reported for subgenera A and C, Coleoptera and Diptera, respectively.

A characteristic property of the EPV genome appears to be the low content of (G + C) residues, on the order of 17–27% (34, 41). The (G + C) content reported for poxviruses of vertebrates is on the order of 32.5–39% (42, 43). The significance of the high (A + T) residue content in terms of DNA transcription is yet to be elucidated.

In DNA–DNA hybridization experiments to determine sequence homologies among EPVs from Lepidoptera, Orthoptera, and vaccinia virus, extensive homology was observed between the lepidopteran viruses AmEPV and EaEPV, whereas another lepidopteran virus from *Oncopera albogutatta* (OaEPV) showed no hybridization with the first two (44). It should be mentioned that the lack of detectable hybridization does not mean the absence of DNA sequence homology. Hybridization under conditions of reduced stringency may show a different result. Also, sequence homologies measured by hybridization have been found to

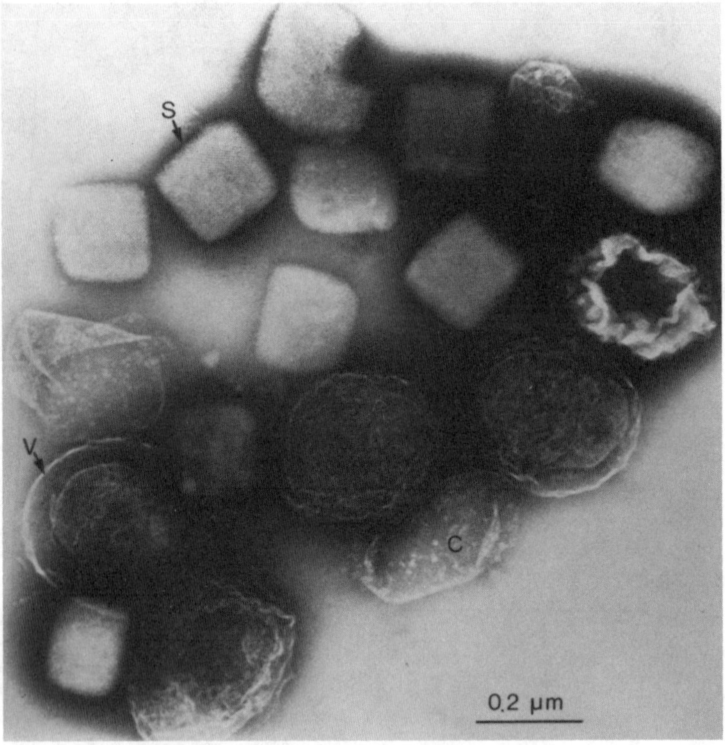


**Figure 6** Microspindles of *Choristoneura biennis* entomopoxvirus. Note the crystalline structure of the spindles (S), viral core (C), and the damaged virions (V).

underestimate actual homologies detected by DNA sequence analysis (45). If one draws an analogy with other occluded viruses, the baculoviruses, where the gene coding for polyhedrin was found to be highly conserved (46), it is reasonable to assume that the spheroidin gene was also conserved among different EPVs. This hypothesis remains to be proven. No hybridization was observed between vaccinia virus DNA and DNAs from EPVs (44).

## REPLICATION

Electron micrograph studies have shown that occlusion bodies ingested by larvae were dissolved in the alkaline environment of the gut and the released virions attached to midgut epithelial cells to initiate infection. The viral envelope fused with membranes of microvilli to facilitate entry of the viral core and lateral bodies to the cell cytoplasm (Figure 7). Further uncoating is probably an enzymatic process following expression of early viral genes. An early sign of viral replication was the formation of a virogenic stroma consisting of an electron-

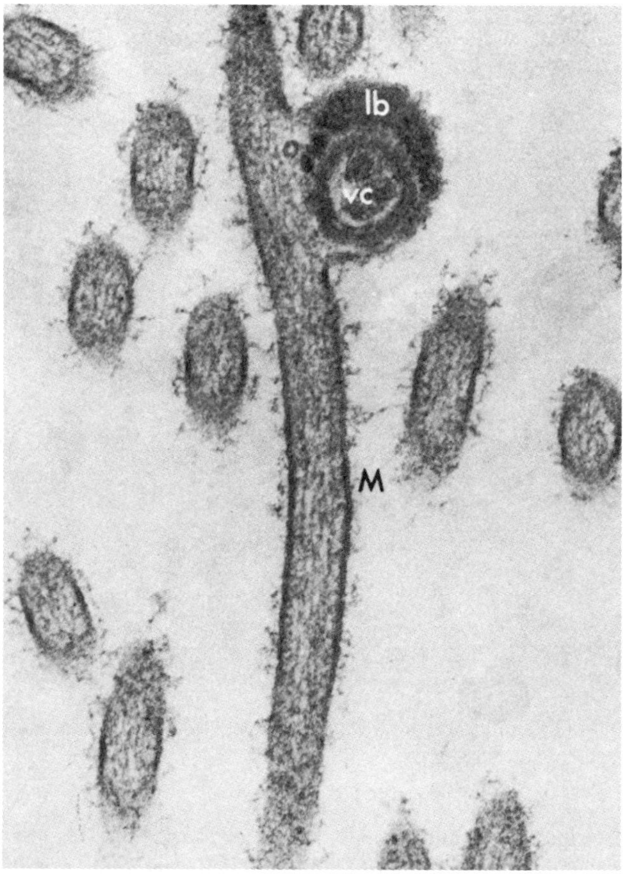


**Figure 7** Poxvirus (EPV) of *Amsacta moorei* Lepidoptera. Entry of the viral core (VC) and the lateral body (lb) into the cytoplasm by fusion of the virus envelope and a microvillus (M) membrane of the midgut epithelium (× 81,000). Courtesy of Dr. A. Granados.

dense and amorphous material surrounded by semicircular structures composed of two unit membranes (viroplasm type I, Figure 8). Eventually, spherical, immature, and electron-dense virus particles were formed within the viroplasm (1, 4, 7, 26). Immature virus particles were also formed in viroplasm type II, which consisted of aggregated granular material interspersed with spherical vesicles (1, 26). MmEPV appeared to replicate mainly in viroplasm type II (47). Within the membrane of the immature particle, the virus underwent differentiation and compression, which led to the formation of the viral core. The

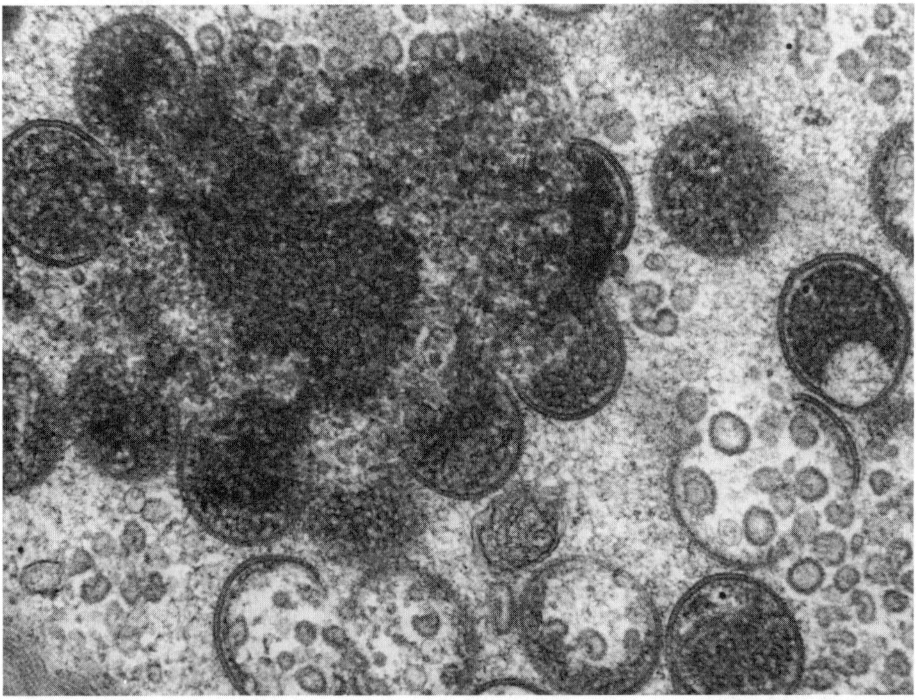

**Figure 8** Viroplasm of *Choristoneura fumiferana* EPV consisting of electron dense amorphous material and immature viral particles (× 58,000).

core developed its familiar rectangular shape surrounded by a three-layered envelope, concomitant with the formation of dense lateral bodies. Only mature virus particles or microspindles were occluded within the spheroid. The process appeared to be specific. Virions that were not destined to be occluded budded through the cytoplasm into the hemocoel. During the process they acquired a second membrane (26, 48, 49). Presumably, nonoccluded EPVs disseminate infection in susceptible larval tissues. In the hemocoel, viropexis was shown to be the mode of viral entry into the cells (1, 49). Occlusion bodies and spindles were rare in hemocytes infected with MmEPV but were found to be plentiful in fat tissue (48). Presumably, hemocytes preferentially allow the replication of the nonoccluded phenotype.

In contrast to other EPVs, where the cytoplasm is the site of virus replication, MsEPV was present in both the cytoplasm and the nucleus (29). This phenomenon was also observed in *L. dispar* cell culture infected with AmEPV, where structures similar to immature virions appeared in the nuclei of infected cells (50).

The replication of EPVs in tissue cultures has not been studied extensively in spite of the availability of several lepidopteran cell lines that are permissive to AmEPV (2). A cell line derived from hemocytes of *E. acrea* designated BTI-EAA was the most commonly used one because of the susceptibility of almost all the cells in the culture to the virus and the relatively fast rate of viral development. The virus replication cycle in tissue culture mimics that observed in larvae. The virions enter the cells by viropexis followed by uncoating and the formation of a virogenic stroma. Extracellular virus reached relatively high titers, on the order of $10^8$–$10^9$ $TCID_{50}$ units per milliliter by 7 days postinfection (51). By 10 days postinfection almost 100% of the cells had produced occlusion bodies in the cytoplasm. The cells were susceptible to infection by nonoccluded virus as well as by alkali-liberated virus. The molecular biology of the replication cycle of EPVs in tissue culture has not been studied. It would be interesting to find out if their replication at the molecular level is similar to that of vertebrate poxviruses or if it has similarity to other occluded DNA viruses.

## Virion-Associated Enzymes

As in the case of poxviruses of vertebrates, several enzymatic activities have been found in the entomopox virion:

1. Nucleotide phosphohydrolase (NPH). It was first found in the virions of EaEPVs and its activity was comparable to that observed with vaccinia virus (52).
2. Neutral and acidic deoxyribonucleases. The optimum pH for the neutral DNase is 7.5 and that for the acidic is 5.0. The activity of the latter was lower than that of vaccinia virus, probably due to partial inactivation when the virus was released from the spheroids at high pH (52).
3. DNA-dependent RNA polymerase. The optimum temperature of this enzyme is 28°C instead of the 37°C required for the activity of the vaccinia enzyme (40). Apparently the enzyme is inactivated if the alkali-liberated virions remain at a high pH environment for prolonged periods. McCarthy and co-workers recommend lowering the pH immediately after dissolution of the spheroids.
4. Polyadenylic acid polymerase. When ATP was supplied as the sole substrate in the RNA polymerase reaction, the RNA product was resistant to RNase and the reaction was not inhibited by actinomycin D (53). These results are consistent with the properties of polyadenylic acid polymerase of vaccinia virus (54, 55).
5. Alkaline protease. The presence of an alkaline protease has been shown to be associated with the matrix protein of baculoviruses (56, 57). A

similar enzyme was found in the spheroidin of EPVs (37, 39). The enzyme quickly degraded spheroidin (120 kD) to a 52 kD protein. Its optimum activity was at pH 10–11, and it was inhibited by $Hg^{2+}$ ions, SDS, soybean trypsin inhibitor, and by heating at 60°C for 10 min. As in the case of baculoviruses, tissue culture–derived spheroids were devoid of the enzyme (36, 58).

## ACKNOWLEDGMENT

We would like to thank Miss Sandra Mantulak for excellent reproductions of the diagrammatic presentations. We are grateful to Dr. R. R. Granados for permission to reproduce illustrations used in this chapter.

## REFERENCES

1. Kurstak, E. and Garzon, S. Entomopoxviruses, in *Altas of Insect and Plant Viruses*, Maramorosch, K., Ed., Academic Press, New York, 1977, p. 29.
2. Granados, R. R. The entomopoxviruses, in *Pathogenesis of Invertebrate Microbial Disease*, Davidson, E. W., Ed., Allanheld Osmun, Montclair, NJ, 1981, p. 101.
3. Arif, B. M. The entomopoxviruses. *Adv. Virus Res.* 29, 195, 1984.
4. Goodwin, R. H. and Filshie, B. K. Morphology and development of an occluded virus from the blacksoil scarab, *Othnonius batesi*. *J. Invertebr. Pathol.* 13, 317, 1969.
5. Matthews, F. E. F. Classification and nomenclature of viruses. *Intervirology* 12, 132, 1979.
6. Roberts, D. W. and Granados, R. R. A poxlike virus from *Amsacta moorei* (Lepidoptera: Arctiidae). *J. Invertebr. Pathol.* 12, 141, 1968.
7. Granados, R. R. and Roberts, W. D. Electron microscopy of a poxlike virus infecting an invertebrate host. *Virology* 40, 230, 1970.
8. Meynadier, G., Fosset, J., Vago, C., Duthoit, J. L., and Bres, N. Une virose à inclusions ovides chez un lepidoptere. *Ann. Epiphyt.* 19, 703, 1968.
9. Weiser, J. A pox-like virus in the midge *Camptochironomus tentans*. *Acta Virol.* 13, 549, 1969.
10. Weiser, J., Tchubianishvili, C., and Zizka, Z. Ultrastucture of the spindle-shaped virus, *Vagoiavirus operophterae*, in the winter moth, *Operophtera brumata* L. *Acta Virol.* 14, 314, 1970.
11. Sutter, G. R. A pox virus of the army cutworm. *J. Invertebr. Pathol.* 19, 375, 1972.
12. Bird, F. T., Sanders, C. J., and Burke, J. M. A newly discovered virus disease of the spruce budworm, *Choristoneura biennis* (Lepidoptera: Tortricidae). *J. Invertebr. Pathol.* 18, 159, 1971.
13. Cunningham, J. C., Burke, J. M., and Arif, B. M. An entomopoxvirus found in populations of the large aspen tortrix, *Choristoneura conflictana* (Lepidoptera: Tortricidae) in Ontario. *Can. Entomol.* 105, 767, 1973.
14. Katagiri, K. A newly discovered entomopoxvirus of *Choristoneura diversana* (Lepidoptera: Tortricidae). *J. Invertebr. Pathol.* 22, 300, 1973.

15. Hurpin, B. and Vago, C. Une maladie à inclusions cytoplasmiques fusiformes chez le coleoptere *Melolontha melolontha* (L.). *Rev. Pathol. Veg. Entomol. Agric. Fr.* 42, 115, 1963.
16. Bergoin, M., Davauchelle, G., Duthoit, J. L., and Vago, C. Étude au microscope électronique des inclusions de la virose à fuseaux des coleopteres. *C. R. Acad. Sci., Ser. D*, 266, 2126, 1968.
17. Hurpin, B. and Robert, P. Sur la spécificité de Vagoiavirus melolonthae. *Entomophaga* 12, 175, 1967.
18. Goodwin, R. H. and Roberts, R. J. Diagnosis and infectivity of entomopoxviruses from three Australian scarab beetle larvae (Coleoptera: Scarabeaeidae). *J. Invertebr. Pathol.* 25, 47, 1975.
19. Vago, C., Monsarrat, P., Duthoit, J. L., Amargier, A., Meynadier, G., and Van Waerebeke, D. A new spindle virosis observed in the stag beetle Coleoptera of Madagascar Lepidoptera *Scarabeidea figulus*. *C. R. Hebd. Séances Acad. Sci.* 266, 1621, 1968.
20. Lipa J. J. and Bartkowski, J. A newly discovered poxlike virus disease of dung beetle, *Geotrupes silvaticus*. *J. Invertebr. Pathol.* 20, 218, 1972.
21. Goodwin, R. H. and Filshie, B. K. Morphology and development of entomopoxviruses from two Australian scarab beetle larvae (Coleoptera: Scarabaeidae). *J. Invertebr. Pathol.* 25, 35, 1975.
22. Katagiri, K., Kushida, T., Kasuga, S., and Ohaba, M. An entomopoxvirus disease of the cupreous chafer, *Anomala cuprea* Hope. *Jpn. J. Appl. Zool.* 19, 243, 1975.
23. Vago, C., Robert, P., Armagier, A., and Duthoit, J. L. Nouvelle virose a spheroides et à fuseaux observée chez le coleoptere *Phyllopertha horticola* L. *Mikroskopie* 25, 378, 1969.
24. Gotz, P., Huger, A. M., and Krieg, A. Über ein insektenpathogens Virus aus der Gruppe der Pickenviren. *Naturwissenschaften* 56, 145, 1969.
25. Huger, A. M., Krieg, A., Einschermann, P., and Gotz, P. Further studies on *Polypoxvirus Chironomi*, an insect virus of the pox group isolated from the midge *Chironomus luridus*. *J. Invertebr. Pathol.* 15, 253, 1970.
26. Stoltz, D. B. and Summers, M. D. Observations on the morphogenesis and structure of a hemocytic poxvirus in the midge *Chironomus attenuatus*. *J. Ultrastruct. Res.* 40, 581, 1972.
27. Federici, B. A., Granados, R. R., Anthony, D. W., and Hazard, E. I. An entomopoxvirus and nonoccluded virus-like particles in larvae of the chironomid *Goeldichironomus holoprasinum*. *J. Invertebr. Pathol.* 23, 117, 1974.
28. Buchatskyi, L. P. Experimental study of a virus disease of densonucleosis type in *Aedes aegypti*. *Mikrobiol. Zh. (Kiev)* 36, 797, 1974.
29. Henry, J. E., Nelson, B. P., and Jutila, J. W. Pathology and development of the grasshopper inclusion body virus in *Melanoplus sanguinipes*. *J. Virol.* 3, 605, 1969.
30. Bergoin, M., Devauchelle, F., and Vago, C. Electron microscopy study of *Melolontha* poxvirus: The fine structure of occluded virions. *Virology* 43, 453, 1971.
31. Bird, F. T. The development of spindle inclusions of *Choristoneura fumiferana* (Lepidoptera: Tortricidae) infected with entomopox virus. *J. Invertebr. Pathol.* 23, 325, 1974.

32. Bergoin, M., Veyrunes, J. C., and Scalla, R. Isolation and amino acid composition of the inclusions of *Melolontha melolontha* poxvirus. *Virology* 40, 760, 1970.
33. Croizier, G. and Veyrunes, J. C. Analyse immunochimique des inclusions de la "Maladie à Fuseaux" du coleoptere *Melolontha melolontha* L. *Ann. Inst. Pasteur* 120, 709, 1971.
34. Arif, B. M. Isolation of an entomopoxvirus and characterization of its DNA. *Virology* 69, 626, 1976.
35. Bilimoria, S. L. and Arif, B. M. Structural polypeptides of *Choristoneura biennis* entomopoxvirus. *Virology* 104, 253, 1980.
36. Langridge, W. H. R. and Roberts, D. W. Structural proteins of *Amsacta moorei*, *Euxoa auxiliaris*, and *Melanoplus sanguinipes* entomopoxviruses. *J. Invertebr. Pathol.* 39, 346, 1982.
37. Mitchell, F. L., Smith, G. E., and Smith, J. W. Characterization of an entomopoxvirus of the lesser cornstalk borer *(Elasmopalpus lignosellus). J. Invertebr. Pathol.* 42, 299, 1983.
38. Jaeger, B. and Langridge, W. H. R. Infection of *Locusta migratoria* with entomopoxvirus from *Arphia conspera* and *Melanoplus sanguinipes* grasshoppers. *J. Invertebr. Pathol.* 43, 374, 1984.
39. Bilimoria, S. L. and Arif, B. M. Subunit protein and alkaline protease of entomopoxvirus spheroids. *Virology* 96, 596, 1979.
40. McCarthy, W. H., Granados, R. R., and Roberts, D. W. Isolation and characterization of entomopox virions from virus-containing inclusions of *Amsacta moorei* (Lepidoptera: Arctiidae). *Virology* 59, 59, 1974.
41. Langridge, W. H. R. and Roberts, D. W. Molecular weight of DNA from four entomopoxviruses determined by electron microscopy. *J. Virol.* 21, 301, 1977.
42. Pfau, C. J. and McCrea, J. F. Release of deoxyribonucleic acid from vaccinia virus by 2-mercaptoethanol and pronase. *Nature* 194, 894, 1962.
43. Yau, T. and Rouhandeh, H. Characterization of monkey pox virus DNA genome. *Biochem. Biophys. Acta* 299, 210, 1973.
44. Langridge, W. H. R. Detection of DNA base sequence homology between entomopoxviruses isolated from *Lepidoptera* and *Orthoptera*. *J. Invetebr. Pathol.* 43, 41, 1984.
45. Yang, R. C. and Wu, R. BK virus DNA sequence: Extent of homology with simian virus 40 DNA. *Proc. Natl. Acad. Sci. U.S.A.* 76, 1179, 1979.
46. Vlak, J. M. and Rhormann, G. F. The nature of polyhedrin, in *Viral Insecticides for Biological Control*, Maramorosch, K. and Sherman, K. E., Eds., Academic Press, New York, 1985, p. 489.
47. Bergoin, M., Devauchelle, G., and Vago, C. Electron microscopy study of the pox-like virus of *Melolontha melolontha* L. (Coleoptera, Scarabaeidae). Virus morphogenesis. *Arch. Gesamte Virusforsch.* 28, 285, 1969.
48. Devauchelle, G., Bergoin, M., and Vago, C. Étude ultrastructurale du cycle de réplication d'un Entomopoxvirus dans les hemocyes de son hôte. *J. Ultrastruct. Res.* 37, 301, 1971.
49. Granados, R. R. Insect poxviruses: Pathology, morphology, and development. *Misc. Publ. Entomol. Soc. Am.* 9, 73, 1973.

50. Quiot, J. M., Bergoin, M., and Vago, C. Développement et pathogenese d'un entomopoxvirus en culture cellulaire de lepidoptere. *C. R. Acad. Sci. Ser. D* 280, 2273, 1975.
51. Granados, R. R. and Naughton, M. Replication of *Amsacta moorei* entomopoxvirus in a continuous cell culture from *Estigmeme acrea*, *Proceedings First International Colloquium on Invertebrate Pathology, Kingston, Ontario*, 1976, p. 113.
52. Pogo, B. G. T., Dales, S., Bergoin, M., and Roberts, D. W. Enzymes associated with an insect poxvirus. *Virology* 43, 306, 1971.
53. McCarthy, W. J., Nesser, C. F., and Roberts, D. W. RNA polymerase activity of *Amsacta moorei* entomopox virions. *Intervirology* 5, 69, 1975.
54. Moss, B., Rosenblum, E. N., and Paoletti, E. Polyadenylate polymerase from vaccinia virions. *Nature, New Biol.* 245, 59, 1973.
55. Sheldon, R. and Kates, J. Mechanism of poly(A) synthesis by vaccinia virus. *J. Virol.* 14, 214, 1974.
56. Eppstein, D. A. and Thoma, J. A. Characterization and serology of the matrix protein from a nuclear polyhedrosis of *Trichoplusia ni* before and after degradation by an endogenous proteinase. *Biochem. J.* 167, 321, 1977.
57. Tweeten, K. A., Bulla, L. A., Jr., and Consigli, R. A. Characterization of an alkaline protease associated with a granulosis virus of *Plodia interpunctella J. Virol.* 26, 702, 1978.
58. Zummer, M. and Faulkner, P. Absence of protease in baculovirus polyhedral bodies propagated *in vitro*. *J. Invertebr. Pathol.* 33, 383, 1979.
59. Weiser, J. and Vago, C. A newly described virus of the winter moth, *Operophtera brumata* Hubner (Lepidoptera: Geometridae). *J. Invertebr. Pathol.* 8, 314, 1966.

# 7

# Invertebrate Iridoviridae

V. K. WARD AND JAMES KALMAKOFF
*University of Otago, Dunedin, New Zealand*

## INTRODUCTION

### General Features

The Iridoviridae are large, icosahedral, nonoccluded dsDNA viruses (1) that infect both vertebrates and invertebrates and are the largest group of nonoccluded insect viruses (2). The family Iridoviridae comprises four genera as determined by the International Committee on Taxonomy of Viruses (3). Two genera infect invertebrates and two infect vertebrates. The two vertebrate genera are *Ranavirus*, for which FV3 is the type species (3), and *Lymphocystivirus*, for which Fish Lymphocystis Disease Virus is the type species. FV3 is the type species for the entire Iridoviridae family and is the model for iridescent virus (IV) replication (2). This virus is thought to be only distantly related to the insect iridescent viruses (4).

The two invertebrate IV genera are the *Iridovirus* genus (small insect iridescent viruses) and the *Chloriridovirus* genus (large insect iridescent viruses). The type species for the *Iridovirus* genus is *Chilo* iridescent virus (CIV) which infects the rice stem borer, *Chilo suppressalis*. This virus was first discovered in Japan in 1964 by Takeishi and subsequently reported and described by Fukaya and Nasu (5), Bellet and Inman (6), and Mitsuhashi and Koyama (7). Virus particles belonging to the *Iridovirus* genus have a typical face-to-face dimension of 130–140 nm compared with those of the *Chloriridovirus* genus, which have a face-to-face dimension of 180–200 nm (8). The type species of the *Chloriridovirus* genus is mosquito iridescent virus type 3 (MIV), which was discovered in the mid-1960s by Clark et al. (9) in *Aedes taeniorhynchus* larvae.

One of the problems with IV taxonomy is the typing scheme by which insect

**Table 1** Recognized Insect Iridescent Viruses

| Iridescent virus type | Host of origin | Iridescent virus type | Host of origin |
|---|---|---|---|
| 1 | *Tipula paludosa* | 17 | *Pterostrictus madidus* |
| 2 | *Sericesthis paludos* | 18 | *Opogonia* sp. |
| 3 | *Aedes taeniorhynchus* | 19 | *Odontria striata* |
| 4 | *Aedes cantans* | 20 | *Simocephalus expinosus* |
| 5 | *Aedes annulipes* | 21 | *Heliothis armigera* |
| 6 | *Chilo suppressalis* | 22 | *Simulium* sp. |
| 7 | *Simulium ornatum* | 23 | *Heteronychus arator* |
| 8 | *Culicoides* sp. | 24 | *Apis cerana* |
| 9 | *Wiseana* spp. | 25 | *Tipula* sp. |
| 10 | *Witlesia sabulosella* | 26 | *"Mayfly"* |
| 11 | *Aedes stimulans* | 27 | *Nereis diversicolor* |
| 12 | *Aedes cantans* | 28 | *Lethocerus columbiae* |
| 13 | *Corethralla brakeleyi* | 29 | *Tenebrio molitor* |
| 14 | *Aedes detritus* | 30 | *Heliothis zea* |
| 15 | *Aedes detritus* | 31 | *Armadillidium vulgare* |
| 16 | *Costelytra zealandica* | 32 | *Porcellio dilatatus* |

*Source:* Reference 2.

IV isolates are recognized. The current scheme is an interim measure, proposed in 1970 by Tinsley and Kelly (10), which is based on the host of origin and the date of isolation. *Tipula* iridescent virus was the first discovered and has been designated insect iridescent virus type 1, while *Chilo* iridescent virus, being the sixth IV to be reported, is insect IV type 6. This scheme provides no real criterion for IV classification and has been overtaken by the number of IV isolates for which no real relationships are known. To date, 32 insect iridescent viruses have been recognized and typed (2) (Table 1). IVs have also been reported in invertebrates other than insects, including terrestrial isopods, nematodes (11), and brachiopods (12). This chapter is limited mainly to recent work on the small insect iridescent virus group.

## Incidence and Distribution

Invertebrate IVs have been found worldwide in a number of different environments. For example, a small iridescent virus has been isolated from *Simulium* spp. in five widespread locations, Czechoslovakia (13), Wales (14), Guatemala (15), North America (16), and New Caledonia (17). The relationships between these viruses has not been studied.

The major common feature of all iridescent viruses is the iridescent, or opalescent, blue-green coloration of heavily infected larvae or purified pellets of the virus. This coloration is caused by closely packed paracrystalline arrays of virus particles within heavily infected tissues causing Bragg reflections of light (18) with the color of the iridescence being determined by the interparticle spacing (19). The production of iridescence in infected larvae does not always occur (20,21), probably because paracrystalline virus arrays do not always form. This has been shown both in vivo (22) and in vitro (23) and may be due to the IV budding continuously from the infected cell (23). Kelly (2) and Stoltz (24) have suggested that iridescence may be a trivial characteristic of these viruses as it has no relation to any functional feature of the virus and is due only to physical structure. A consequence of the use of iridescence for taxonomy has been that the *Chloriridovirus* and *Iridovirus* genera, which are probably genetically unrelated, have both been called insect iridescent viruses (2), whereas other viruses have not been included in the Iridoviridae because they do not show iridescence (25,26).

It is not known how IVs infect their hosts. An oral route of entry is the most obvious, but it is generally agreed that the oral route is not infectious in most cases (2). Recent studies have indicated that per os infection may be more common than first thought (27,28), with infection rates in excess of 40% being obtained. Stoltz and Summers (29) showed that for a mosquito IV (MIV), most of the MIV particles are degraded shortly after entry into the midgut and that the virus did not seem capable of penetrating the peritrophic membrane. Carter (30) found that treatment of viral particles with gut fluids had no effect on the infectivity of IV injected into the hemocoel. These results indicate that the peritrophic membrane may play an important role in the resistance of insects to IV infection.

Other possible routes of infection include spiracles, transovarial infection, and contamination of insect eggs. These have all been considered unlikely routes of infection for the small insect iridescent viruses (8,31). Unlike the small insect IVs, large insect IVs appear to be capable of transovarial transmission. Studies by Fukuda and Clark (32) and Linley and Nielsen (33,34) indicated that the IV of the mosquito *Aedes taeniorhyncus* (RMIV) was transmitted transovarially, and Woodard and Chapman (35) found that larvae infected late in their life cycle often passed the virus to their offspring, which became infected and later died.

Infected cadavers have also been proposed as a source of IV infection, especially among cannibalistic insects, since cannibalism would cause a massive uptake of concentrated virus (36). Wounding of insects through cannibalistic behavior may also facilitate entry for IV present in the environment and would be analogous to the intrahemocoelic infection route often used in the laboratory. Parasites may also play a role in the infection of susceptible insects with an iridescent virus.

## Composition and Structure

Most of the research on the insect iridescent viruses has been with the iridovirus type species CIV or with TIV, which was the first insect iridescent virus to be discovered. The general features of insect IVs are 11–18% DNA (1), the presence of an internal lipid membrane (37,38), an electron-dense central core, and an icosahedral outer capsid. In the case of CIV the DNA has been shown to possess 15% repeat sequences (39) and is cyclically permuted with a terminal redundancy of 12% (40). No sequence homology has been found between CIV and other Iridoviridae (39).

The central core of insect IVs consists of DNA packed in a nucleosomallike structure with six DNA binding proteins in the case of CIV (41). The core is highly hydrated (42) and in the vertebrate iridescent virus FV3 the DNA–protein complex appears to be a long, convoluted filament (43). A lipid membrane surrounds the inner core (2,37) and makes up approximately 9% of the virion (44). Protein complexes extend through the single unit membrane, attaching the core polypeptides to the capsid polypeptides; this may be an explanation for the high stability of these viruses (41). This finding is further supported by the studies of Klump and colleagues (45), who observed that core DNA appeared to be attached to capsid proteins. The lipid membrane has been shown to be approximately 4 nm thick in IV type 29 (2) and has a different composition to host cell lipid (38,44).

The lipid membrane is closely associated with the outer shell (37). This shell is icosahedral in shape (46) and in the case of TIV has been shown to be a lattice of hexagonally packed units (47). The lattice of TIV is composed of 1472 triangular, and in some cases pentagonal, subunit aggregates, which are cylindrical in shape with a central hole (48,49). In 1972 Willison and Cocking (50) found surface fibers at the periphery of virus particles, and Stoltz (37) noted that the fibrillar fringe was composed of one fibril per subunit but may not be present on all IVs. Studies by Kelly (2) using x-ray scattering and hydrodynamic studies indicated the presence of a second outer protein shell in IV type 29. Little is known about this second outer shell.

Small insect IVs have been shown to produce at least 20 polypeptides (51–53). The number of proteins found in IV particles varies with the conditions under which they are analyzed. For CIV, Kelly and Tinsley (54) identified 19 polypeptides in a range of 10–213 kD with the major polypeptide being 65 kD, while Barray and Devauchelle (53) obtained 16 polypeptides by SDS/$\beta$-mercaptoethanol treatment, 26 by SDS/urea, and more than 50 by 2D gel electrophoresis. These results indicate that some proteins in CIV are oligomeric and possess disulfide linkages between the peptide chains. The predominant IV polypeptide accounts for approximately 40% of the total mass of the virus and surface labeling indicates that this protein is located on the surface of the viral particle (55). Tissue culture studies combined with pulse-labeling experiments have indicated that CIV infection causes the production of 28 virus-specific polypeptides, of which 19 are structural.

Numerous enzymatic activities have been associated with insect IVs. These include an RNA polymerase (51), a nucleotide phosphohydrolase, a protein kinase (56), and an alkaline protease (57,58). The function of the protein kinase is unknown as no particle proteins are phosphorylated despite this enzyme being capable of phosphorylating some low molecular weight polypeptides in vitro (56). The protein kinase is found primarily in the nuclei of infected cells associated with newly synthesized viral proteins during CIV infection. This indicates a nuclear involvement for insect IV replication. It is possible that the enzyme may have some regulatory role in the host cell. Very little is known about the nucleotide phosphohydrolase other than that it has a requirement for $Mg^{2+}$ ions (56) and the function of the alkaline protease is not understood. The virion-associated protease is thought to be a larval contaminant rather than of viral origin as it is not found in tissue culture–grown virus (57,58). The alkaline protease may have some role is viral uncoating.

## Replication

Little is known about the cellular aspects of insect IV infection. The primary site of viral replication in the insect is not known (2), although all tissues become infected eventually; thus it may be that no particular tissue is required for primary replication. Viral uptake into cells is by viropexis (59) and incorporation into lysosomes (23,60). Inside the lysosome the virus particles disintegrate and are presumably uncoated (61).

Many of the cellular events of IV replication have been determined by CIV infection of tissue culture cell lines including *Choristoneura fumiferana* (62) and *Trichoplusia ni* cells (63). The range of cell lines susceptible to IV infection appears to be related to the in vivo host range of the virus, with cell type, temperature, multiplicity of infection, and growth conditions all affecting virus multiplication (62–64).

The gross pathological effects of IV infection include syncytia formation. This is caused by some virus component and not viral multiplication as UV-treated virus still causes syncytia formation, but heat treatment or antibodies prevent this from occurring (65). Host cell RNA, DNA, and protein sythesis are inhibited by viral components (66,67) and there is an increase in cellular DNA polymerase, thymidine kinase, and RNA polymerase levels. CIV has been found to increase phosphorylation of proteins and to cause the production of new phosphoproteins of viral origin early in infection (68), which may involve the protein kinase. CIV infection also causes the disaggregation of polyribosomes in both permissive and nonpermissive cell lines even when UV-irradiated (69).

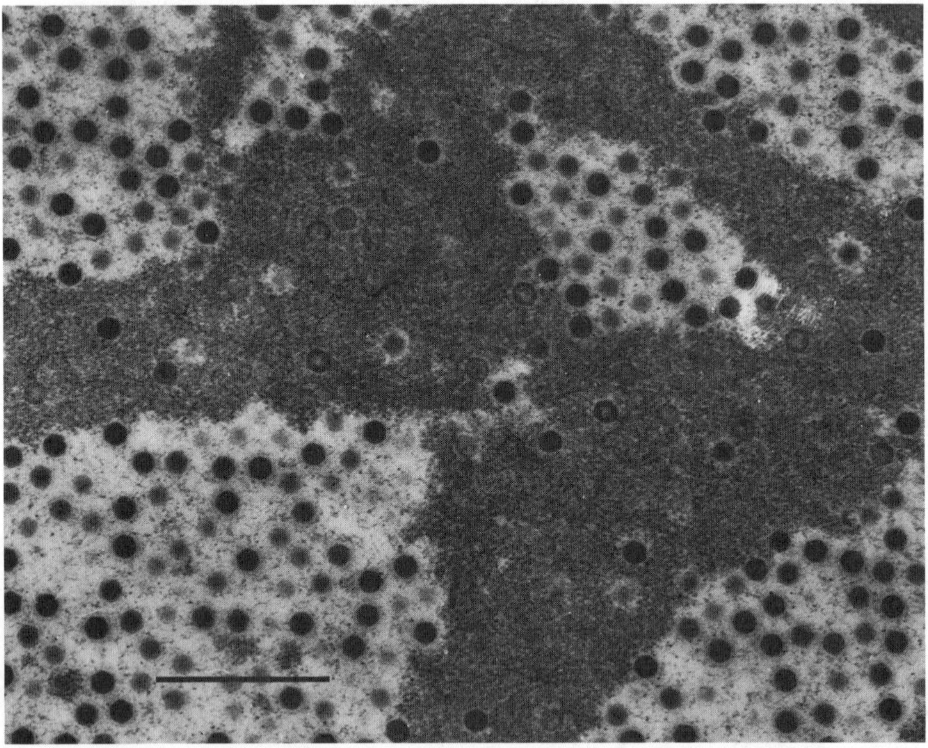

**Figure 1** Cellular virogenic stroma showing viral particles at various stages of assembly. Bar - 1μm. (Courtesy Dr. B. S. Batson, Microbiology Department, University of Otago, Dunedin, New Zealand.)

Studies of TIV in hemocytes of the waxmoth *Galleria mellonella* indicate that an electron-dense virogenic stroma is formed in response to IV infection (70) (Figure 1). In FV3 the viroplasm has been reported to involve cytoskeletal elements (71). This involvement of cytoskeletal elements does not appear to occur in TIV as no detectable changes in microfilaments, microtubules, or intermediate filaments were noted during infection with TIV (72). Virogenic stroma occur in the cytoplasm and are composed of fine fibrils that contain viral DNA (73) and proteins (74).

The molecular aspects of insect iridescent virus replication have been little studied. Therefore, the vertebrate iridescent virus FV3 is used as the model for IV replication (2). How relevant the replication strategy of this virus may be for insect iridescent viruses is not known. FV3 replication has been reviewed by

Murti et al. (75). The genome is cyclically permuted and terminally redundant, which are common features of insect IVs that have been investigated for these properties (40,76). Iridescent viruses are the only eukaryotic viruses that are cyclically permuted and terminally redundant.

For replication a functional host cell nucleus is required (77) with initial viral nucleic acid synthesis being in the nucleus. After synthesis in the nucleus the DNA migrates to the cytoplasm, where DNA synthesis resumes. This switch to the cytoplasm is associated with a viral protein (73), and it is not known if this event occurs with insect iridescent virus replication as no definite nuclear involvement has been found for insect IV replication. In the cytoplasm the viral DNA recombines to form long concatemers, which are required for the regeneration of DNA ends (79). Early FV3 mRNA synthesis also occurs in the nucleus and employs a viral modified RNA polymerase. Bladon et al. (80) noted that IV viroplasms have proteins related to nuclear matrix proteins and that as viroplasms increase, the detectable amount of nuclear protein decreases. This would indicate there is at least some nuclear requirement for insect IV replication.

Protein production occurs in two phases. The first set of proteins is produced within five hours of infection of tissue culture cells and the second phase proteins are produced at eight hours postinfection. The late proteins appear to be associated with DNA replication and require the presence of the early proteins for production (81). IV structural proteins do not appear to be glycosylated or phosphorylated.

There are different views regarding the assembly of virions and the encapsidation of the viral genome. Yule and Lee (70) proposed that the capsid was formed face by face then filled with the DNA-containing fibrils of the virogenic stroma through a hole left in the capsid. Devaucheile et al. (23) propose that as the shell enlarges, it sequesters a portion of the viroplasm and forms immature particles (82). In accordance with the observations of Devauchelle, Stoltz and Summers (83) suggested that the inner membrane segregates viral genomes from the viroplasm during assembly in a mechanism similar to poxvirus assembly. Once assembled the IV particle may leave the cell by budding (70,73), total cell lysis (30), or extrusion in vacuoles (63,84).

## DETECTION AND IDENTIFICATION OF IRIDESCENT VIRUSES

In the past iridescent viruses have been detected by the observation of iridescence in infected larvae (17,85,86). As previously noted, the usefulness of iridescence is limited since it does not always occur (20,21). For this reason it is necessary to develop a method for the detection of IVs which is sensitive and accurate and does not depend on the iridescence of infected larvae.

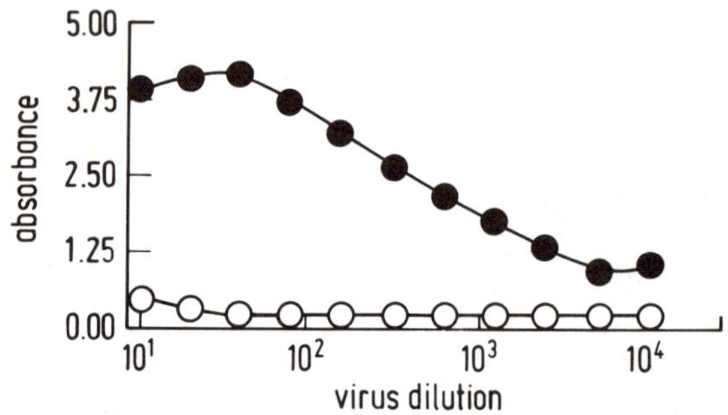


**Figure 2** ELISA assay of mosquito iridescent virus (untyped)–infected larval material. ●, test sample (antibodies to MIV); ○, control (normal rabbit serum).

Serology is the most commonly used alternative procedure to detect the presence of IVs. Serological methods include serum neutralization, complement fixation, indirect ELISA, sandwich ELISA, latex agglutination, immunoperoxidase staining, and radioimmunoassay (20,31,87–90). The various merits of some of these techniques have been summarized by Kelly et al. (20), who noted that serum neutralization was impractical for screening because of bacterial contamination of tritrated larvae and complement fixation, which is relatively insensitive when larval extracts are used, is not quantitative. Latex agglutination was found to be 50 times less sensitive than ELISA, while radioimmunoassay (RIA) is even less sensitive and is technically demanding. The ELISA assay, apparently the best serological technique available, has the disadvantage of being semiquantitative.

Serological methods are technically very simple and require little specialized equipment. An example of using an ELISA technique to detect a mosquito iridescent virus is presented in Figure 2. Purified mosquito iridescent virus (12) was used to prepare polyvalent antiserum by injection into rabbits. Samples of field-collected mosquito larvae were triturated in a small volume of water then fixed into microtiter wells by drying. These wells were reacted with rabbit antiserum followed by HRPO-labeled sheep antirabbit antibody and read in an ELISA reader. An estimate of the amount of virus present could be made with reference to a standard curve.

Antigenic comparisons between iridescent viruses can be made using RIA. Figure 3 shows the antigenic comparisons of viruses isolated from *Wiseana* spp. (WIV) and *Costelytra zealandica* (CzIV), two pasture pests that share the same soil environment. Using CzIV as the antigen, a comparison of the homologous and WIV antisera was made. There is at least a 100-fold difference in the dilution end points of the respective antisera. Further analysis can be made using their

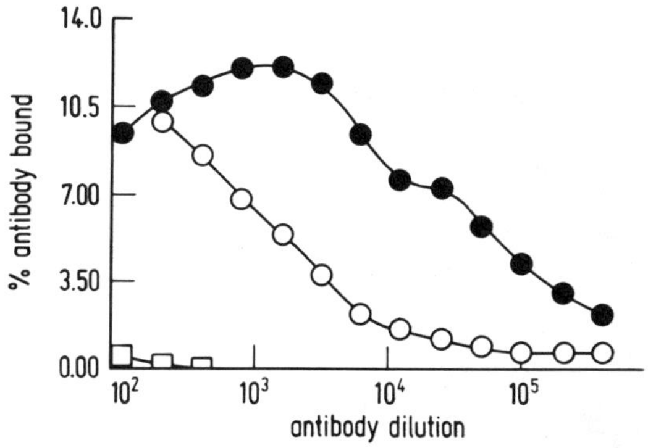


**Figure 3** Radioimmunoassay for the detection and comparison of iridescent viruses. CzIV fixed to microtiter wells was used as the antigen. ○, WIV (type 9) antiserum; ●, CzIV (type 16) antiserum; □, MIV (untyped) antiserum.

relative antibody binding affinities to whole virus or to isolated coat proteins. By using precipitin end points and cross-absorption tests (91), it has been shown that CIV is antigenically unrelated to other iridescent viruses tested. There are distinct viral antigens among the iridescent viruses which are not related to the host used for propagation. However, many iridescent viruses are propagated in experimental laboratory hosts such as *Galleria mellonella*, and it is possible that there has been cross-contamination between iridescent viruses. For example, there are antigenic differences in TIV propagated in *Lymantria dispar* or *Tipula paludosa* when compared to TIV propagated in *Galleria mellonella* (91).

All of these techniques, when used to detect virus, involve the partial purification of samples prior to detection (20). Sera of known specificity are required and antiserum production may be affected by differences in inoculation procedure, animals used, and the timing of serum collection. Problems may also arise from using impure virus preparations during antiserum production and, in addition, the sensitivity of the assay is affected by the presence of insect material. Interpretation, reproducibility, and standardization are problems and in many cases IVs are serologically cross-reactive (88,91,92). Despite these problems, serology does play an important role in the detection of insect viruses and will probably continue to do so, especially with the development of monoclonal antibody techniques.

The advent of molecular biology techniques, in particular DNA–DNA hybridization, has led to the development of new methods for the detection of viruses. Viruses that have been detected by nucleic acid probes include rotaviruses (93), sugarcane mosaic virus (94), polyoma virus (95), avocado sunblotch virus (96), tobacco mosaic virus (97), and a nuclear polyhedrosis virus (98). The

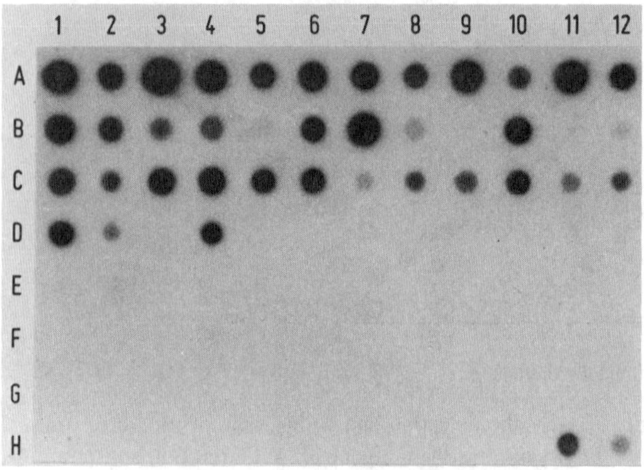


**Figure 4** Dot-blot assay for the detection of WIV (type 9) in a number of insect hosts (see Table 2). A1–B8 = WIV-infected *Heliothis armigera larvae*; B9–D4 = WIV-infected *Laspeyresia pomonella* larvae; D5–D9 = uninfected *L. pomonella* larvae; D10–E7 = uninfected *Mithimna separata* larvae; E8–G3 = WIV-infected *M. separata* larvae; G4–G8 = uninfected *Agrotis ipsilon* larvae; G9–H10 = WIV-infected *A. ipsilon* larvae; H11–H12 = WIV positive controls.

use of DNA–DNA dot-blot hybridization has been shown to be more sensitive than ELISA (97), virus isolation, electron microscopy (95), light microscopy, and radioimmunoassay (98).

In our laboratory we have developed a dot-blot assay for the detection of insect IV type 9 from *Wiseana* spp. larvae. This assay is simple to use and has proven to be very sensitive. The following is a brief description of the method. Individual insect larvae were homogenized in phosphate buffer and a 5 $\mu$l sample was applied to GeneScreen hybridization membrane (New England Nuclear Boston, MA.) using a Bio-Rad dot-blot apparatus. The membrane-bound samples were placed onto a pad of filter paper soaked with 0.5 *M* NaOH, 1.5 *M* NaCl to lyse any virus particles present in the sample and to denature the DNA. After 30 min at room temperature the membrane was removed from the pad and allowed to dry. Released DNA was bound to the membrane by exposure to germicidal UV light for 5 min (99) and the membrane was washed to remove protein material capable of interfering with the assay. The presence of viral DNA was determined by hybridization using nick-translated WIV DNA followed by autoradiography.

Figure 4 shows a dot-blot assay of some different insect species that were infected with WIV in a host range study. These results and the use of virus isolation to detect IV infection are summarized in Table 2. None of the larvae displayed any indication of iridescence at any stage of these infection studies.

**Table 2** Host Range Analysis of *Wiseana* spp. Iridescent Virus

| Host tested | WIV |
|---|---|
| Galleria mellonella (waxmoth) | + |
| Heliothis armigera (tomato fruitworm) | + |
| Planotortrix excessana (green-headed leafroller) | + |
| Laspeyresia pomonella (codling moth) | + |
| Epiphyas postvittana (light brown apple moth) | + |
| Bombyx mori (silkworm) | + |
| Odontria spp. (coleopteran beetle) | + |
| Wiseana spp. (New Zealand grass caterpillar) | + |
| Ctenopseustis obliquana (brown-headed leafroller) | − |
| Mithimna separata (cosmopolitan armyworm) | − |
| Agrotis ipsilon (greasy cutworm) | − |
| Costelytra zealandica (New Zealand grass grub) | − |

*Key*: +, virus growth detected by dot-blot hybridization (example seen in Figure 4); −, no virus growth detected.

Dot-blot analysis indicated that 7 of the 11 species of insect inoculated by intrahemocoelic injection were infected, while the use of virus isolation to detect IV infection indicated that only 5 of the 7 dot-blot positive species were susceptible. The presence of virus particles in infected insects was confirmed by electron microscopy in those cases in which virus was isolated.

The purification of virus or of viral DNA was found to be unnecessary since good results were obtained directly from the insect homogenates. Advantages of the assay are that no discernible cross-reaction to other *Wiseana* spp. viruses was found and removal of the probe and rehybridization with a new probe allows for the detection of multiple types of virus in the same sample. The major disadvantages with this system lie in the requirement for expensive and short-lived radiochemicals; however, the development of biotinylated probes should extend the shelf life and lower the hazards involved with probe handling (100).

An important criterion for the study of any virus is the accurate determination and identification of that virus. The identification of IVs by serological methods is useful but may not be completely accurate and is unlikely to have the sensitivity to separate closely related strains of the same species of IV. Most insect IVs show some serological cross-reactivity, with only IV type 6 from *Chilo suppressalis*, IV type 24 from *Apis cerana*, and MIV being completely unrelated to other insect IVs (2). In addition, if a virus is to be used as a biocontrol agent it is essential to be able to determine any changes in the virus during its production or use. For precise identification some constant and distinguishing feature is required.

Protein profiles produced by polyacrylamide gel electrophoresis (PAGE) have been used to separate viruses and show potential for the separation of IVs. The problems with PAGE lie in the variability between laboratories and a wide choice of conditions for the running of polyacrylamide gels. This places limits on the usefulness of protein profiles for the precise identification of IVs. Despite these problems PAGE has been used to good effect. Analysis of the polypeptides from different insect IVs indicates a degree of similarity between IV isolates, although distinct patterns do exist (101).

The use of restriction endonucleases to separate closely related strains of insect baculoviruses is well documented (102–104), although there are few reports of the use of restriction enzymes for the identification of insect IVs (105). There has been no indication of whether small differences between IV strains are detectable or if variant strains exist. Restriction endonucleases have the potential for universal use as a standard procedure for the identification and separation of IVs.

We have investigated the use of restriction enzymes for the study of WIV and CzIV which were isolated from the lepidopteran host *Wiseana* spp. (type 9) and the coleopteran insect pest *Costelytra zealandica* (type 16). The patterns produced by electrophoresis of the DNA of these viruses after digestion with the restriction enzyme EcoRI are shown in Figure 5, as are the EcoRI restriction patterns for CIV, TIV, and IV type 22. It can be clearly seen that distinct patterns exist for these viruses.

Further investigation of the restriction patterns produced for WIV using the restriction enzymes BamHI, EcoRI, and PstI indicated that a number of submolar bands were present, indicated by an asterisk in Figure 6. These bands are thought to be due to a mixed population of virus. Evidence for this has come from passage of WIV through larvae of the waxmoth *Galleria mellonella* and subsequent analysis by restriction endonuclease digestion of passaged WIV DNA. It was found that the highest molecular weight EcoRI-* fragment was lost upon repeated passage of the virus, which indicates selection against this strain of virus. Other submolar fragments (*) appear to be homologous to molar fragments. For example, hybridization studies of the 5.2 kb BamHI-* fragment suggest it is a subfragment of BamHI-L.

These results show that at least three populations of WIV exist in this isolate as determined by the use of restriction enzymes. These three populations are (1) the major population present, (2) the population containing the EcoRI-* fragment lost on serial passage through *Galleria mellonella* larvae, and (3) the population(s) containing the submolar fragments found to be fractions of molar fragments. This demonstrates that restriction analysis can detect specific strains of virus within a species and thus should be useful for both species and strain identification.

Tajbakhsh et al. (106) have reported a heterogeneous composition of bulk-

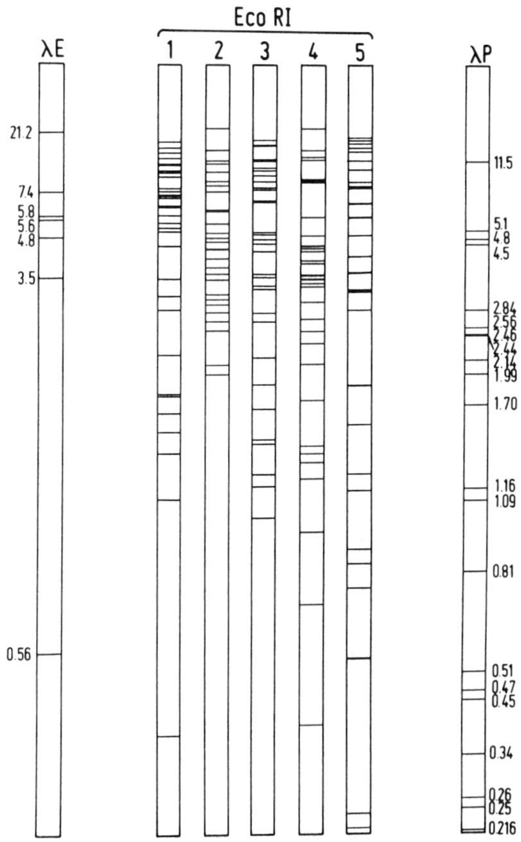


**Figure 5** Comparison of EcoRI restriction patterns for five different insect IVs. (Drawn from data obtained in Refs. 76,105–107,111.) λE = EcoRI lambda DNA standard; λP = PstI lambda DNA standard; 1 = CIV (type 6); 2 = CzIV (type 16); 3 = SIV (type 22); 4 = TIV (type 1); 5 = WIV (type 9).

purified TIV DNA with submolar restriction fragments being detected. This group found that TIV particles contained four species of DNA when virus was isolated from larval homogenates. In addition to the major genome molecule designated L, there were three different small molecules of DNA found in the DNA preparation, which were designated S1, S2, and S3. All of these fragments were related to L and possessed a common sequence. These fragments have been shown to be of viral origin, not larval. Tajbakhsh et al. also postulate that these small molecules may be derived from the partially filled capsids commonly isolated from larval homogenates. This may provide another explanation for the

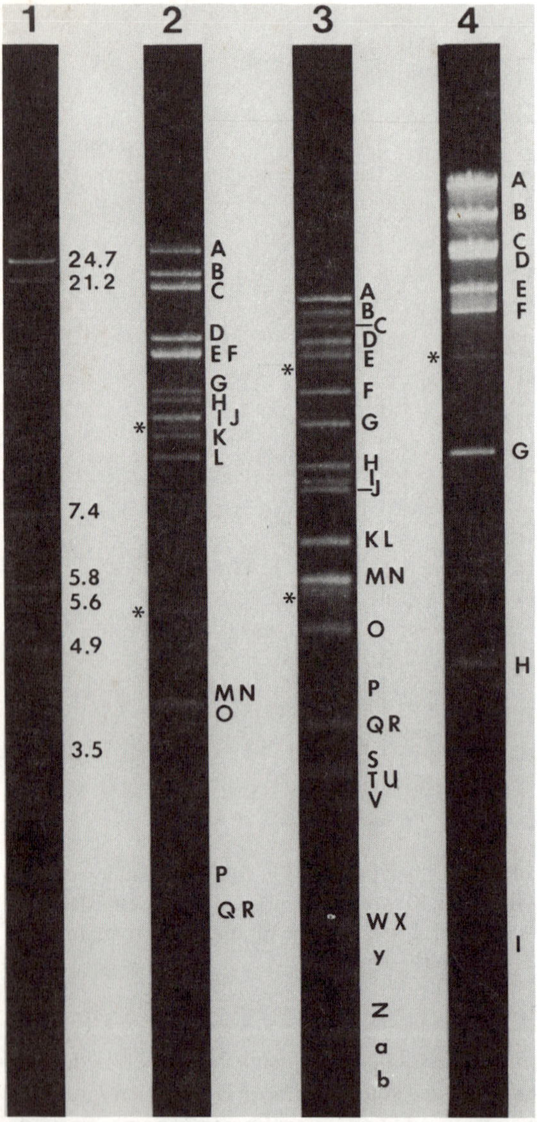


**Figure 6** Restriction endonuclease digestion of WIV DNA and subsequent separation by electrophoresis on 0.6% agarose (76). 1 = EcoRI-digested lambda DNA standard; 2 = BamHI-digested WIV DNA; 3 = EcoRI-digested WIV DNA; 4 = PstI-digest WIV DNA; * = submolar fragments.

presence of the submolar fragments found in WIV. It was also noted that methylation of the genome was not responsible for the presence of submolar fragments.

Due to the current problems associated with the identification of IVs the taxonomy of insect IVs is very confusing with no definite criteria having been established for the separation of IVs. Restriction endonuclease analysis has the advantage of being a defined procedure that uses a standardized set of commercially available reagents. The use of restriction analysis in conjunction with hybridization studies should provide a useful basis for the study of IV taxonomy, a study which is long overdue.

## IRIDESCENT VIRUS MOLECULAR BIOLOGY

The genome of insect iridescent viruses consists of single-copy, double-stranded linear DNA of approximately 110–155 MD for small iridescent viruses (6,107) and 160–185 MD for large IVs (8). The genomes of the vertebrate iridescent viruses FV3 and FLDV have been well characterized, as has the genome of the small insect iridescent virus CIV. The terminal redundancies of these viruses vary from 4–6% for FV3 (75), to 12% for CIV (40), to 50% for FLDV (108).

Studies by Murti et al. (109) indicated that the terminally redundant sequences present in FV3 are limited to a portion of the genome in a manner similar to that of bacteriophage P22 (110). The termini of these linear genomes are variable; however, the variable sequences at the termini come from only 25% of the genome. Approximately 75% of the sequences in the genome may never be at the termini of the genome. Studies of the genome of CIV indicate that this limiting of terminally redundant sequences may not occur. End-labeling of CIV DNA followed by digestion with restriction enzymes showed that all restriction endonuclease–generating DNA fragments were labeled (40), hence all sequences must have been at the termini of different linear genome molecules. End digestion with exonucleases and subsequent digestion with restriction enzymes also indicated that all parts of the genome were capable of functioning as genome termini (40). It is not known if this difference between FV3 and CIV is due to inherent differences between vertebrate and invertebrate IVs or if vertebrate and invertebrate IVs may show either type of terminal redundancy.

To further understand the molecular biology of insect iridescent viruses we have undertaken the physical mapping of the genome of insect IV type 9 (WIV) (76). The reasons and possible uses of physical mapping have been summarized for CIV by Soltau et al. (111):

> The construction of the physical map of the CIV genome is the basis for the analysis of further structural and functional properties of this interesting genome, e.g. the determination of DNA origin(s) of replication, packaging

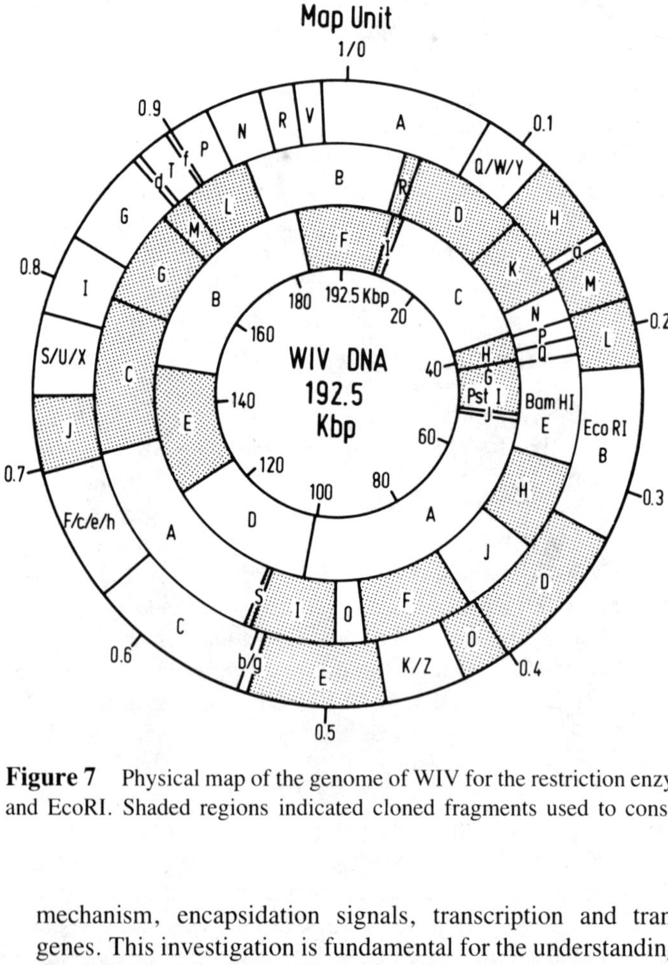


**Figure 7** Physical map of the genome of WIV for the restriction enzymes PstI, BamHI, and EcoRI. Shaded regions indicated cloned fragments used to construct the map.

mechanism, encapsidation signals, transcription and translation of viral genes. This investigation is fundamental for the understanding of the virus life cycle, host range, virus host interaction and pathogenicity of CIV.

We have mapped the genome of WIV using the restriction enzymes BamHI, EcoRI, and PstI. Restriction fragments were cloned in the plasmid vectors pUC8 and pBR328 using *Escherichia coli* strains JM83 and HB101, respectively. Map construction was by multiple enzyme digest and hybridization of the cloned restriction fragments to Southern blots of digested WIV DNA. The physical map was analyzed by linear modeling using an SAS (Statistical Analysis System, NC) computer program (112). The results of the study are presented in Figure 7. The construction of this physical map indicated that the genome of WIV, like other IVs, is cyclically permuted and hence probably terminally redundant. Because of

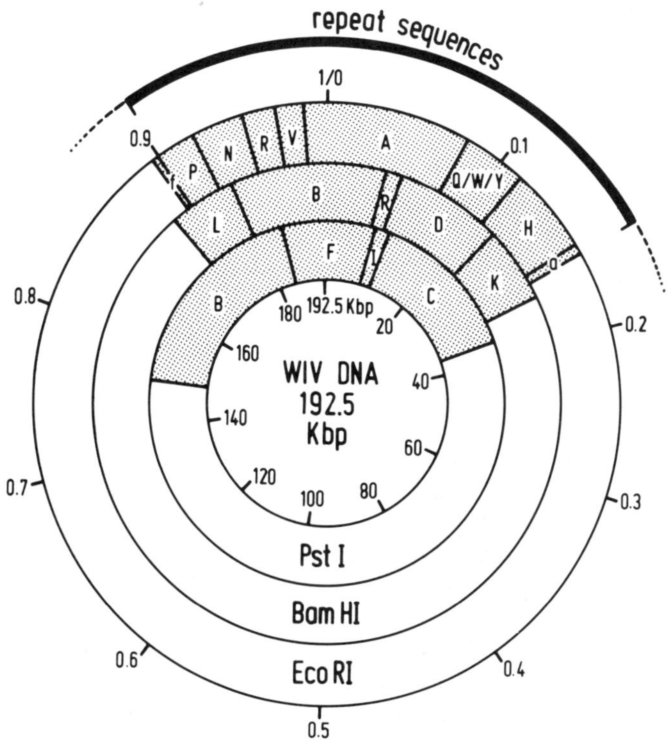


**Figure 8** Physical map for the region of the WIV genome found to contain repeat sequences.

the cyclic permutation of these genomes, the map is presented in a circular form. The restriction digest patterns from which this map was derived are presented in Figure 6.

Hybridization studies also indicated that the genome of WIV contains repetitive DNA sequences as some restriction fragments were found to hybridize to many more fragments than was expected. EcoRI-N that is cut once by BamHI and not at all by PstI was found to hybridize to BamHI B, D, L, and R, PstI B, F, I, and C, and EcoRI, A, H, M/N, Q/R, V, W/X, Y, a, and e/f. Hybridization of BamHI-D also implicated EcoRI-P as containing a repeat sequence(s). A summary of these hybridization results is presented in Figure 8, where all fragments thought to possess repeat sequences are shown. This indicates that the portion of the genome which appears to possess repeat sequences is an approximately 48 kb (25%) region of the genome from EcoRI-f to somewhere between the BamHI-K/N cut site and the EcoRI-a/H cut site.

Repeat sequences have been identified in a number of viruses and are implicated in many viral functions. Repeats at either end of the dsDNA genome of adenoviruses have been identified as the origin of replication with either repeat sequence being capable of acting as the origin (113). Vaccinia has a 30-fold repetition of a 70 bp sequence within a 10 kb region at each end of the genome (114), which is associated with early mRNA production (115). Vaccinia has also been found to contain a mixture of different repeats (116). Tandem repeats in polyoma virus have been associated with early gene expression and control of host genes, while the complexity of repeats in various herpesviruses have been associated with host range in different cell types; that is, the more complex a repeat region is, the wider the host range (117). The origin of replication of HSV1 is also within a repeat region (118). It is possible that any or some of the repeats in WIV may function as the origin of replication. Hybridization results indicated that it is likely that more than one type of repeat sequence may be present in this region of the genome since not every fragment detected was capable of hybridizing to all other fragments found to possess repeat sequences. It is possible that the mixture of repeat sequences found in WIV is the mechanism for the apparently wide host range of IVs. The presence of repeat sequences in one region of the genome indicates that some important functions are likely to occur in this region.

It would be interesting to investigate other insect IVs to determine if the presence of repeat sequences is common and if a mixture of repeats appears to be present. Preliminary results suggest the genome of CzIV also contains a number of repeat sequences (119). Studies of TIV have indicated that several repeat sequences are present in the genome of this virus (106).

The only other insect IV reported to have been physically mapped is CIV (111). As for WIV, FLDV, and FV3, the physical map of this genome has been presented as a circle. Figure 9 is the physical map of the CIV genome for the restriction enzymes EcoRI, NcoI, PvuII, SphI, Asp718, BamHI, SalI, SmaI, and ApaI. The restriction patterns for BamHI, EcoRI, ApaI, NcoI, and SalI are seen in Figure 10. The presence of repeat sequences was also noted in CIV. These repeats were not found to be clustered in one region of the genome but were located in EcoRI fragments C and H, which are well separated on the physical map (Figure 9). This may be another feature in which CIV appears to be different from other insect IVs. Comparison of the restriction patterns and physical maps of CIV and WIV indicate little similarity between these viruses.

McMillan (107) compared the genomes of CzIV and WIV by Southern hybridization using different stringency conditions. Hybridization of $^{32}$P[dCTP]-labeled CzIV DNA to Southern blots of restriction endonuclease–digested WIV and CzIV DNA under conditions of 68% homology required for hybridization (20% formamide, 42°C) indicated that a large proportion of these genomes is highly homologous. Six WIV DNA fragments produced by the enzymes BamHI,

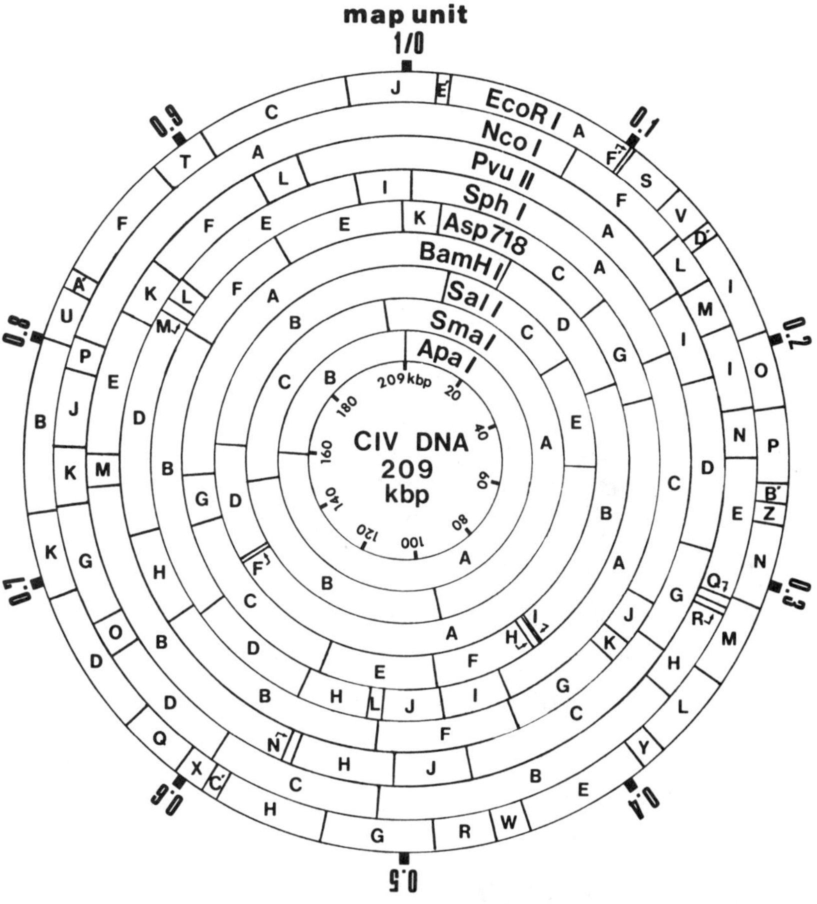


**Figure 9** Physical map of the CIV genome for the restriction endonucleases EcoRI, NcoI, PvuII, SphI, Asp718, BamHI, SalI, SmaI, and ApaI. (Courtesy Professor G. Darai, Institut für Medizinische Virologie, Universität Heidelberg.)

EcoRI, and PstI were found not to show any homology at this stringency. The results were identical when the degree of homology required for hybridization was increased to 84% (50% formamide, 42°C). The six WIV fragments found not to hybridize to CzIV DNA were PstI-G and H, BamHI-M, N, and P, and EcoRI-L (EcoRI-L was determined nonhomologous by hybridization of cloned EcoRI-L to CzIV DNA). When we compared these bands to the physical map for WIV (Figure 7), it was found that five of the fragments came from the same region of the physical map.

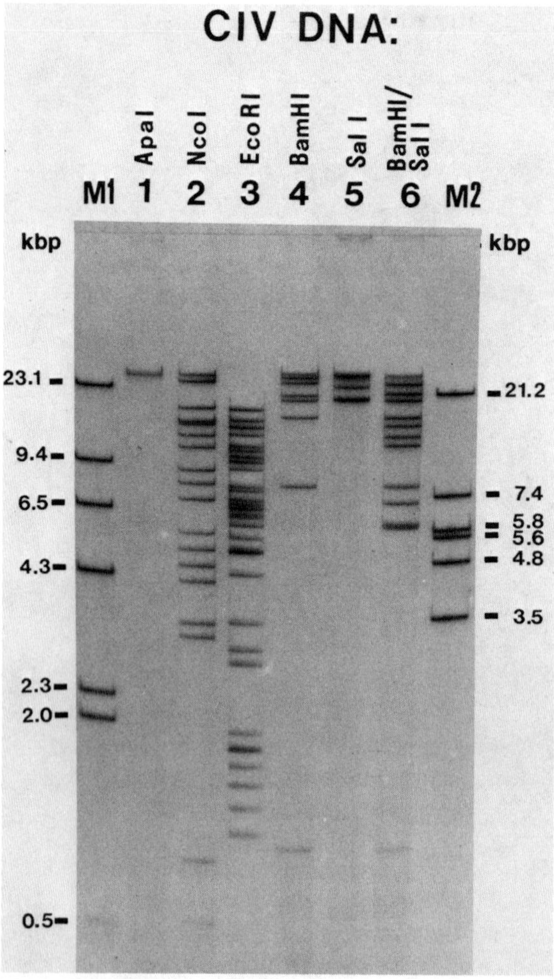


**Figure 10** Restriction endonuclease digestion of CIV DNA for the restriction enzymes ApaI, NcoI, EcoRI, BamHI, SalI, and BamHI/SalI. (Courtesy Professor G. Darai.)

Comparison of the size of the CzIV and WIV genomes shows a difference of 22.8 kb (107). The major area of low or nonhomology is at least 15.38 kb and the BamHI fragment M is approximately 4 kb, thus indicating a low of nonhomologous area of at least 19.4 kb. The 3.4 kb difference unaccounted for (between 22.8 and 19.4 kb) is probably in the regions flanking the fragments which show little or no homology. It would be interesting to know if this difference in size between these two genomes is due to deletions or insertions at

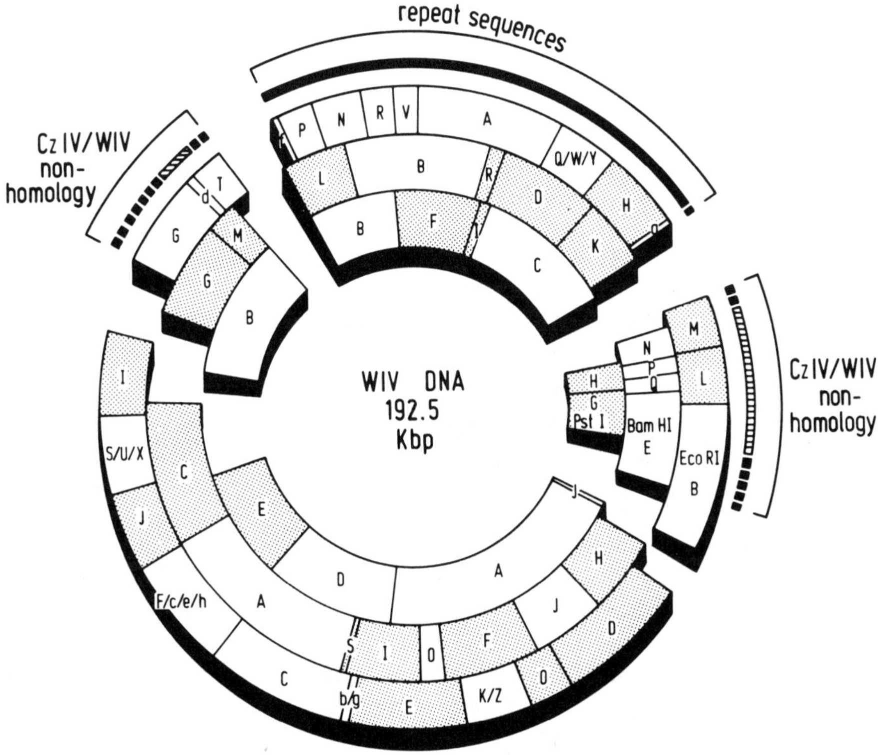


**Figure 11** Location of WIV repeat sequences and CzIV nonhomology regions on the physical map of WIV. Broken lines indicate possible range of regions of nonhomology and repeat sequences.

the two regions demonstrated by these results. One further observation is that these regions occur at either end of the section of the WIV genome which appears to contain repeat sequences. The results of these molecular biology studies are summarized in Figure 11.

## ECOLOGY OF IRIDESCENT VIRUSES

A fundamental question concerning IVs is how they infect their hosts. Early reports of some IVs indicated infection rates of 70–90% (120, 121), and it is hard to imagine wounding, cannibalism, or parasitic inoculation causing infection levels as high as this in a field population. Preliminary studies in this laboratory have indicated that per os infection does occur and may be at levels of up to

100% when a large amount of virus is used. The per os infection of *Galleria mellonella* larvae with WIV produced no iridescence in any larvae despite larvae becoming obviously sick and not developing. The use of a DNA–DNA dot-blot hybridization assay indicated that 100% of the larvae were infected with WIV; subsequent virus isolation confirmed this result. Sikorowski and Tyson (28) indicated infection rates in *Heliothis* of nearly 50% with iridescence as the criterion for infection and subsequent staining and light microscopy for confirmation of infection. The development of infected larvae was found to be much slower than uninfected larvae, being up to 89 days compared with the normal maximum development period of 21 days. Infected larvae were often deformed and all obviously infected larvae died in the larval (95%) or pupal (5) stages. Per os studies by Sieburth and Carner (27) with the IV of *Anticarsi gemmatalis* indicated that this IV is lethal to its host by per os infection, although large inocula were required. $LD_{50}$ values of up to 3 $\mu$g of virus per milligram of insect were obtained. Mortality was variable at all inoculum concentrations, especially at low levels, and it was found to be virtually impossible to raise the infection levels to 100% mortality. These authors also found that all six larval instars of *A. gemmatalis* could be infected per os and that the time to death was dependent on larval instar. Older larvae lived for a shorter period after infection than early instar larvae, which is unusual for viruses. Most larvae died prior to pupation, even when pupation was only a matter of days after inoculation and no extensive infection had occurred. This implies that massive viral proliferation is not necessary for IVs to cause insect death, which may explain the lack of iridescence in many infected larvae.

Iridescent viruses have not been considered as potential biocontrol agents because they were thought not to cause the death of their hosts. As indicated earlier, IVs are capable of killing their host in significant numbers and heavily infected larvae may pupate but do not progress to adults. These results demonstrate that IVs do have potential as biocontrol agents and are worthy of further investigation.

The natural ecology of IVs may be complex. IVs are found in multiple infections with a number of insect pathogens including bacteria (122), nematodes (121,123), an NPV, a fungus (27), and a picornavirus (124). In the case of IV infection of *A. gemmatalis* larvae it was found that there was activation of a latent NPV in 10% of the infected larvae (27). The virus isolated from terrestrial isopods, "wood-lice," may be transmitted by nematodes (11), since it has been observed that a higher percentage of IV-infected isopods are parasitized by nematodes. This suggests that nematodes may transmit virus passively by surface contamination or actively, since the virus replicates within the nematode. Furthermore, the virus from isopods can infect soil-dwelling coleopterans such as *Phyllophaga anxia*.

Another possible interaction is the virus from the *Chlorella*-like algae (125). This virus infects the protozoan *Paramecium bursaria* and the green hydra,

*Hydra viridis*. The *Chlorella* virus has similar properties to the *Chloriridovirus* genus (mosquito iridescent viruses). Both are icosahedral and approximately 180–200 nm, have dsDNA, 5–10% lipid, and assemble in the cytoplasm. Recently, a mosquito iridescent virus was isolated from *Aedes subalbirostris* and *A. antipodeus* larvae in freshwater pools in South Otago, New Zealand (12). In the same pools were found Daphnia *(Cerodaphnia dubia)* infected with a similar (the same?) virus. It is possible that the iridescent virus could have an algal, mosquito and/or Daphnia, and protozoan cycle.

## CONCLUSIONS

The iridescent viruses have many unique features with regard to their genomic structure, molecular biology, and ecological interactions. This makes this group of viruses worthy of further study. The major areas that require investigation are their taxonomy and their mode of infection under natural conditions. The typing of iridescent viruses based on iridescence, host of origin, and date of isolation is no longer suitable since they provide no information as to their taxonomic relationships. It is possible that within the Iridoviridae there are quite distinct groups of viruses, as their serology and DNA homology would suggest. The use of restriction enzymes, physical mapping of the genome, and DNA homology has the potential of resolving this problem.

Recent studies indicate that iridescent viruses are capable of per os infection and cause the death of their host. Possibly the major problem of many early studies was the reliance on iridescence as a means of detecting infection. The formation of iridescence is not a definite consequence of infection, and heavy infection may not be necessary for causing the death of the host. The dot-blot assay has indicated that iridescence is not a suitable criterion for virus detection. The dot-blot assay is suitable for determining routes of infection, secondary hosts, virus transmission, and identification.

Iridescent viruses have not been seriously considered as candidates for development as biological control agents. With recent information on their molecular biology and genetic mapping there is potential for strain selection and genetic manipulation. Along with their attractive iridescence they may become the Cinderella of the insect virus world.

## ACKNOWLEDGMENTS

The authors wish to acknowledge the assistance Dr. G. Darai, Heidelberg University, who supplied preprints and figures on the molecular biology of CIV. In addition, the CzIV work of Nigel McMillan is acknowledged. The WIV research was made possible through a grant from the Southland Frozen Meats Company Ltd., New Zealand.

## REFERENCES


1. Vaughn, J. L. Virus and Rickettsial diseases, in *Insect Diseases*, vol. 1, Cantwell, G. E., Ed., Marcel Dekker, New York, 1979, p. 49.
2. Kelly, D. C. Insect iridescent viruses. *Curr. Topics Microbiol. Immunol.* 116, 23, 1985.
3. Brown, F. The classification and nomenclature of viruses: Summary of results of meetings of the International Committee on Taxonomy of Viruses in Sendai, September 1984. *Intervirology* 25, 141, 1986.
4. Kelly, D. C. and Avery, R. J. Frog virus 3 deoxyribonucleic acid. *J. Gen. Virol.* 24, 339, 1974.
5. Fukaya, M. and Nasu, S. A *Chilo* iridescent virus (CIV) from the rice stem borer *Chilo suppressalis* Walker (Lepidoptera, Pyralidae). *Appl. Ent. Zool.* 1, 69, 1966.
6. Bellet, A. J. D. and Inman, R. B. Some properties of deoxyribonucleic acid preparations from *Chilo, Sericesthis,* and *Tipula* iridescent viruses. *J. Mol. Biol.* 25, 424, 1967.
7. Mitsuhashi, J. and Koyama, K. Stability of *Chilo* iridescent virus. *Jpn. J. Appl. Ent. Zool.* 11, 177, 1967.
8. Kelly, D. C. Non-occluded viruses, in *Pathogenesis of Invertebrate Microbial Diseases*, Davidson, E. W., Ed., Allanheld Osmun, Montclair, NJ, 1981, chap. 2.
9. Clark, T. B., Kellen, W. R., and Lum, P. T. M. A mosquito iridescent virus (MIV) from *Aedes taeniorhynchus* (Weidemann). *J. Invertebr. Pathol.* 7, 519, 1965.
10. Tinsley, T. W. and Kelly, D. C. An interim nomenclature system for the iridescent group of viruses. *J. Invertebr. Pathol.* 12, 66, 1970.
11. Hess, R. T. and Poinar, G. O., Jr. Iridoviruses infecting terrestrial isopods and nematodes. *Curr. Topics Microbiol. Immunol.* 116, 49, 1985.
12. Anderson, J. P. T. The Biology and Host Range of a Mosquito Iridescent Virus, Dip. Sci. thesis, University of Otago, Dunedin, New Zealand, 1983.
13. Weiser, J. Iridescent virus from the blackfly *Simulium ornatum* Meigen in Czechoslovakia. *J. Invertebr. Pathol.* 12, 36, 1968.
14. Batson, B. S., Johnston, M. R. L., Arnold, M. K., and Kelly, D. C. An iridescent virus from *Simulium* sp. (Diptera Simuliidae) in Wales. *J. Invertebr. Pathol.* 27, 133, 1976.
15. Takaoka, H. Pathogens of blackfly larvae in Guatemala and their influence on natural populations of 3 species of onchocerciasis vectors. *Am. J. Trop. Med. Hyg.* 29, 467, 1980.
16. Avery, S. W. and Bauer, L. Iridescent virus from *Prosimulium* collected in Maine. *J. Invertebr. Pathol.* 43, 430, 1984.
17. Batson, B. S. A small iridescent virus *(Iridovirus)* from *Simulium neornatipes* Dumbleton (Diptera: Simuliidae) in the South Pacific island of New Caledonia. *J. Invertebr. Pathol.* 48, 384, 1986.
18. Smith, K. M. *Insect Virology*, Academic Press, New York, 1967, p. 80.
19. Williams, R. C. and Smith, K. M. A crystallizable insect virus. *Nature* 179, 119, 1957.

20. Kelly, D. C., Edwards, M. L., and Robertson, J. S. The use of enzyme-linked immunosorbent assay to detect and discriminate between small insect iridescent viruses. *Ann. Appl. Biol.* 90, 369, 1978.
21. Ward, V. K. Characterisation of Two *Wiseana* spp. Viruses, Ph.D. diss., University of Otago, Dunedin, New Zealand, 1987.
22. Stoltz, D. B. A Virus Disease in *Chironomus plumosus*, Ph.D. diss., McMaster University, Hamilton, Ontario, Canada.
23. Devauchelle, G., Stoltz, D. B., and Darcy-Tripier, F. Comparative ultrastructure of Iridoviridae. *Curr. Topics Mircrobiol. Immunol.* 116, 1, 1985.
24. Stoltz, D. B. The structure of icosahedral cytoplasmic deoxyriboviruses. *J. Ultrastruct. Res.* 37, 219, 1971.
25. Schuster, A. M., Burbank, D. E., Meister, B., Skrdla, M. P., Meints, R. H., Hattman, S., Swinton, D., and van Etten, J. L. Characterization of viruses infecting a eukaryotic *Chlorella*-like green alga. *Virology* 150, 170, 1986.
26. Meints, R. H., Lee, K., and van Etten, J. L. Assembly site of the virus PBCV-1 in a *Chlorella*-like green alga: Ultrastructural studies. *Virology* 154, 240, 1986.
27. Sieburth, P. J. and Carner, G. R. Infectivity of an iridescent virus for larvae of *Anticarsia gemmatalis* (Lepidoptera: Noctuidae). *J. Invertebr. Pathol.* 49, 49, 1987.
28. Sikorowski, P. P. and Tyson, G. E. *Per os* transmission of iridescent virus of *Heliothis zea* (Lepidoptera: Noctuidae). *J. Invertebr. Pathol.* 44, 97, 1984.
29. Stoltz, D. B. and Summers, M. D. Pathway of infection of mosquito iridescent virus. *J. Virol.* 8, 900, 1971.
30. Carter, J. B. A Virus Disease of Leatherjackets, Ph.D. diss., University of Glasgow, Scotland, 1971.
31. Carter, J. B. The mode of transmission of *Tipula* iridescent virus. I. Source of infection. *J. Invertebr. Pathol.* 21, 123, 1973.
32. Fukuda, T. and Clark, T. B. Transmission of the mosquito iridescent virus (RMIV) by adult mosquitos of *Aedes taeniorhynchus* and their progeny. *J. Invertebr. Pathol.* 25, 275, 1975.
33. Linley, J. R. and Nielsen, H. T. Transmission of a mosquito iridescent virus in *Aedes taeniorhynchus*. I. Laboratory experiments. *J. Invertebr. Pathol.* 12, 7, 1968.
34. Linley, J. R. and Nielsen, H. T. Transmission of a mosquito iridescent virus in *Aedes taeniorhynchus*. II. Experiments related to transmission in nature. *J. Invertebr. Pathol.* 12, 17, 1968.
35. Woodard, D. B. and Chapman, H. C. Laboratory studies with the mosquito iridescent virus (MIV). *J. Invertebr. Pathol.* 11, 296, 1968.
36. Carter, J. B. The mode of transmission of *Tipula* iridescent virus. II. Route of infection. *J. Invertebr. Pathol.* 21, 136, 1973.
37. Stoltz, D. B. The structure of icosahedral cytoplasmic deoxyriboviruses. An alternative model. *J. Ultrastruct. Res.* 43, 58, 1973.
38. Balange-Orange, N. and Devauchelle, G. Lipid composition of an iridescent virus type 6 (CIV). *Arch. Virol.* 73, 363, 1982.
39. Kelly, D. C. and Avery, R. J. The DNA content of four small iridescent viruses:

Genome size, redundancy and homology determined by renaturation kinetics. *Virology* 57, 425, 1974.
40. Delius, H., Darai, G., and Flügel, R. M. DNA analysis of insect iridescent virus 6: Evidence for circular permutation and terminal redundance. *J. Virol.* 49, 609, 1984.
41. Cerutti, M. and Devauchelle, G. Characterisation and localisation of CIV polypeptides. *Virology* 145, 123, 1985.
42. Cuillel, M., Tripier, F., Braunwald, J., and Jacrot, B. A low resolution structure of frog virus 3. *Virology* 99, 277, 1979.
43. Tripier-Darcy, F. and Nermut, M. V. Cryodecapage d'un virus, le FV3 (frog virus 3): Structure fine de la capside et organisation interne. *Biol. Cell* 48, 17a, 1983.
44. Kelly, D. C. and Vance, D. E. The lipid content of two iridescent viruses. *J. Gen. Virol.* 21, 417, 1973.
45. Klump, H., Beaumais, J., and Devauchelle, G. Structural and thermodynamic investigation of the *Chilo* iridescent virus (iridovirus type 6). *Arch. Virol.* 75, 269, 1983.
46. Williams, R. C. and Smith, K. M. The polyhedral form of the *Tipula* iridescent virus. *Biochem. Biophys.* Acta 28, 464, 1958.
47. Wrigley, N. G. An electron microscope study of the structure of *Sericesthis* iridescent virus (SIV). *J. Gen. Virol.* 5, 123, 1969.
48. Wrigley, N. G. An electron microscope study of the structure of *Tipula* iridescent virus. *J. Gen. Virol.* 6, 169, 1970.
49. Manyakov, V. F. Fine structure of the iridescent virus type 1 capsid. *J. Gen. Virol.* 36, 73, 1977.
50. Willison, J. H. M. and Cocking, E. C. Frozen fractured viruses: A study of virus structure using freeze etching. *J. Microsc.* 95, 397, 1972.
51. Kelly, D. C. and Tinsley, T. W. Ribonucleic acid polymerase activity associated with particles of iridescent virus types 2 and 6. *J. Invert. Pathol.* 22, 199, 1973.
52. Carey, G. P., Lescott, T., Robertson, J. S., Spencer, L. K., and Kelly, D. C. Three African isolates of small iridescent viruses. *Virology* 85, 307, 1978.
53. Barray, S. and Devauchelle, G. Etude des polypeptides de structure du virus iridescent de *Chilo suppressalis* (Iridovirus type 6). *Can. J. Microbiol.* 25, 841, 1979.
54. Kelly, D. C. and Tinsley, T. W. The proteins of iridescent virus types 2 and 6. *J. Invertebr. Pathol.* 19, 273, 1972.
55. Moore, N. F. and Kelly, D. C. A comparative study of the polypeptides of three iridescent viruses by N-terminal analysis, amino acid analysis, and surface labelling. *J. Invertebr. Pathol.* 36, 415, 1980.
56. Monnier, C. and Devauchelle, G. Enzyme activities associated with an invertebrate iridovirus: Protein kinase activity associated with iridescent virus type 6 (*Chilo* iridescent virus). *J. Virol.* 35, 444, 1980.
57. Farara, T. and Attias, J. Alkaline protease activity associated with iridescent virus type 6. *FEMS Microbiol. Lett.* 20, 351, 1983.
58. Farara, T. and Attias, J. Further characterisation of an alkaline protease activity associated with iridescent virus type 6. *Arch. Virol.* 87, 307, 1986.

59. Webb, S. R., Paschke, J. D., Wagner, G. W., and Campbell, W. R. Pathology of mosquito irridescent virus of *Aedes taeniorhynchus* in cell cultures of *Aedes aegypti*. *J. Invertebr. Pathol.* 27, 27, 1976.
60. Younghusband, H. B. and Lee, P. E. Virus cell studies of *Tipula* iridescent virus in TIV hemocytes. *Virology* 38, 247, 1969.
61. Mathieson, W. B. and Lee, P. E. Cytology and autoradiography of *Tipula* iridescent virus infection of insect suspension cell cultures. *J. Ultrastruct. Res.* 74, 59, 1981.
62. Cerutti, M., Guérillon, J., Arella, J., and Devauchelle, G. La replication de l'iridovirus de type 6 (CIV) dans différentes lignées cellulaires. *C. R. Séances Acad. Sci.* ser. 3, 292, 797, 1981.
63. McIntosh, A. H. and Kimura, M. Replication of the insect *Chilo* iridescent virus (CIV) in a poikilothermic vertebrate cell line. *Intervirology* 4, 257, 1974.
64. Kelly, D. C. and Robertson, J. S. Icosahedral cytoplasmic deoxyriboviruses. *J. Gen. Virol. (Suppl.)* 20, 17, 1973.
65. Cerutti, M. and Devauchelle, G. Cell fusion induced by an invertebrate virus. *Arch. Virol.* 61, 149, 1979.
66. Cerutti, M. and Devauchelle, G. Inhibition of macromolecular synthesis in cells infected with an invertebrate virus (iridovirus type 6 or CIV). *Arch. Virol.* 63, 297, 1980.
67. Cerutti, M. and Devauchelle, G. Isolation and reconsitution of *Chilo* iridescent virus membranes. *Arch. Virol.* 74, 145, 1982.
68. Petit, F. and Devauchelle, G. Modifications in the phosphorylation of ribosomal proteins and ribosome-associated proteins in invertebrate cells infected with iridovirus type 6. *Ann. Inst. Pasteur Virol.* 137, 205, 1986.
69. Petit, F. and Devauchelle, G. Isolation of polysomes from permissive and nonpermissive invertebrate cell lines infected with *Chilo* iridescent virus. *Med. Microbiol. Immunol.* 174, 67, 1985.
70. Yule, B. G. and Lee, P. E. A cytological and immunological study of *Tipula* iridescent virus-infected *Galleria mellonella* larval hemocytes. *Virology* 51, 409, 1973.
71. Murti, K. G. and Goorha, R. Interaction of frog virus 3 with the cytoskeleton. I. Altered organisation of microtubules, intermediate filaments and microfilaments. *J. Cell. Biol.* 96, 1248, 1983.
72. Seagull, R. W., Lee, P. E., and Frosch, M. Comparison of microtubules and microfilaments in *Tipula* iuridescent virus-infected and uninfected cells. *Can. J. Biochem. Cell Biol.* 63, 543, 1985.
73. Younghusband, H. B. and Lee, P. E. Cytochemistry and autoradiography of *Tipula* iridescent virus-infected *Galleria mellonella* larval hemocytes. *Virology* 51, 409, 1970.
74. Chinchar, V. G., Goorha, R., and Granoff, A. Early proteins are required for the formation of frog virus 3 assembly sites. *Virology* 135, 148, 1984.
75. Murti, K. G., Goorha, R., and Granoff, A. An unusual replication strategy of an animal iridovirus. *Adv. Virus Res.* 30, 1, 1985.
76. Ward, V. K. and Kalmakoff, J. Physical mapping of the DNA genome of insect iridescent virus type 9 from *Wiseana* spp. larvae. *Virology*, 160, 507, 1987.

77. Goorha, R., Willis, D. B., and Granoff, A. Macromolecular synthesis in cells infected by frog virus 3. VI. Frog virus 3 replication is dependent on the cell nucleus. *J. Virol.* 21, 802, 1977.
78. Goorha, R. and Dixit, P. A temperature sensitive (TS) mutant frog virus 3 (FV3) is defective in second stage DNA replication. *Virology* 136, 186, 1984.
79. Watson, J. D. Origin of concatemeric T7 DNA. *Nature* 239, 197, 1971.
80. Bladon, T., Frosch, M., Sabour, P. M., and Lee, P. E. Association of nuclear matrix proteins with cytoplasmic assembly sites of *Tipula* iridescent virus. *Virology* 155, 524, 1986.
81. Devauchelle, G., Attias, J., Monnier, C., Barray, S., Cerutti, M., Guerillon, J., and Orange-Balange, N. *Chilo* iridescent virus. *Curr. Topics Microbiol.* Immunol. 116, 37, 1985.
82. Devauchelle, G. Ultrastructural characterization of an iridovirus from the marine worm *Nereis diversicolor* (O. F. Müller). *Virology* 81, 237, 1977.
83. Stoltz, D. B. and Summers, M. D. Observations on the morphogenesis and structure of a hemocytic poxvirus in the midge *Chironomus attenuatus*. *J. Ultrastruct. Res.* 40, 581, 1972.
84. Hukuhara, T. and Hashimoto, Y. Multiplication of *Tipula* and *Chilo* iridescent viruses in cells of *Anthereae eucalypti*. *J. Invertebr. Pathol.* 9, 278, 1967.
85. Fowler, M. and Robertson, J. S. Iridescent virus infection in field populations of *Wiseana cervinata* (Lepidoptera: Hepialidae) and *Witlesia* sp. (Lepidoptera: Pyralidae) in New Zealand. *J. Invertebr. Pathol.* 19, 151, 1972.
86. Xeros, N. A second disease of the leatherjacket, *Tipula paludosa*. *Nature* 174, 562, 1954.
87. Cole, A. and Morris, T. J. A new iridovirus of two species of terrestrial isopods, *Armadillium vulgare* and *Porcellio scaber*. *Intervirology* 14, 21, 1980.
88. Elliot, R. M., Lescott, T., and Kelly, D. C. Serological relationships of iridescent virus type 25 recently isolated from *Tipula* sp. with two other iridescent viruses (types 2 and 22). *Virology* 81, 309, 1977.
89. Charpentier, G., Bordeleau-Cloutier, P., Massicotte-Rousseau, C., Guerillon, J., and Devauchelle, G. *Chilo suppressalis* virus induced antigens in insect cells and tissue localisation and kinetics of appearance. *Ann. Inst. Pasteur Virol.* 137E, 37, 1986.
90. Bilimoria, S. L., Parkinson, A. J., and Kalmakoff, J. Comparative study of $^{125}I$ and ($^{3}H$) acetate labelled antibodies in detecting iridescent viruses. *Appl. Microbiol.* 28, 133, 1974.
91. Kalmakoff, J., Moore, S., and Pottinger, R. P. An iridescent virus from the grass grub *Costelytra zealandica*: Serological study. *J. Invertebr. Pathol.* 20, 70, 1972.
92. Kalmakoff, J. and Robertson, J. S. Serological relationship of *Wiseana* iridescent virus (WIV) to other iridescent viruses. *Proc. Univ. Otago Med. Sch.* 48, 16, 1970.
93. Lin, M., Imai, M., Bellamy, A. R., Ikegami, N., Furuichi, Y., Summers, D., Nuss, D. L., and Deibel, R. Diagnosis of rotavirus infection with cloned cDNA copies of viral genome segments. *J. Virol.* 55, 509, 1985.
94. French, R. C., Simon, S. H., and Derrick, K. S. Detection of sugarcane mosaic virus by dot-blot hybridisation. *Phytopathology*, 74, 859, 1984.

95. Gibson, P. E., Gardner, S. D., and Porter, A. A. Detection of human poloma virus DNA in urine specimens by hybridot assay. *Arch. Virol.* 84, 233, 1985.
96. Barker, J. M., McInnes, J. L., Murphy, P. J., and Symons, R. H. Dot-blot procedure with [$^{32}$P] DNA probes for the sensitive detection of avocado sublotch and other viroids in plants. *J. Virol. Meth.* 10, 87, 1985.
97. Sela, I., Reichman, M., and Weissbach, A. Comparison of dot molecular hybridisation and enzyme linked immunosorbent assay for detecting tobacco mosaic virus in plant tissues and protoplasts. *Phytopathology*, 74, 385, 1984.
98. Ward, V. K., FLeming, S. B., and Kalmakoff, J. Comparison of a DNA-DNA dot-blot hybridisation assay with light microscopy and radioimmunoassay for the detection of a nuclear polyhedrosis virus. *J. Virol. Meth.* 15, 65, 1987.
99. Reed, K. C. and Mann, D. A. Rapid transfer of DNA from agarose gels to nylon membranes. *Nucl. Acids Res.* 13, 7207, 1985.
100. Kulski, J. K. and Norval, M. Nucleic acid probes in diagnosis of viral diseases of man. *Arch. Virol.* 83, 3, 1985.
101. Kelly, D. C., Ayers, M. D., Lescott, T., Robertson, J. S., and Happ, G. M. A small iridescent virus (type 29) isolated from *Tenebrio molitor*: A comparison of its proteins and antigens with six other iridescent viruses. *J. Gen. Virol.* 42, 95, 1979.
102. Lee, H. H. and Miller, L. K. Isolation of genotypic variants of *Autographa californica* nuclear polyhedrosis virus. *J. Virol.* 27, 754, 1978.
103. Kislev, N. and Edelman, M. DNA restriction pattern differences from geographic isolates of *Spodoptera littoralis* nuclear polyhedrosis virus. *Virology* 119, 219, 1982.
104. Miller, L. K. and Dawes, K. P. Restriction endonuclease analysis to distinguish two closely related nuclear polyhedrosis viruses: *Autographa californica* MNPV and *Trichoplusia ni* MNPV. *Appl. Environ. Microbiol.* 35, 1206, 1978.
105. Hibbon, J. A. and Kelly, D. C. Iridescent virus type 22 DNA. *Arch. Virol.* 68, 9, 1981.
106. Tajbakhsh, S., Dove, M. J., Lee, P. E., and Seligy, V. L. DNA components of *Tipula* iridescent virus. *Biochem. Cell Biol.* 64, 495, 1986.
107. McMillan, N. A. J. Some Properties of the DNA of *Costelytra zealandica* Iridescent Virus, B. Sc. (Hons.) thesis, University of Otago, Dunedin, New Zealand, 1986.
108. Darai, G., Anders, K., Koch, H. G., Delius, H., Gelderblom, H., Samalecos, C., and Flügel, R. M. Analysis of the genome of fish lymphocystis disease virus isolated directly from epidermal tumours of pleuronectes. *Virology* 126, 466, 1983.
109. Murti, K. G., Goorha, R., and Granoff, A. Structure of the frog virus 3 genome: Size and arrangement of nucleotide sequences as determined by electron microscopy. *Virology* 116, 275, 1982.
110. Tye, B., Huberman, J. A., and Botstein, D. Non-random circular permutation of phage P22 DNA. *J. Mol. Biol.* 85, 501, 1974.
111. Soltau, J. B., Fischer, M., Schnitzler, P., Scholz, J., and Darai, G. Characterisation of the genome of insect iridescent virus type 6 by physical mapping. *J. Gen. Virol.*, 68, 2717, 1987.

112. Holford, T. R., Brown, S. E., and Knudson, D. L. Estimation of DNA fragment size and generation of DNA restriction endonuclease maps using linear models. *J. Virol. Meth.* 10, 117, 1985.
113. Stillman, B. W., Topp, W. C., and Engler, J. A. Conserved sequences at the origin of adenovirus DNA replication. *J. Virol.* 44, 530, 1982.
114. Wittek, R. and Moss, B. Tandem repeats within the inverted terminal repetition of vaccinia virus DNA. *Cell* 21, 277, 1980.
115. Venkatesan, S., Gershowitz, A., and Moss, B. Complete nucleotide sequences of two adjacent early vaccinia genes located within the inverted terminal repetition. *J. Virol.* 44, 637, 1982.
116. Pickup, D. J., Bastia, D., Stone, H. O., and Joklik, W. K. Sequence of terminal regions of cowpox virus DNA: Arrangement of repeated and unique sequence elements. *Proc. Natl. Acad. Sci. U.S.A.* 79, 7112, 1982.
117. Roizman, B. The structure and isomerisation of herpes simplex virus genomes. *Cell* 16, 481, 1979.
118. Stow, N. D. Localisation of an origin of DNA replication within the TRs/IRs repeated region of the herpes simplex virus type 1 genome. *EMBO J.* 1, 863, 1982.
119. McMillan, N. A. J. Personal communication, 1987.
120. Ricou, G. Production de *Tipula paludosa* Meig. en prairie en function de l'humiditie du sol. *Rev. Ecol. Biol. Sol.* 12, 69, 1975.
121. Chapman, H. C., Clark, T. B., Anthony, D. W., and Glenn, F. E. An iridescent virus from the larvae of *Corethralla brakelyi* (Diptera: Chaoboridae) in Louisiana. *J. Invertebr. Pathol.* 18, 284, 1971.
122. Carter, J. B., Green, E. I., and Kirkham, A. J. A *Tipula paludosa* population with a high incidence of 2 pathogens. *J. Invertebr. Pathol.* 42, 312, 1983.
123. Stadelbacher, E. A., Adams, J. R., Faust, R. M., and Tompkins, G. J. An iridescent virus of the Bollworm *Heliothis zea* (Lepidoptera: Noctuidae). *J. Invertebr. Pathol.* 32, 71, 1978.
124. Wagner, G. W., Paschke, J. D., Campbell, W. R., and Webb, S. R. Proteins of two strains of mosquito iridescent virus. *Intervirology* 3, 97, 1974.
125. van Etten, J. L., Meints, R. H., Kuczmarski, D., Burbank, D. E., and Lee, K. Viruses of symbiotic *Chlorella* like algae isolated from *Paramecium bursaria* and *Hydra viridis*. *Proc. Natl. Acad. Sci. U.S.A.* 79, 3867, 1982.

# 8

# Nodaviridae of Invertebrates

DONALD A. HENDRY
*Rhodes University, Grahamstown, South Africa*

## INTRODUCTION

The Nodaviridae are a family of small, isometric insect-pathogenic riboviruses, each possessing a bipartite messenger sense genome encapsidated within a single virion (1). This family represents the simplest class of multipartite viruses. The only other known bipartite viruses are those of plants; these, however, have their genome parts encapsidated in separate virions (2).

The family name is derived from the type member Nodamura virus, which was named after the village in Japan where it was first isolated (3). The natural host of Nodamura virus is thought to be the mosquito *Culex tritaeniorhynchus*, but it can also infect mice and possibly pigs (3,4). This ability to infect both insects and mammals aroused much interest in Nodamura virus in the 1970s, during which period the unique physicochemical properties of this virus became evident (5–9). The focus has now largely shifted to other nodaviruses because of their vigorous growth in tissue culture (10,11) and consequent amenability to molecular analysis. This has resulted in a number of significant advances: the elucidation of the respective roles of the two genomic RNAs (9,12–15); the determination of the complete sequence of a nodavirus genome (16–18); the first derivation of infectious transcript RNA from cloned cDNA copies of insect (or animal) virus RNA (19); and the determination of the atomic structure of a nodavirus at 3.0 Å resolution (20).

Several review articles dealing with the Nodaviridae have been published (21–24).

# CLASSIFICATION

## Historical Perspective

Nodamura virus was classified initially as an "arbovirus" on the basis of its original isolation from mosquitoes, its ability to multiply in various insects, and its transmissibility by mosquitoes to susceptible mice (3,4). However, unlike most arboviruses, Nodamura virus is resistant to both ether and chloroform (3). When it became apparent that Nodamura virus resembles the vertebrate enteroviruses in a number of respects—morphology, resistance to acid, and bouyant density in cesium chloride—it was regarded as a picornavirus (5). However, Nodamura virus is not stabilized against heat by magnesium ions (5), which effectively protect enteroviruses. Also, exhaustive serological testing reveals no relatedness between Nodamura virus and human, murine, or swine picornaviruses (3). When it was shown that Nodamura virus has a divided genome (6) and that the two RNAs are encapsidated together (8), it became evident that it belongs to a unique virus family.

## Members of the Family

Unenveloped icosahedral insect-pathogenic viruses of about 30 nm diameter (Figure 1), having a bipartite single-stranded RNA genome encapsidated in a single virion, are defined as nodaviruses (1). Six viruses have been classified as members of this family by the International Committee on Taxonomy of Viruses (ICTV), and these are listed in Table 1. All have been isolated from insects, but none are particularly host-specific.

## Diagnosis and Identification

The diagnosis of a nodaviral infection is most conveniently done by serology, using immunodiffusion, immune electron microscopy (IEM), or enzyme-linked immunosorbent assay (ELISA), as the known members of this family are serologically related (see Table 2). Alternatively, or as confirmation of the serological diagnosis, nodaviruses can be diagnosed using a combination of the virus particle diameter (about 30 nm), the size of the major coat protein (about 40 kD), and the presence of two genomic RNA species of $0.5 \times 10^6$ and $1 \times 10^6$ daltons (1).

Individual nodaviruses are distinguishable serologically (25–28) and, in certain cases, by plaque size (11). Thus antiserum to BBV reacts with FHV (26,28), MwV (28), BoV (27), and weakly with GMV (25) but not with NV (21). Antiserum to NV, while not reacting with BBV, FHV, or MwV (28), nevertheless reacts with BoV (27) (see Table 2). Virus identification can be done using properties such as host range, particle density, size of capsid protein, and electric charge. Only NV is lethal for day-old mice (3) and does not plaque on *Drosophila* monolayers (29), in addition to being the most atypical nodavirus serological-

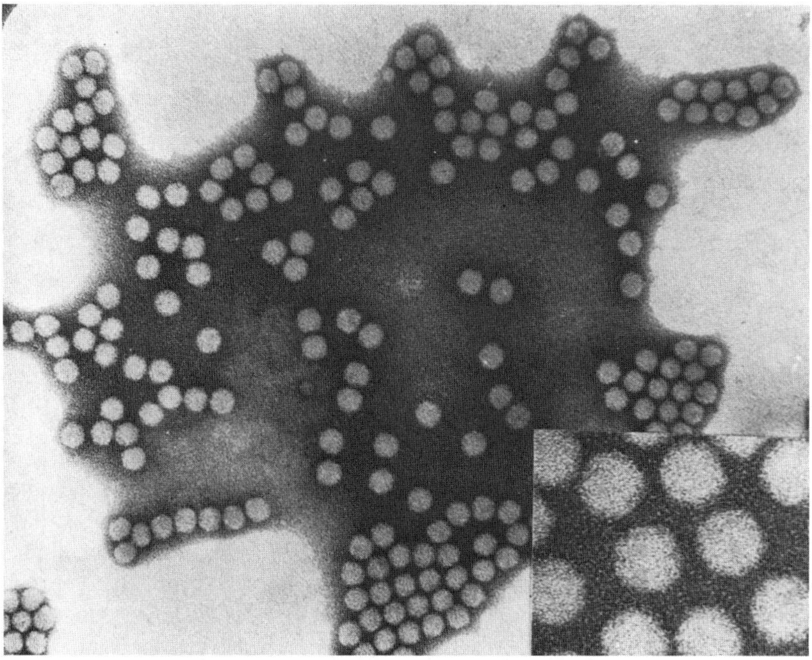

**Figure 1** Particles of FHV, a typical nodavirus, stained with 1% uranyl acetate (×115,000; insert, ×300,000). (Reprinted with permission from D. M. Blackhurst, M.Sc. thesis, Rhodes University, Grahamstown, 1987.)

ly. Black beetle virus has the lowest density of all the known members of this family (see Table 3 and ref 28), and migrates cathodically in Tris buffer at pH 8.4 (26, 28). Flock House virus has the highest density and largest coat protein of the nodaviruses and apparently contains no minor proteins; GMV has the lowest density and both the smallest coat proteins and RNA1 (see Table 3 and Ref. 26 and 25).

**Table 1** The Family Nodaviridae

| Virus | Origin | Natural host | Reference |
|---|---|---|---|
| Nodamura virus (NV) | Japan | *Culex tritaeniorhynchus* | 3 |
| Black beetle virus (BBV) | New Zealand | *Heteronychus arator* | 39 |
| Flock House virus (FHV) | New Zealand | *Costelytra zealandica* | 34 |
| Boolarra virus (BoV) | Australia | *Oncopera intricoides* | 27 |
| Gypsy moth virus (GMV) | Papua New Guinea | *Lymantria ninayi* | 25 |
| Manawatu virus (MwV) | New Zealand | *Costelytra zealandica* | 28 |

**Table 2** Serological Relatedness of the Nodaviridae[a]

| | Antiserum | | | | |
|---|---|---|---|---|---|
| Virus | NV | BBV | FHV | BoV | GMV |
| NV | ++ | – | – | * | * |
| BBV | – | ++ | + | + | + |
| FHV | – | ++ | ++ | + | * |
| BoV | + | ++ | * | ++ | * |
| GMV | * | + | * | * | ++ |
| MwV | – | + | ++ | – | * |

*Key*: ++, strong reaction; +, weak reaction; –, no reaction; *, not tested.
[a]The information is extracted from Refs. 21, 25–28.

Boolarra virus occupies a serological position midway between NV and the other nodaviruses, sharing antigens with both NV and BBV (27). In addition, only BoV has been reported to be unstable in 1% SDS (29). Manawatu virus contains a single capsid protein of the same size as BBV but differs from BBV in both electrophoretic mobility and density (28). The use of virus density as a criterion for identification should be approached with caution, however, as comparative determinations on all the members have not been performed in one and the same laboratory.

The W17 strain of BBV, developed by serial passage of the original virus isolate in *Drosophila* cells (11) and used subsequently for molecular analyses (18,19,64,65), is serologically indistinguishable from FHV but easily distinguishable from BBV (30). However, FHV and BBV W17 are not identical, the latter making smaller plaques (31).

## PATHOBIOLOGY, PROPAGATION, AND ASSAY

### Epidemiology

Very little is known regarding the epidemiology of nodaviruses, as each virus was obtained after a single isolation from insects collected in one locality. Only in the case of FHV has the natural distribution of the virus been examined (31a).

Female *Culex tritaeniorhynchus* (Diptera: Culicidae) mosquitoes, collected in a pig-baited trap at Nodamura near Tokyo on July 23, 1956, were the source of NV (3). Little is known concerning its prevalence in mosquitoes, but transmission to swine probably occurred over a fairly wide radius during 1956 and 1957, as pigs less than a year old from different areas contained neutralizing antibody to NV when bled in September 1957. Also, large populations of *C. tri-*

*taeniorhynchus* were recorded during June and July of 1956 and 1957 (4). Surprisingly, about 3% of cattle examined in southern Japan in 1974 contained complement-fixing antibodies to NV (32). No neutralizing antibodies to NV were found in either bird or human sera collected near Tokyo during 1957–1958 (4).

The source of BBV was a black beetle, *Heteronychus arator* Fabricius (Coleoptera: Scarabaeidae), collected near Helensville in New Zealand in April 1974 (33). This insect has become a major pest of improved pasture in that country, high-density populations reaching 140/m$^2$ occurring in kikuyu grasslands. Of the diseased or dying insects collected, 25% of the third instar larvae and 7% of the adults contained BBV, the remaining mortalities being due to other pathogens (33).

Flock House virus was isolated from third instar larvae of the New Zealand grass grub, *Costelytra zealandica* (White) (Coleoptera: Scarabaeidae), collected from a pasture at Flock House in New Zealand (34). The distribution of FHV in New Zealand was limited to pastures around Flock House, as indicated by a survey of over 3000 larvae of *C. zealandica* collected from both North and South Islands between February and October 1980 (31a). Using immunoosmophoresis, FHV could be detected in larvae but not in pupae or adults. The virus occurred at a low level in the insect population, the incidence of FHV in larvae collected from Flock House reaching a maximum of 12.5%, with the virus playing only a minor role in overall grass grub mortality.

One larva of *C. zealandica*, collected from the pasture where FHV was isolated, was used for the isolation of MwV, named after the region in North Island where it was found (28).

Dead or moribund larvae of the underground grassgrub, *Oncopera intricoides* (Walker) (Lepidoptera: Hepialidae), found in improved pastures near Boolarra in Australia in 1977, were the source of BoV (27), while GMV was isolated from dead or diseased larvae of *Lymantria ninayi* (Lepidoptera: Lymantriidae) found in Papua New Guinea (35).

## Host Range and Pathogenesis

The Nodaviridae are all insect-pathogenic viruses, having been isolated from lepidopteran, coleopteran, or dipteran hosts (see Table 1). However, none are particularly host-specific, and they can be transmitted to other insects. The type member NV is the only member not confined to the class Insecta, being also pathogenic for mammals (3).

The natural host for NV is thought to be the mosquito *Culex tritaeniorhynchus*, from which it was first isolated by inoculation of a suspension of female mosquitoes into suckling mice (3). The virus can, however, be transmitted from diseased to healthy mice by a different mosquito, *Aedes aegypti*, within which it appears to multiply (3). The virus also multiplies in the mosquitoes *Culex tarsalis* (3), *Aedes albopictus*, and *Toxorhynchites amboinensis*, but not in *Culex quinquefasciatus* (36), when inoculated by injection. Shortly before death, the mosquitoes manifest loss of balance, inability to fly, and eventual paralysis; head squashes reveal large amounts of virus antigen in the brain. Mosquitoes infected orally, or by immersion of larvae in virus suspension, also contain virus antigen in the brain but are not killed. Transovarial transmission of the virus does not occur (36).

NV can also replicate in other arthropods. Virus multiplication, as judged by serial passage, occurs after virus is injected into the hemocoel of the soft tick, *Ornithodorus savignyi* (Acarina: Argasidae), and of larvae of both the Indian meal moth, *Plodia interpunctella* (Lepidoptera: Pyralidae), and the black carpet beetle, *Attagenus piceus* (Coleoptera: Dermestidae). No multiplication occurs in the bedbug, *Cimex lectularius* (Hemiptera: Cimicidae), or the German roach, *Blattella germanica* (Blattodia: Blattellidae) (3). The virus also replicates in the honeybee, *Apis mellifera* (Hymenoptera: Apidae), and larvae of the wax moth, *Galleria mellonella* (Lepidoptera: Galleriidae). Injected bees die after about 7 days, the anterior two pairs of legs first becoming paralyzed (37). Wax moth larvae become paralyzed in the five or six last segments 4–6 days after injection. The paralysis spreads and leads to death 15–20 days postinjection (37,38).

Intracranial, subcutaneous, or intraperitoneal injection of NV kills 1–3 day old mice within 5–15 days, causing extensor paralysis of the hind limbs 3 hours to 3 days before death. Lesions are most pronounced in muscle and central nervous system (4; see "Virus-Infected Tissue" below). Antibody formation without disease occurs after injection of pigs, adult mice, and rabbits. No illness occurs after inoculation of baby chicks, weanling mice and rabbits, weanling guinea pigs and Syrian hamsters, or embryonated chicken eggs (4).

The replication of the other nodaviruses is restricted to insects (however, see FHV below). The primary natural host for BBV is the black beetle, *Heteronychus arator*. The virus causes infected third instar larvae and prepupae to be flaccid and translucent compared with the creamy white appearance of healthy larvae. The virus can be transmitted by injection to *Pseudaletia separata* (Lepidoptera: Noctuidae), *G. mellonella*, *Aphodius tasmaniae* (Coleoptera: Aphodiinae), and *Pericoptus truncatus* (Coleoptera: Scarabaeidae) (33). After injection with BBV, day-old mice develop no symptoms and no virus replication can be detected (39).

The grass grub, *Costelytra zealandica*, is the natural host of MwV (28) and of FHV, which also replicates in *G. mellonella* (26, 34). The original host of GMV is *Lymantria ninayi*, but the virus can be transmitted to the gypsy moth *Lymantria dispar* by feeding or injection. Injected *L. dispar* larvae die a minimum of 8 days after injection, hanging in a flaccid condition from the foliage by their prolegs (35). The underground grassgrub, *Oncopera intricoides*, was the source of BoV, although larvae of *G. mellonella* can be infected with BoV and die about 10 days postinjection. No deaths occur among mice injected with BoV (27).

A recent surprising observation is that once the barrier of the infectious process is breached, FHV can replicate in a host that is taxonomically distant from insects (73). If leaves or plant protoplasts of various species (barley, chenopodium, cowpea, or tobacco) are inoculated with the RNA of FHV and then homogenized 2–14 days later, homogenates contain up to $3.2 \times 10^5$ PFU per milligram of leaf tissue when plaque-assayed on *D. melanogaster* cells. The plaques are due to FH virions, as the infectivity is sensitive to anti-FHV serum but not to ribonuclease. No symptoms are produced, and no systemic spread of FHV occurs. These results suggest that it is the delivery and release mechanisms that restrict the host range of FHV.

## Cytopathology and Virus Morphogenesis

### Virus-Infected Tissue

Studies on nodavirus morphogenesis and nodavirus-induced cytopathology in infected animals have been made on tissue samples of mice infected with NV (4,5) and of wax moths infected with NV (38) and BBV (39). Histologically, mice infected with NV suffer necrosis of spinal cord neurons and brown fat as well as inflammation and degeneration of paravertebral and limb skeletal muscles (4). These histopathological changes are similar to those caused in mice by a variety of viral agents, such as Coxsackie and Sindbis viruses (4). Histological examination of NV-infected wax moth larvae indicates that cellular lesions occur mainly in the cytoplasm of cells of the muscle, the nerve ganglia, and the salivary and moulting glands. Dense cytoplasmic inclusions are formed, and muscle fibers lose their cross-striations (38). In contrast to the muscle cell tropism of NV in both organisms, BBV shows a tropism for epithelial midgut cells in the wax moth (39).

Electron microscopic examination of tissue from NV-infected mice (5) or from NV- or BBV-infected wax moths (38,39) reveals a number of cytopathological similarities. In all cases, the lesions are confined to the cytoplasm and consist of the loss of internal organization and the accumulation of dispersed or paracrystallized virions (Figure 2). In the case of NV-infected mouse or wax moth muscle cells. the virus causes cytoplasmic degeneration, disorganization of the muscular fibrils, and, at the advanced stage of infection, almost complete replacement of the sarcolemma by virus crystalloids (Figure 3). The virus particles exist dispersed between the fibrils, but later in the infection also become aggregated into mainly two types of cytoplasmic inclusions consisting of either dense masses with a smooth granular texture or masses of irregular reticular strands. These structures ("viroplasm") are indistinguishable from those seen in picornavirus-infected cells (40,41). Apart from dispersed virions and those in

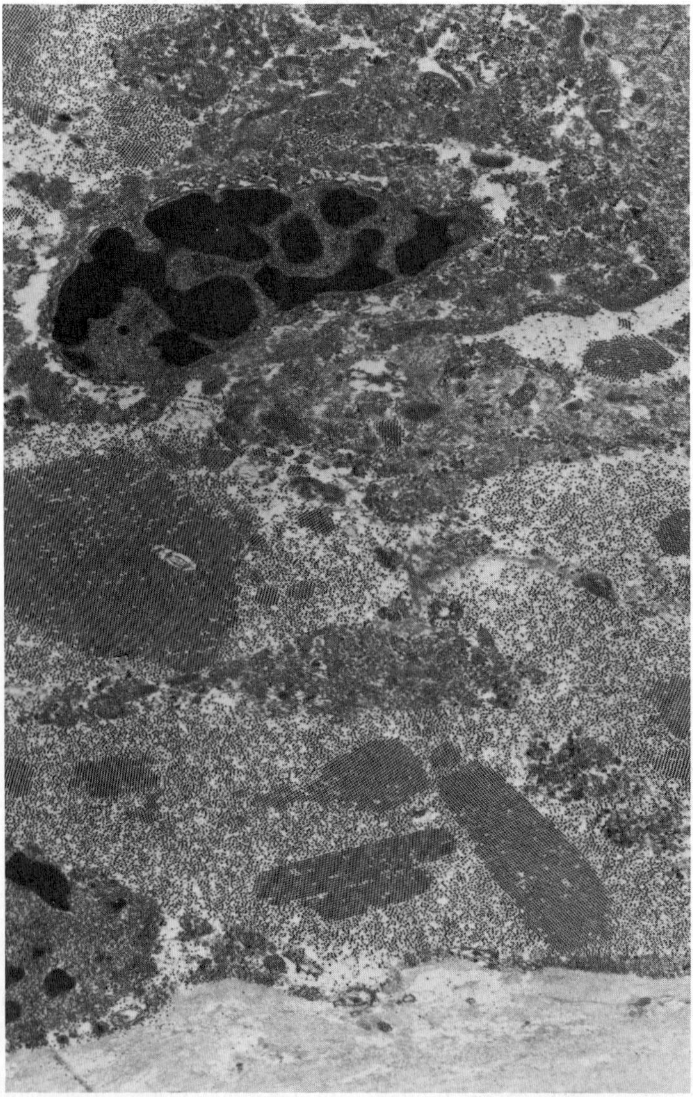

**Figure 2** Hypodermal cells of *Galleria mellonella* heavily infected with NV. A large number of virus particles and paracystalline bodies are present in the lower cell near the cuticle. In the second cell, virus particles are still in the stage of assembly around the virogenic stroma. Two nuclei show condensation of the chromatin (×18,600). (From Ref. 38.)

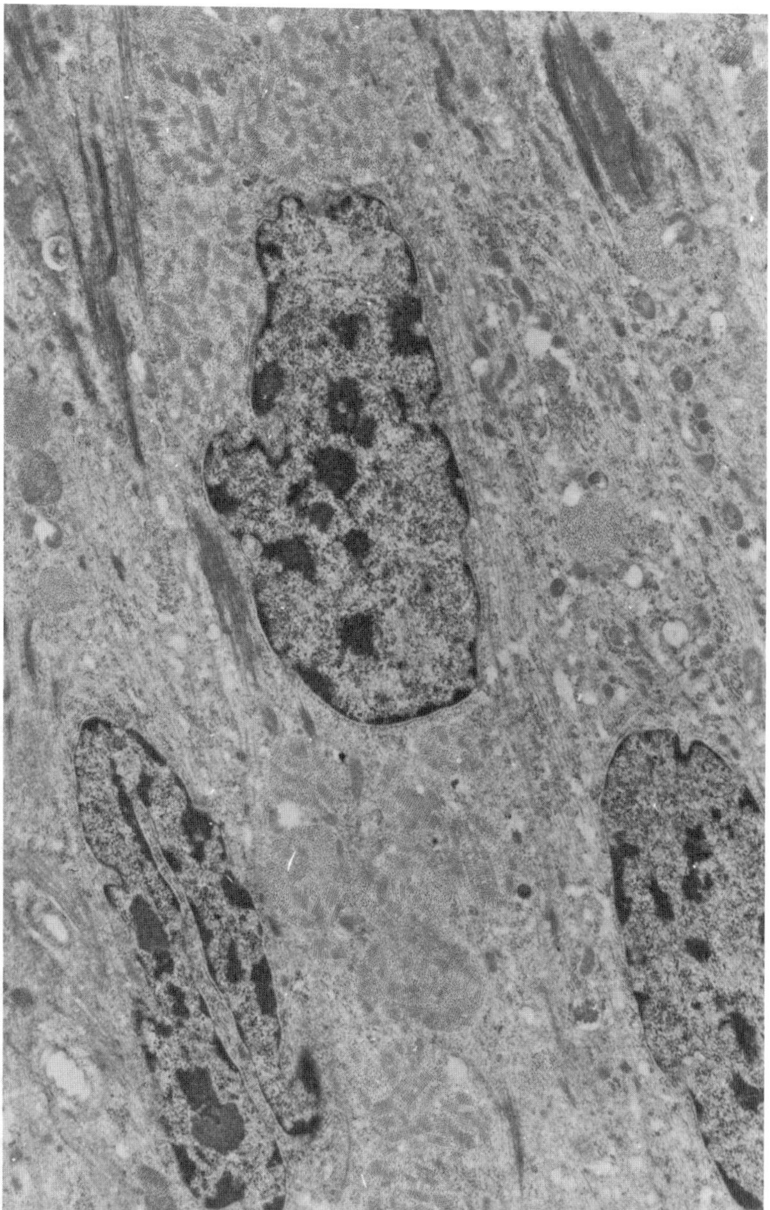

**Figure 3** Muscle tissue of *Galleria mellonella* at an advanced stage of infection with NV. The sarcoplasm is almost completely replaced by virus particles, some in paracrystalline arrays (×14,900). (From Ref. 38).

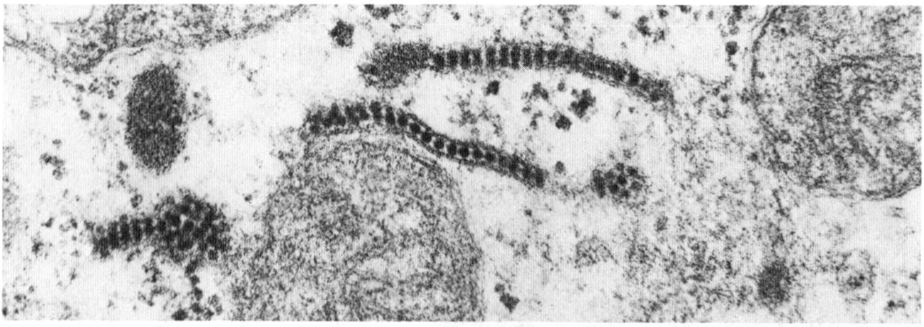

**Figure 4** Nodamura virus in stringlike arrangements between two unit membranes in infected *Galleria mellonella* cells (×57,700). (From Ref. 38.)

crystalline arrays, the virions can also occur in strings (Figure 4) between two smooth membranes, a phenomenon also observed in BoV-infected *Drosophila* cells (42). These stringlike arrangements could be the first step in the formation of virus-filled vesicles (see below), or they could be a mechanism for the continuous release of virus from infected cells.

Virus particles also occur in the cytoplasm in membrane-bound vesicles that are released intact on a large scale; in the mouse, muscle macrophages and liver Kupfer cells contain large masses of these membrane-bound viral structures, but no free virions. This phenomenon ("cortical shedding") has also been observed with poliovirus-infected cells (41) and is postulated to be an autophagic defense mechanism (43).

In NV-infected wax moth tissue, virions are also seen in the hypodermis, adipose tissue, trachea, and hemocytes (38).

Virus-Infected Cultured Cells

The cytopathic effect of nodaviral infection on cultured cells has been examined in BBV- and FHV-infected *Heteronychus arator* and *Drosophila melanogaster* cells (44) and in BoV-infected *Drosophila* cells (42). This approach permits sequential studies of virus morphogenesis in synchronously infected cells.

The cytopathic effect usually consists of the infected cells rounding up and aggregating (Figure 5), the extent depending on the infecting virus (44). The viruses replicate in the cytoplasm and, except for FHV in the *Heteronychus* cells (44), form crystalline arrays (42) (Figure 6). Cellular changes in BoV-infected *Drosophila* cells are first visible 24 hr postinfection (p.i.) as large, electron-dense areas (virogenic stroma) (Figure 7), which by 48 hr p.i. contain arrays of 16 nm electron-lucent particles. The accompanying cellular changes consist of the formation of multilamellar membrane stacks and vesicular bodies (Figure 7). By 72 hr p.i. the arrays in the virogenic stroma contain, in addition to the smaller

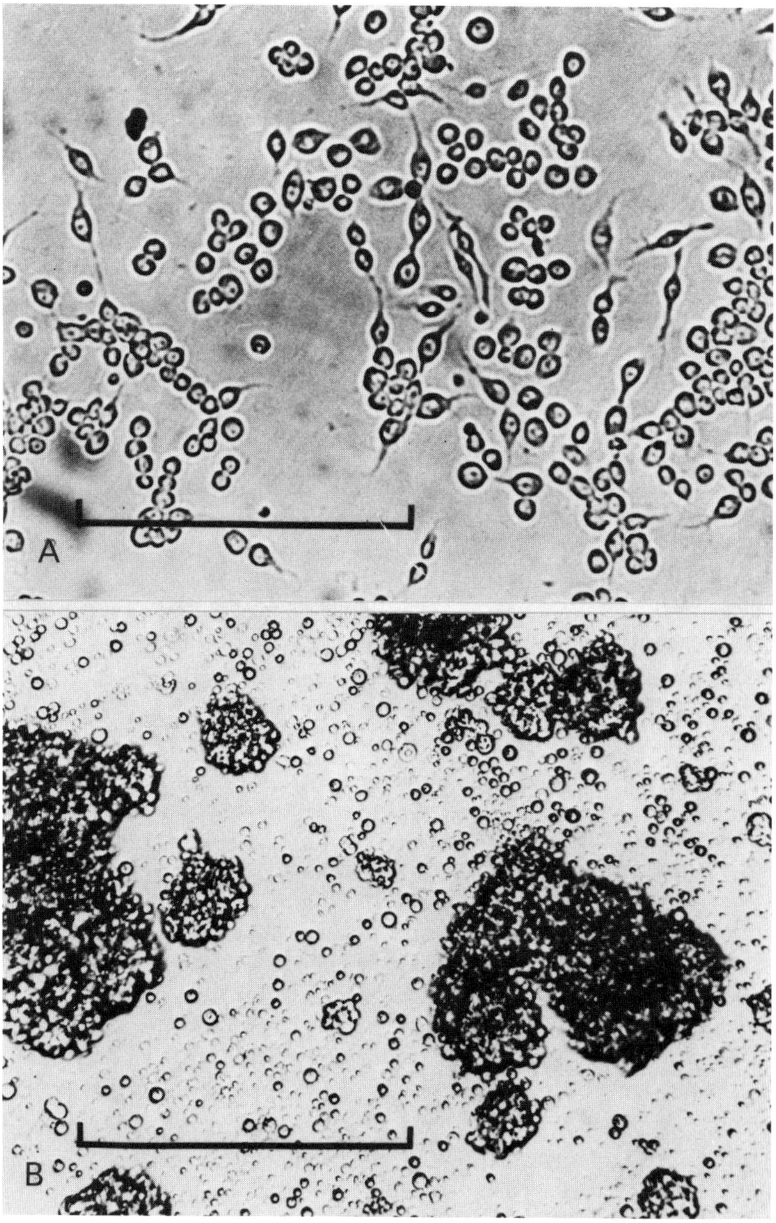


**Figure 5** CPE of FHV in *Drosophila* line 1 cells. (A) Uninfected cells. (B) FHV-infected cells. Bars = 100 $\mu$m. (From Ref. 26.)

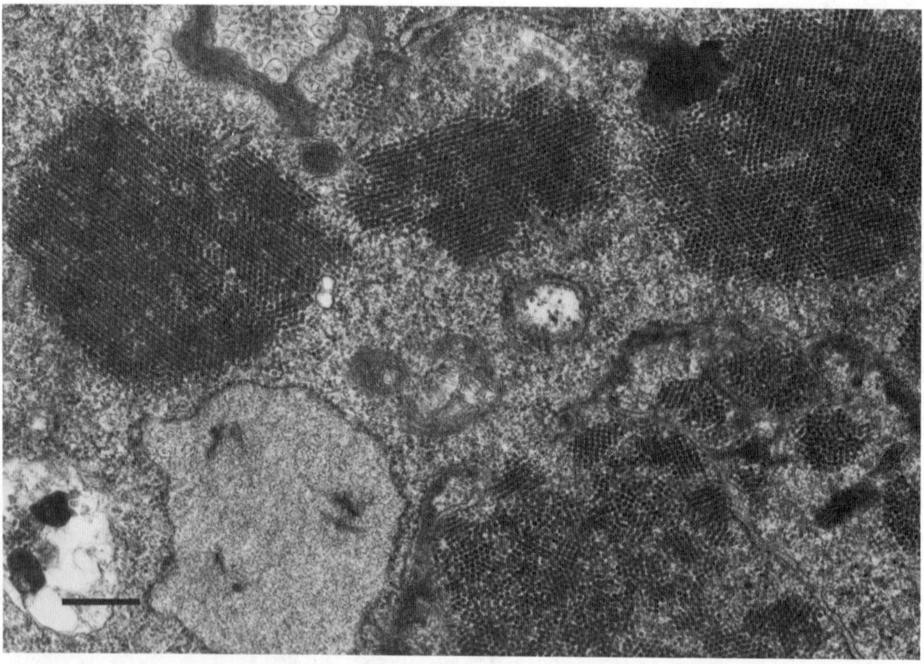

**Figure 6** *Drosophila* cell infected with BoV, showing numerous paracrystalline arrays. Bar = 400 nm. (From Ref. 42.)

electron-lucent particles, also 30 nm electron-dense ones, which by 96 hr p.i. are tightly packed and the only ones visible in the stroma. By 120 hr p.i. virus strings in membranous vesicles are visible, while at 144 hr p.i. the cytoplasm is packed with large paracrystalline arrays and virus particles in vesicular bodies (Figure 8). Both the virus strings and the virus-containing vesicles are also features of nodavirus-infected tissue.

## Propagation in Whole Organisms

Prior to the development of suitable tissue culture systems, nodaviruses had to be propagated in whole organisms. This is still the method employed for the propagation of NV, for which a high-yielding tissue culture system has yet to be found.

The propagation of NV is accomplished by the intraperitoneal or intracranial injection of virus suspension into neonatal mice (5,6). Once paralysis has set in, the hind limb muscles are used for virus purification (9). Muscle tissue produces

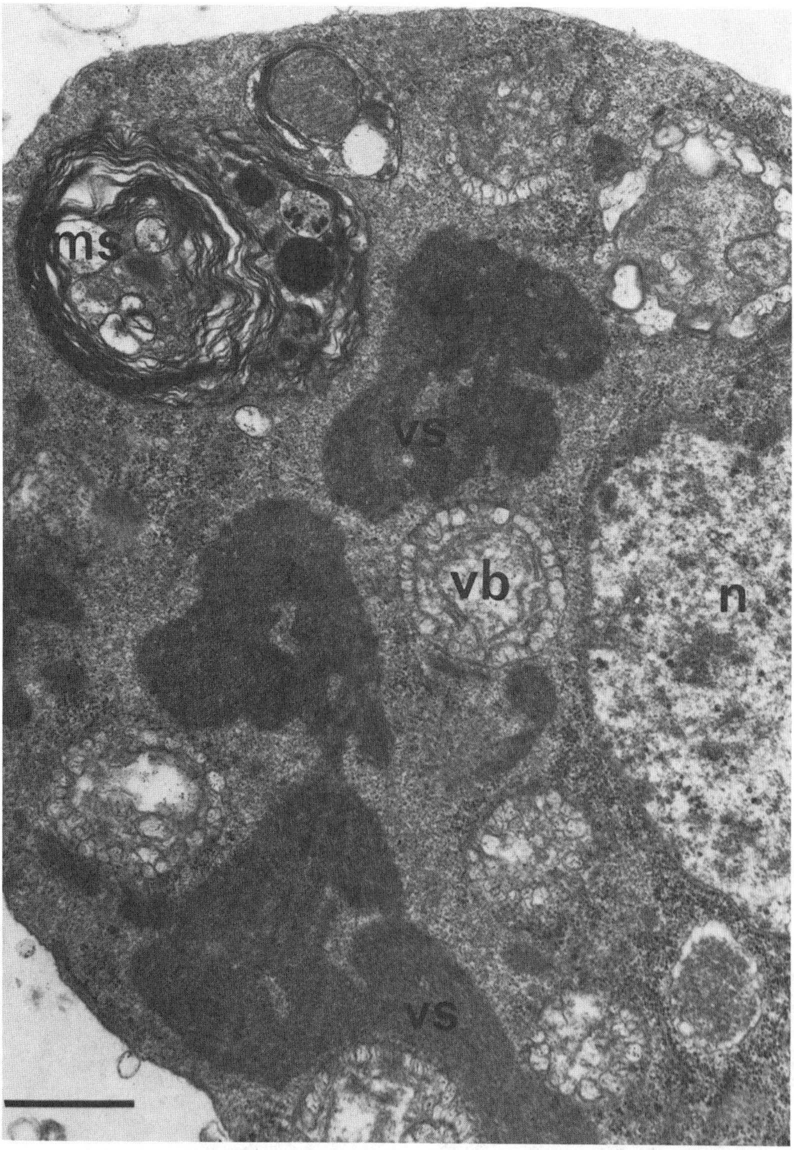


**Figure 7** A BoV-infected *Drosophila* cell at an early stage of infection (24 hr p.i.). The cell contains virogenic stroma (vs), vesicular bodies (vb), and multilamellar membrane stacks (ms). The nucleus (n) is intact. Bar = 1 $\mu$m. (From Ref. 42.)

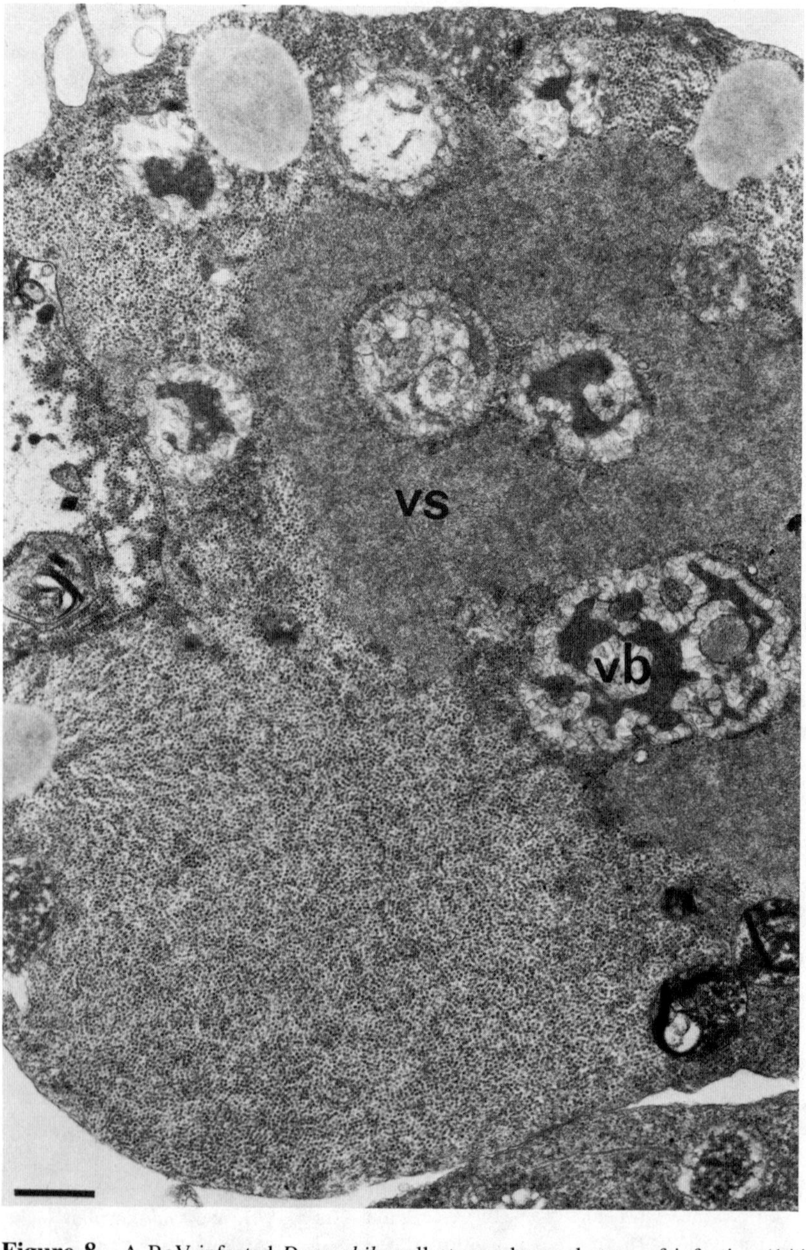


**Figure 8** A BoV-infected *Drosophila* cell at an advanced stage of infection (144 hr p.i.). Paracrystalline array occupies a large portion of the cell. Vesicular bodies (vb) are surrounded by virogenic stroma (vs). Bar = 0.5. μm. (From Ref. 42.)

a virus yield that is about $10^4$-fold the $LD_{50}$ of the inoculum, while the yield from brain tissue is about 1% that from muscle (6). Adult bees *(Apis mellifera)* and larvae of the wax moth *(Galleria mellonella)* also support NV replication (37).

Moth larvae can be used for the propagation of other nodaviruses. Thus larvae of *G. mellonella* also support the replication of BBV (39), GMV (35), and BoV (27), while larvae of *Lymantria dispar* can be used to propagate GMV (35). Larvae are either injected intrahemocoelically or droplet-fed with crude or purified virus suspensions, which can contain penicillin, and are then reared at 25–30°C for 10–30 days. Injected larvae usually die sooner than those fed with the virus (35).

## Propagation in Cultured Cells

Although NV grows in BHK-21 hamster and mosquito (*Aedes albopictus* and *A. aegypti*) cells (45), multiplication occurs at such low levels that virus-specific protein synthesis is difficult to detect (9). A major impetus to the study of nodaviruses came with the demonstration (10) that BBV replicates to very high concentrations in cultured *Drosophila melanogaster* line 1 cells (WR strain) (46). It is possible to purify milligram quantities of virus a few days after infecting a culture seeded with $10^8$ cells, by which stage the virus can account for up to 20% of the total cell-associated protein (10). This is accomplished without cell lysis, as the infected cells remain intact for four days or more after infection. This long period of synthesis probably contributes to the high yield of virus. One immediate effect of this vigorous virus growth was that, for the first time, nodavirus-specific polypeptides could be distinguished relatively easily from the background of host proteins (12).

Cultured *Drosophila* cells also support the replication of FHV (26), BoV (27), and MwV (28). In the case of FHV, a burst size of 250,000 particles per cell has been calculated (26). Both BBV and FHV also replicate in cultured cells of the black beetle, *Heteronychus arator* (47), although the infections are not as productive as in *Drosophila* cells (44).

In the absence of a plaque assay for any of the nodaviruses, the first growth curve was obtained using an enzyme-linked immunosorbent assay (ELISA) to detect BBV production in *Drosophila* cells (48). Intracellular production of BBV is detectable as little as 2 hr p.i. Release of virus from the cells is a gradual process starting 6–8 hr p.i. and continuing to increase, without cell lysis, for at least 48 hr. A tissue culture dilution end point ($TCID_{50}$) assay, based on the cytopathic rounding and clumping of infected cells, was used to measure the replication of FHV in both *Drosophila* and *H. arator* cells (44). In *Drosophila* cells, extracellular FHV is detectable a few hours after infection, increasing rapidly for a further 40 hr, and then leveling off. In *H. arator* cells, replication is slower and the yield is lower. Virus yield in the *Drosophila* cells is about double that in *H. arator* cells, measured as about 100 and 50 pg virus per cell, respectively (44), in spite of the latter cell type being derived from the same family as the virus host.

The synthesis of BBV-specific proteins in *Drosophila* cells is detectable after infection at a multiplicity as low as 10 particles per cell and is maximal at 300 particles per cell, above which multiplicity no increase in viral protein synthesis is observed (12). However, BBV replication is severely curtailed by the presence of persistently infecting defective-interfering particles that develop if virus is passaged serially at high multiplicities (13). This is prevented by passaging serially at low multiplicity (100 particles per cell) followed by a single passage at high multiplicity (10,000 particles per cell) when a high virus yield is required (14). In this regard, a New Zealand subline of *Drosophila* cells containing an endogenous BBV-related nodavirus supports only very low levels of BBV replication. Although the identity of this endogenous virus is unknown, its coat protein differs electrophoretically from that of BBV and [as found with both FHV (26) and MwV (28)] it exhibits no evidence of minor proteins (10).

## Assay Systems

Initial nodavirus infectivity assays were performed using whole insects. Adult honeybees *(Apis mellifera)* and wax moth larvae *(Galleria mellonella)* can be used in $LD_{50}$ assays for NV (45). Both insects are killed by the virus, bees within two weeks after injection and moth larvae within four. Tests with wax moths are also less consistent than those with bees, but they are useful when bees are unavailable or unsuitable for experiments. Suckling mice are used in a similar fashion for NV assays (3–6,45). Both BBV (10) and BoV (27) can be assayed in wax moth larvae.

The first plaque assay for a nodavirus employed *Drosophila* cell monolayers to titrate BBV (49). An improved plaque assay for BBV, providing reliability and convenience, was developed using a rapidly growing strain (BBV-W17) of the virus (11). This modification revealed that cell density of the monolayer is critical for plaque formation. Thus a strictly controlled number of cells (usually $4 \times 10^4$ cells per 60 mm tissue culture dish) is used for each assay. This assay reveals a linear relationship between PFU and the number of virus particles and also permits the titration of other slower-growing nodaviruses. Thus wild-type BBV, BBV-W17, and FHV can be distinguished on the basis of plaque size (11). All the nodaviruses except NV form plaques on *Drosophila* cell monolayers (29).

Other methods that have been used to assay nodaviruses are an indirect fluorescent antibody technique (36), enzyme-linked immunosorbent assay (48), immuno-osmophoresis (26), and scoring for cytopathic effect in infected insect cell cultures (44).

# PHYSICOCHEMICAL PROPERTIES

## The Virion

The physicochemical properties of the six characterized members of the Nodaviridae are summarized in Table 3. The most authoritative values are those reported for BBV by Hosur et al. (20), which are derived from their structural data and from the known nucleotide and amino acid sequences of the genomic RNAs and the capsid proteins, respectively (16,18).

### Particle Dimensions

Nodavirus virions are unenveloped particles of 29–31 nm diameter. The particle diameters listed in Table 3 were all, except for that of BBV, obtained from electron micrographs of negatively stained virus preparations. The diameter of 31.2 nm given for BBV is that obtained from small-angle x-ray scattering (50), whereas a value of 29 nm was originally obtained by electron microscopy (34).

### Physical Properties

The most reliable particle weight for a nodavirus is $9.4 \times 10^6$, as listed for the BBV virion in Table 3. It is based (20) on the established structure and on the accurate molecular weights now available for the protein and nucleic acid components of this virus and is a refinement of the value of $8 \times 10^6$ originally reported (8,50). The previous estimate (21) of $5 \times 10^6$ is now accepted as being too low. The sedimentation coefficients, which range between 135 and 142 S, are all comparative values obtained by centrifuging virus in sucrose density gradients together with markers of known $s_{20,w}$. The virus densities, determined by centrifuging virus in gradients of cesium chloride, vary between 1.30 and 1.37 g/ml, depending on the species and the procedure employed (28).

### Chemical Composition

Experimental determinations of the nucleic acid content were done using colorimetric methods in the case of NV (8) and BBV (39) and a spectrophotometric method in the case of BoV (27). All of these indicated an RNA content in excess of 20%. In the case of BBV, the RNA content of 16% was calculated from the particle structure and the molecular weights of the protein and nucleic acid components (20). This is now the accepted figure for the RNA content of nodaviruses, and the reason for the higher values obtained experimentally is unclear. From these figures, a total complement of $1.51 \times 10^6$ daltons of RNA and $7.89 \times 10^6$ daltons of protein can be calculated for BBV.

The virus particles probably contain no lipid, as at least one nodavirus has been shown to be resistant to organic solvents (3). The presence of carbohydrate has not been investigated.

**Table 3** Physicochemical Properties of the Nodaviridae

| Physicochemical property | NV | BBV | FHV | BoV | GMV | MwV |
|---|---|---|---|---|---|---|
| Particle size (nm) | 29 | 31.2 | 29 | 30 | 30 | * |
| Molecular weight of virion (× $10^6$) | 8 | 9.4 | * | * | * | * |
| $s_{20,w}$ of virion | 135 | 137 | 142 | 140 | * | 142 |
| Density (g/ml)[a] | 1.34 | 1.335 | 1.351 | 1.34 | 1.30 | (1.366) |
| | (1.352) | (1.372) | | | | |
| % RNA | 20.5 | 16 | * | 21.2 | * | * |
| Proteins (MW × $10^3$)[b] | **40** | **39** | **43** | **38.0** | **31** | **40** |
| | 43 | 44 | — | 40.6 | 34 | — |
| | (38) | 5 | — | — | — | — |
| RNAs (MW × $10^6$) | 1.15 | 1.04 | 1.04 | 1.03 | 0.95 | 1.1 |
| | 0.46 | 0.47 | 0.46 | 0.47 | 0.47 | 0.46 |
| $s_{20,w}$ of RNAs | 22 | 22 | 22 | 22 | * | * |
| | 15 | 15 | 15 | 17 | * | * |
| Composition of 22 S RNA | | | | | | |
| A | 22.4 | 29.0 | * | * | * | * |
| C | 27.0 | 24.1 | * | * | * | * |
| G | 27.6 | 22.8 | * | * | * | * |
| U | 23.0 | 24.1 | * | * | * | * |
| Composition of 15 S RNA | | | | | | |
| A | 24.8 | 26.4 | * | * | * | * |
| C | 28.2 | 23.8 | * | * | * | * |
| G | 22.8 | 24.7 | * | * | * | * |
| U | 24.2 | 25.2 | * | * | * | * |
| $E_{260nm}$ at 1 mg/ml | * | 4.15 | * | * | * | * |
| $E_{260nm}/E_{280nm}$ | 1.61 | 1.56 | * | * | * | * |
| Reference | 5,6,8 | 13,20,39 | 18,26,34 | 27 | 25,35 | 28 |

*Key:* *, not determined; –, not found.

[a]Values in parentheses are from Ref. 28.

[b]The molecular weight of the major protein is in boldface type.

### Stability

Particles of NV are resistant to both ether and chloroform (3), probably reflecting an absence of lipid. However, the resistance of NV to sodium deoxycholate is variable, no inactivation occurring if virus is tested in suspension of mouse brain, but irregular and unpredictable inactivation occurring in suspension of mouse muscle (51). No other nodaviruses have been examined for their resistance to lipid solvents.

The majority of nodaviruses are unstable in CsCl. In alkaline CsCl gradients, freshly purified NV gives a single infectious peak at 1.34 g/ml (5,6), which, however, represents only about 1% overall recovery of infectivity (6). Only between pH 7.5 and 8.5 is this single peak observed, two peaks being found if the gradient is below pH 7.5 or above pH 8.5 (8). Purified NV that has been stored at 4°C for a few days in Tris-HCl yields a number of peaks in CsCl, none of which are infectious, the major proportion being in two peaks of about 1.5 and 1.6 g/ml (6). This loss of both infectivity and particle integrity is due to purified NV particles being unstable in the presence of $Cl^-$ ions at all pH values above 4 (8). No heterogeneity is observed if NV is sedimented in $Cs_2SO_4$ gradients (8) or if the virus is first fixed with 0.05% formaldehyde (6). This latter observation indicates that the denser peaks obtained in the presence of $Cl^-$ ions represent virions that have lost protein during the centrifugation.

An instability in CsCl is also observed with BBV (39) and GMV (35) provided the purified virions are stored frozen prior to centrifugation. Freshly purified BBV, or virus treated with 0.05% formaldehyde, shows no instability in CsCl (39). Particles of FHV are unaffected by CsCl (26).

Only NV and FHV of the nodaviruses have been examined for pH and heat stability. Incubation of NV for 3 hr at pH 3.7 has no effect on the infectivity titer, while heating virus at 50°C for 2 hr causes a reduction in infectivity of about 3 $\log_{10}$ provided the protecting effect of contaminating tissue is eliminated by dilution (5). The presence of molar $MgCl_2$ does not protect NV against heat (5). The infectivity of FHV is unaffected by incubation for 15 min at pH values from 2.5 to 8.4 and by 10 min heating at 40°C (26). Heating of FHV for 10 min at 53°C or at 58°C results in a 90% and a 99% loss of infectivity, respectively (26). Acidification to pH 3 has no effect on BBV (28).

All the nodaviruses except Boolarra virus are stable in 1% sodium dodecyl sulfate (29), a property utilized in the purification procedures developed for NV (6) and for BBV (12).

## The Viral Genome

### General Description

The characteristic feature of the nodaviral genome is that it consists of two RNA strands, a 15 and a 22 S species, that are encapsidated within a single virion (1).

Both RNAs, which are messenger-sense and single-stranded, are required for infectivity (6,14). The two RNAs of BBV contain no significant common sequences (18,52). Neither of the two RNAs of NV (7), FHV (26), or BBV (16,18) is polyadenylated. Both of the RNAs of BBV have $m^7$ GpppGp cap structures at their 5' ends and have unidentified moieties (possibly proteins) at the 3' ends that prevent polyadenylation or ligation (16,18). Replicative forms of each of the two RNAs of both NV (8) and BBV (53) are found in infected cells, indicating that the RNAs have independent genetic functions. They can, however, by gentle treatment, be isolated from NV virions as a single infectious 27 S complex, presumably linked by labile, noncovalent bonds (8; also see "RNA Encapsidation and Virion Maturation" below).

The 15 S RNA

The smaller nodaviral RNA (RNA2) is a 15 S species with a molecular weight of about $0.47 \times 10^6$, there being remarkably little variation in the molecular weight estimates of this component obtained from the different nodaviruses (see Table 3). Although the RNA2 of GMV migrates ahead of that of BBV on composite agarose–acrylamide gels, it was nevertheless estimated to have a molecular weight of $0.47 \times 10^6$ (25); a molecular weight of $0.5 \times 10^6$ was assumed for BBV RNA2. Sequencing studies (16) show that the 15 S RNA of BBV contains 1399 bases (Figure 15), giving it a molecular weight of $0.47 \times 10^6$. This figure agrees with estimates obtained for BBV RNA2 by denaturing gel electrophoresis (39) and by oligonucleotide fingerprinting (52) and with those obtained for the RNA2 of the other nodaviruses.

The 22 S RNA

The larger nodaviral RNA (RNA1) is a 22 S species with a molecular weight of about $1.1 \times 10^6$ (see Table 3). The RNA1 of GMV appears to be smaller than that of the other members, the RNA1 of this virus migrating slightly faster than that of BBV on composite agarose–acrylamide gels and having an estimated molecular weight of $0.95 \times 10^6$ (25). The RNA1 of BBV consists of 3106 bases (Figure 14) (54). The molecular weight calculated from this sequence is $1.04 \times 10^6$, which is in reasonable agreement with values obtained experimentally (39,52).

## The Capsid Proteins

The nodaviral capsid, using BBV as the prototype (12,20), consists of a major and two minor protein species (see Table 3). The major protein of about 39 kD has been designated beta ($\beta$), while one of the minor proteins (gamma, $\gamma$) is about 4.5 kD. Beta and gamma are derived from the cleavage of a 44 kD precursor (alpha, $\alpha$), the other minor capsid protein. The amino acid sequence of

**Table 4** Amino Acid Composition of BBV Capsid Proteins

| | Number of residues[a] | | |
|---|---|---|---|
| Amino acid | Protein alpha | Protein beta | Protein gamma |
| Ala | 44 | 38 | 6 |
| Asx | 0 | 0 | 0 |
| Cys | 5 | 5 | 0 |
| Asp | 14 | 14 | 0 |
| Glu | 11 | 9 | 2 |
| Phe | 24 | 21 | 3 |
| Gly | 30 | 24 | 6 |
| His | 2 | 2 | 0 |
| Ile | 20 | 16 | 4 |
| Lys | 11 | 9 | 2 |
| Leu | 24 | 20 | 4 |
| Met | 10 | 8 | 2 |
| Asn | 28 | 27 | 1 |
| Pro | 31 | 29 | 2 |
| Gln | 18 | 18 | 0 |
| Arg | 27 | 26 | 1 |
| Ser | 32 | 24 | 8 |
| Thr | 29 | 29 | 0 |
| Val | 36 | 34 | 2 |
| Trp | 5 | 4 | 1 |
| Tyr | 6 | 6 | 0 |
| Glx | 0 | 0 | 0 |
| Total number | 407 | 363 | 44 |
| Molecular weight | 43,838 | 39,400 | 4,456 |

[a]The number of residues of each amino acid is derived from the nucleotide sequence of the open reading frame on BBV RNA2 (16) and from the position of the cleavage site on the coat protein precursor alpha (20) (also see Figure 15).

precursor protein $\alpha$ (16), showing the position of the cleavage between the asparagine residue at position 363 and the alanine residue at position 364, which generates $\beta$ and $\gamma$ (20), is given in Figure 15. The amino acid compositions of proteins alpha, beta, and gamma are given in Table 4.

### Major Protein

The molecular weight of the major nodaviral polypeptide ($\beta$), as estimated by SDS-polyacrylamide gel electrophoresis, varies from 31 kD in the case of GMV to 43 kD for FHV (see Table 3). The major protein of GMV is conspicuously smaller than that of the other nodaviruses (25). A value of 39 kD has been calculated for BBV protein $\beta$ from its primary structure (16) (see Figure 15). This value has been accepted as being the size of the major nodaviral polypeptide, and confirmation of the other values differing from this will have to await the sequencing of the appropriate viral genomes.

Minor Proteins

There is considerable discrepancy regarding the reported number and sizes of the minor nodaviral polypeptides (see Table 3). Only the 43 kD precursor (and an unidentified protein of 38 kD) was detected in the case of NV (8), while BBV contained minor proteins of 45 kD (precursor $\alpha$ ?) and an unidentified protein of 33 kD (39). Both GMV (52) and BoV (27) appeared to have only one minor protein, of 34 and 41 kD, respectively, with the GMV protein again being the smallest of all the nodaviral $\alpha$ proteins. No minor proteins were detected in the case of either FHV (26) or MwV (28).

Only in the case of BBV have both minor proteins, $\alpha$ and $\gamma$, as well as their relationship, been demonstrated (12) and subsequently confirmed by sequencing (16,18) and structural (20) analyses. These unequivocal demonstrations of capsid geometry and of the sizes of the BBV capsid polypeptides have led to this virus becoming the prototype of nodaviral capsid structure.

## STRUCTURE

### Particle Morphology

Electron micrographs show negatively stained nodavirions to be rather featureless, roughly spherical particles of about 30 nm diameter (Figure 1). Triangular facets are visible in the case of NV, although capsomeres are not resolved (5). In crystalline arrays, seen in thin sections of infected cells, Nodamura virions have an opaque core and a lucent surface layer (5), whereas BoV occurs as both electron-dense and electron-lucent particles (42). Negatively stained BBV reveals a hexagonal symmetry expected for an icosahedron (50).

### Crystallographic Fine Structure

The structure of the BBV capsid has been determined at 3.0 Å resolution by x-ray diffraction analysis of rhombic dodecahedral virus crystals (20,50). The protein shell is built of 180 identical protomers arranged according to $T = 3$ icosahedral symmetry, each protomer consisting of 407 amino acids (16) (see Table 4). [The term protomer refers to the coat protein precursor $\alpha$ or its two cleavage products $\beta + \gamma$. The BBV capsid consists of about 75–90% $\beta + \gamma$, the rest being uncleaved $\alpha$ (20); see "The Capsid Proteins" above.] These protomers are grouped into 60 triangular units, each forming the icosahedral asymmetric unit typical of a T = 3 lattice (55).

### Protomer Structure

The three identical polypeptide chains comprising the asymmetric unit have been traced at 3.7 Å resolution, and a model derived from the primary structure could be fitted by computer graphics to the 3.0 Å electron density map (20). The basic folding pattern of each polypeptide chain is very similar to that seen with other plant and animal virus coat proteins (56–59). Each consists of an eight stranded antiparallel $\beta$ barrel (Figure 9A), comprising residues 92 to 310. This region of each polypeptide cooperatively forms the contiguous shell of the assembled BBV capsid. Three large insertions, situated between three pairs of adjacent $\beta$ barrel strands within one polypeptide, extend outwards above the contiguous shell. A further chain excursion at the end of the eighth strand of the $\beta$ barrel extends below the contiguous surface, forming a prominent internal feature not seen on other spherical viruses.

Both the amino- (residues 1 to 64) and the carboxy-terminal (residues 345 to 407) regions are invisible in the electron density map and are therefore disordered. Since the visible amino terminus (Ala 65) lies internally, the invisible terminal region probably adapts in a random fashion to the various conformations of the RNA. This region contains 17 basic amino acids (16) capable of neutralizing nucleic acid phosphate residues. The visible carboxy terminus at Leu 344 is also internal. This internal carboxy terminus, a feature unique to the nodaviruses, is probably related to the presence of a protomer cleavage site being situated within the neighboring invisible region (see Figure 15).

### The Icosahedral Asymmetric Unit

Each of the 60 icosahedral asymmetric units consists of three protomers arranged with approximate threefold symmetry (Figure 9B). The asymmetric unit shows a prominent terraced pyramid which rises to a peak (Figure 9C), formed cooperatively by the large insertions on each of the three neighboring protomers. This is clearly visible in a cross section of the shell (Figure 9C), which shows the pyramid extending from145 to 166 Å radius. The central layer, consisting of the eight $\beta$ barrel strands and forming the contiguous shell by bonding between neighboring asymmetric units, lies at 120 to 145 Å radius. The unique internal protrusion extends down to 100 Å.

### Quaternary Structure

The assembled BBV capsid forms a rhombic triicontahedron (Figure 9D). Each asymmetric unit bonds to three neighboring units, one joint occurring across an icosahedral twofold axis and the other two across quasi-twofold axes (Figure 10B). Bonding across a twofold axis forms a dihedral angle of 180° between the

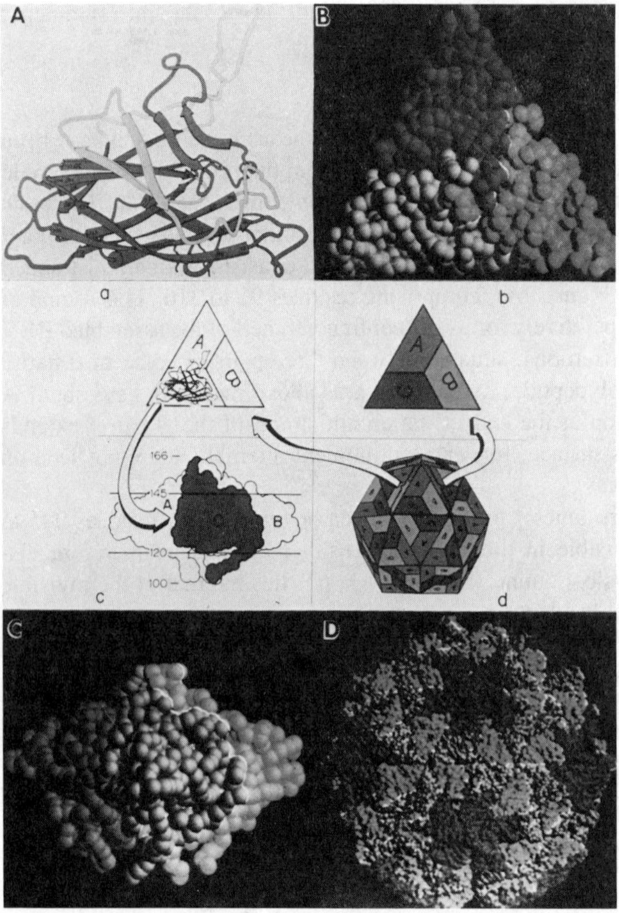


**Figure 9** Organization of BBV established from the 3.0 Å electron density map. The central portion of the figure (lowercase letters) provides expository information for the corresponding outer portion (uppercase letters). Each of the identical capsid subunits (A) consists of an eight-stranded antiparallel $\beta$ barrel. Although the chain consists of 407 amino acids, only those from residue 65–344 are visible. The subunits are organized into trimeric clusters that make up the icosahedral asymmetric unit (b,B). The three major loops of each subunit are terraced to form a prominent pyramid on the virion surface, clearly evident in a side view of the trimeric cluster (c,C). The intact virion (d,D) consists of 60 trimeric clusters arranged with icosahedral symmetry. (From Ref. 20.)

two asymmetric units, resulting in their lying in the same plane and forming a flat, diamond-shaped surface (Figure 10C). By contrast, a dihedral angle of 136° is found at a quasi-twofold axis (Figure 10D), this angle being governed by the presence or absence of an extended protein chain ("arm") inserted into the hydrophobic groove between neighboring asymmetric units. The icosahedral twofold joint is prevented from bending by the insertion of the chain, whereas at a quasi-twofold joint the chain is absent and the hydrophobic surfaces interact directly. The inserted arm [a feature common to a number of T = 3 viruses (57,58)] probably corresponds in the case of BBV to a predominantly hydrophobic region of the polypeptide between residues 376 and 407. This difference in the angles between the asymmetric units at the twofold and quasi-twofold axes largely determines the shape of the BBV shell.

## MOLECULAR BIOLOGY

The first indications that the Nodaviridae have a unique molecular biology were obtained from investigations done using NV (6–9). A severe limitation, however, was the fact that this virus does not replicate in common lines of cultured animal cells (45). A more tractable virus emerged with the demonstration that BBV grows vigorously in cultured cells of *Drosophila melanogaster* (10), as noted previously. The exceptional virus yield provided the basis for significant advances in our understanding of nodaviral molecular biology: the determination of the sizes and roles of BBV-encoded proteins (12–14,48,49,60) and of BBV RNAs (12–14,52,53), the demonstration of a viral replicase (61,62), the elucidation of BBV translational control (15), the sequencing of the entire BBV genome (16–18), and the production of infectious RNA derived by transcription of cloned BBV cDNA (19).

### The Synthesis of Virus-Encoded Proteins

Our knowledge of the nodaviral proteins synthesized during infection comes mainly from experiments done using BBV-infected *Drosophila* cells. In summary, the synthesis of viral polypeptides commences as early as 2 hr p.i. and continues for up to 48 hr p.i. (12,49). In all, five virus-specific polypeptides (including a putative replicase, a coat protein precursor, and two coat proteins) are produced as a result of infection. These are synthesized at different rates and at different times during the infection cycle (12). Although infection causes very little cell lysis, host protein synthesis declines significantly from about 12 hr p.i. on (12), eventually being almost totally suppressed (53).

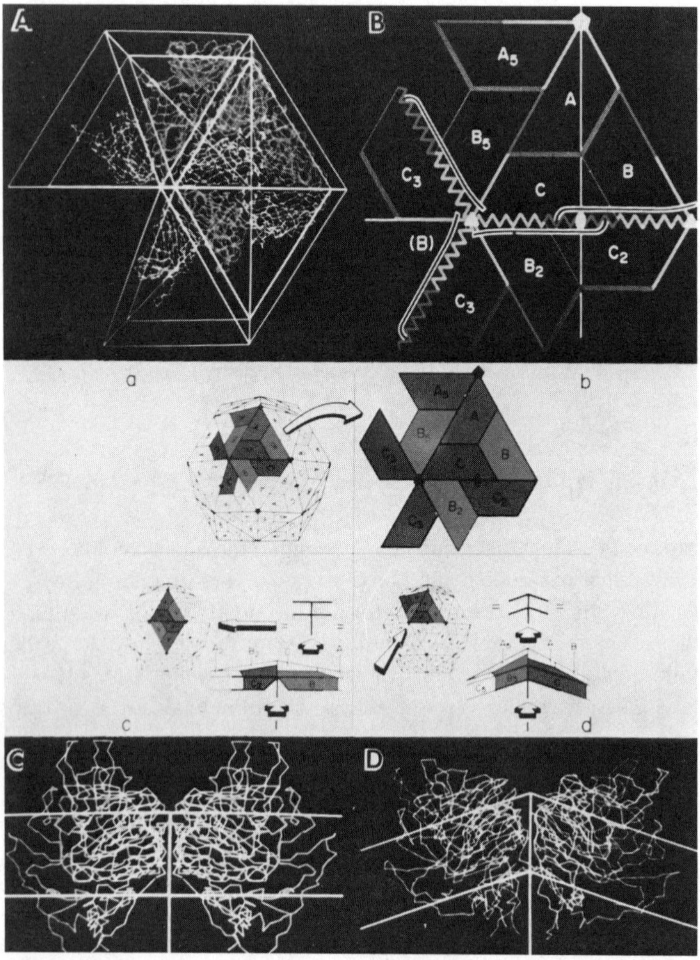


**Figure 10** Differences in bonding interactions at the icosahedral and quasi-symmetry axes of BBV. (A) and (B) show the asymmetric unit and its symmetry relationships to adjacent subunits. Icosahedral symmetry axes are indicated in (B) by white symbols: pentamer (fivefold axis), triangle (threefold), and ellipse (twofold). The difference in dimeric bonding interactions at twofold joints [sawtooth line in (B)] and at quasi-twofold joints [smooth line in (B)] is illustrated in (c,C) and (d,D), respectively. The bend contact at the quasi-twofold joint provides the curvature required to form a spherical shell. (From Ref. 20.)

## Cell-Free Translation of Nodaviral RNAs

In the absence of a susceptible cell system in which readily detectable levels of viral polypeptides could be produced, early studies on the translational strategy of nodaviral RNAs were done in cell-free extracts.

### *Efficiency as Messenger RNAs*

Nodaviral RNAs are very active messengers in cell-free systems, stimulating an incorporation of radiolabeled amino acids into acid-insoluble material that is many times the level found in the corresponding RNA-free lysates (Table 5). The RNA of BBV, for example, in reticulocyte lysate directs incorporation that is 1.6 times that directed by encephalomyocarditis virus (EMCV) RNA, an exceptionally active viral message (12); in *Drosophila* cell lysate, the incorporation is 12 times that directed by EMCV RNA (63). If the RNAs of BBV are separated, in reticulocyte lysate RNA1 stimulates about 1.5 times as much incorporation as RNA2 at saturating levels (12). Conversely, in *Drosophila* cell

**Table 5** Cell-Free Translation of Nodaviral RNAs

| RNA | Lysate type | Stimulation[a] | Saturating RNA concentration (μg RNA/ml lysate) | Reference |
|---|---|---|---|---|
| NV RNA1 | Wheat germ | — | 90 | 9 |
| | HeLa cell | 2–4 | 40 | 9 |
| NV RNA2 | Wheat germ | 10–20 | 60 | 9 |
| | HeLa cell | 2–4 | 20 | 9 |
| BBV RNA1 | *Drosophila* | 18 | 240 | 63 |
| | Reticulocyte | ~70 | 100 | 12 |
| BBV RNA2 | *Drosophila* | 27 | 240 | 63 |
| | Reticulocyte | ~50 | 100 | 12 |
| BBV RNAs 1 + 2 | *Drosophila* | 30 | — | 63 |
| | Wheat germ | — | 60 | 63 |
| | Reticulocyte | — | 80 | 63 |
| | Reticulocyte | 60–75 | 100 | 12 |
| | Reticulocyte | 5 | — | 60 |
| GMV RNAs 1 + 2 | Reticulocyte | 15 | — | 25 |

[a]The figures represent the multiplicative factor by which the indicated RNA stimulates the incorporation of radiolabeled amino acids into acid-insoluble material, relative to the incorporation measured in the corresponding RNA-free lysate.

lysate, the stimulation is 18-fold for RNA1 and 27-fold for RNA2 (63). Translational discrimination of nodaviral RNAs will be discussed further.

*Proteins Synthesized In Vitro*

The first nodaviral RNAs to be examined in cell-free translational systems were those of NV (9). Each RNA directs the synthesis of a single protein representing about 90% of the theoretical coding information. In both wheat embryo and HeLa cell extracts, RNA1 catalyzes the synthesis of a 105 kD protein, presumed to be the viral polymerase, whereas RNA2 directs the synthesis of a protein of 43 kD. Their tryptic fingerprints indicate that the two proteins have no sequences in common, but that the 43 kD protein is indistinguishable from the major 40 kD capsid protein of NV. The 43 kD protein, assumed to be the precursor of the mature coat protein, is not processed to a 40 kD product in either of the two lysates.

The foregoing cell-free translational strategy is also applicable to BBV RNA, whether in lysates of *Drosophila melanogaster* cells (63) or of rabbit reticulocytes (12,13,15,60) (Figure 11). Purified BBV RNA1 directs the synthesis of a large protein (designated A) for which molecular weight values of 104, 110, and 120 kD have been estimated (12,60,63). A number of minor bands are routinely observed in addition to A, presumably representing incomplete forms of this protein generated by premature termination of translation (12). Purified RNA2 directs the synthesis of the coat protein precursor $\alpha$, for which molecular weight estimates of 47, 46, and slightly more than 40 kD have been obtained (12,63,60). (The molecular weights of A and $\alpha$ of BBV are now known from sequencing studies to be 112,290 and 43,837, respectively.) Cell-free translation of RNA2 shows a relative absence of minor bands, suggesting that premature termination occurs less frequently with RNA2 than with RNA1 (12). Translation of equal amounts of BBV RNAs 1 and 2 results in nearly equal synthesis of both products A and $\alpha$ (Figure 11), although $\alpha$ first appears within 5–10 min, and A only within 20 min, after the addition of RNA (12).

If reticulocyte lysate is stimulated with GMV RNA, three GMV-specific proteins (A, $\alpha$, and $\beta$) are observed, with molecular weights of 100,000, 34,000, and 31,000, respectively. Protein $\alpha$ acts as a precursor to $\beta$ in the lysate. All the GMV-specified proteins have lower molecular weights than their counterparts synthesized in response to BBV RNA (25).

If the BBV RNAs to be used for cell-free translation are extracted from infected cells instead of virions, an additional third protein (B) of 10 kD is produced (13). The messenger for protein B is a 9 S subgenomic RNA that is not found in purified virions.

*Cell-Free Cleavage of the Coat Protein Precursor*

Although some workers (9,12,30) find that cleavage of $\alpha$ to $\beta$ and $\gamma$ does not occur in the lysate, in vitro cleavage has been reported for protein $\alpha$ of BBV (60)

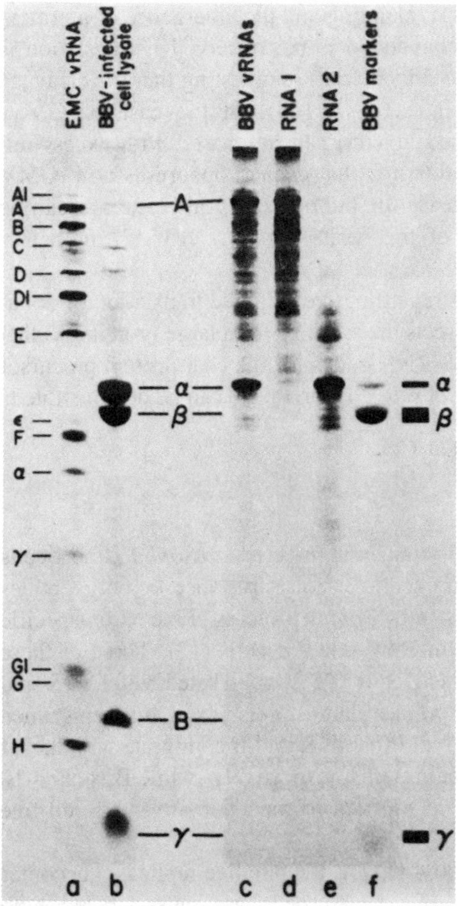


**Figure 11** Electrophoretic comparison of the in vitro and in vivo [$^{35}$S]methionine-labeled translation products of BBV RNA. Reticulocyte lysates were programmed with encephalomyocarditis RNA (lane a), total BBV virion RNA (lane c), BBV RNA1 (lane d), and BBV RNA2 (lane e). Lane b contains lysate of BBV-infected *Drosophila* cells, and lane f virion protein markers. (From Ref. 12.)

FHV (64,65), and GMV (25). The addition of excess unlabeled methionine to lysate after 90 min of translation causes the [$^{35}$S]methionine-labeled BBV capsid precursor to be chased over a 24 hr period into two proteins, one that comigrates with capsid protein and a second of about 7 kD (60). Similarly, if continued translation of either unfractionated FHV RNA or purified RNA2 is inhibited after 30 min by the addition of cycloheximide to the reticulocyte lysate, further incubation results in the concentration of $\alpha$ declining and the emergence of a protein that comigrates with $\beta$ (65). This conversion is not observed if translation is stopped by the addition of pancreatic ribonuclease, suggesting that the cleavage is dependent on the presence of intact RNA (64). In the case of GMV, [$^{35}$S]methionine-labeled $\alpha$ synthesized in vitro can be chased with excess unlabeled methionine into a smaller protein $\beta$ that coelectrophoreses with GMV capsid protein. Of a number of protease inhibitors tested, this cleavage can be partially inhibited by pretreatment of the lysate with 0.1 m*M* phenylmethyl sulfonyl fluoride only (25).

Thus a consistent picture emerges regarding the cell-free translation of nodaviral virion RNAs. The 22 S RNA directs the synthesis of a large protein (A) that is probably the polymerase. The 15 S RNA codes for the coat protein precursor $\alpha$. In some cases, in vitro processing of $\alpha$ to coat protein $\beta$ can be demonstrated.

## Viral Proteins Synthesized in Infected Cells

### *Detection of Viral Polypeptides*

The exceptionally high yield of BBV obtained in infected *Drosophila* line 1 cells (10) permitted the first detection of intracellular nodavirus-specific polypeptides against the large background of host *Drosophila* proteins. Five virus-specific proteins (Figure 11) are synthesized in BBV-infected cells (12). Three of these ($\alpha$, $\beta$, and B) are detected if infected cells are simply lysed with SDS and subjected to electrophoresis on SDS–polyacrylamide gels which are then stained with an appropriate protein-specific dye. Proteins $\alpha$ and $\beta$ comigrate with the 44 kD minor and 40 kD major BBV coat proteins, respectively, while B is an 8–10 kD protein (12, 49) found neither in purified virions nor after the cell-free translation of virion RNAs.

A fourth BBV-specific protein of 104 kD (A, the putative replicase) becomes evident, in addition to $\alpha$ and B, if infected cells are pulse-labeled with [$^{35}$S]methionine at various times after infection (12,49). These three proteins are detected by electrophoresis of the pulsed samples followed by autoradiography (12). The major coat protein $\beta$ is not detected by pulse-labeling because the cleavage that generates $\beta$ is slow relative to the short labeling periods used in this type of experiment.

The coat protein $\beta$, as well as a fifth BBV-specific protein of 5 kD ($\gamma$), are detected if the flow of radioactivity is examined by pulse-labeling infected cells during maximal coat protein synthesis (i.e., about 12 hr p.i.) and then transferring the cells to a medium containing an excess of unlabeled methionine (12). Protein $\alpha$ is seen to decrease steadily with an accompanying increase in protein $\beta$ and emergence of $\gamma$. Precursor $\alpha$ is thus cleaved to generate capsid proteins $\beta$ and $\gamma$. This latter protein is usually not detected in purified virions because of its small size, but it can be demonstrated if heavily labeled virions are electrophoresed in SDS–polyacrylamide gels of sufficiently high acrylamide concentration.

A similar pattern of five virus-induced proteins is detected in BoV-infected *Drosophila* cells (66).

*Rates of Protein Synthesis*

Unlike the situation observed in cell-free systems, these five BBV-specific proteins are not all synthesized at comparable rates in infected cells. If autoradiograms of pulsed BBV-infected cell samples are quantified by densitometry (Figure 12), protein A is seen to be the least abundant protein (12). Its synthesis peaks 4–6 hr p.i. and then declines rapidly. Protein B is synthesized in greater amounts than A, peaking about 8–10 hr p.i. The synthesis of the coat protein precursor greatly exceeds that of both A and B, still increasing at 12 hr and continuing for at least 48 hr p.i. Regarding the temporal synthesis of capsid

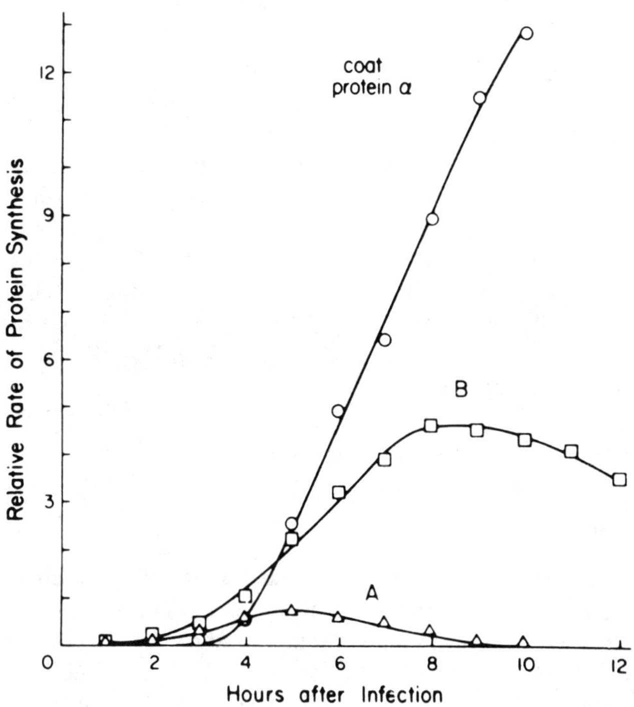


**Figure 12** Relative rates of synthesis of the major BBV-induced polypeptides in *Drosophila* cells. Autoradiograms similar to that in Figure 11 were quantified by densitometry. (From Ref. 12.)

proteins $\beta$ and $\gamma$, $\beta$ does not appear in infected cells until about 18 hr p.i. It accumulates to the same concentration as $\alpha$ by 48 hr, by which time $\gamma$ is also clearly evident. Pulse-chase analysis of infected cells indicates that the precursor $\alpha$ has a half-life of about 1.5 days (12).

This study of the BBV-specific proteins produced in vivo thus demonstrates four important aspects of the nodaviral infection cycle: (1) five virus-specific proteins are produced in virus-infected cells; (2) protein $\alpha$, the primary translation product of RNA2, is cleaved intracellularly; (3) protein B, not synthesized in cell-free extracts, is produced in large amounts in infected cells; and (4) the syntheses of proteins A, B, and $\alpha$ are under temporal and quantitative control not paralleled in cell-free systems (12). This pattern is consistent with the idea that RNA1 codes for an early protein (A, a replicase element), whereas RNA2 codes for a late protein, the capsid precursor $\alpha$ (12).

*Synthesis of Protein B off a Subgenomic RNA*

*Drosophila* cells infected with BBV synthesize abundant amounts of a protein B of about 10 kD (12,49) (Figure 11, lane b). This protein is not derived from proteins A or $\alpha$ and is not produced when cell-free extracts are programmed with either of the two viral RNAs (12,49,60). However, if RNA extracted from infected cells when synthesis of B is at a maximum [i.e., about 9 hr p.i. (12)] is used for cell-free translation, three proteins are synthesized (13). These comigrate with proteins A, $\alpha$, and B from infected cells, which indicates that an active messenger for B (as well as for A and $\alpha$) is present in BBV-infected cells. Polysomes isolated from actinomycin-treated cells between 7 and 15 hr after infection contain four size classes of mRNA: 22, 15, 11, and 9 S. The 9 S RNA programs the synthesis of protein B in cell-free extracts. The 22 and 15 S RNAs are indistinguishable from virion RNAs 1 and 2, respectively, as judged by electrophoretic mobility and in vitro translation activity, while the 11 S RNA has no demonstrable in vitro messenger activity (13).

Protein B is therefore a virus-encoded protein synthesized off a 9 S subgenomic RNA (RNA3) that has an estimated molecular weight of $0.15 \times 10^6$ (13). The origin of RNA3 is discussed later.

*Preferential Translation of RNA2*

In BBV-infected *Drosophila* cells, proteins A, B, and $\alpha$ are synthesized in markedly different amounts (12). In particular, the synthesis of protein A peaks very early in the infection cycle, and the amount synthesized is far less than that of coat protein $\alpha$. However, this low rate of intracellular synthesis of protein A is not due to the lack of competent RNA1 messenger in the cell: RNAs 1 and 2 are made in roughly equimolar proportions throughout the infection cycle (13,53), and RNA extracted from these infected cells directs vigorous, cell-free synthesis of protein A (13). As the two virion RNAs are translated in vitro with the same relative frequencies (12), the increase in protein A synthesis seen in vitro is not due to the preferential translation of RNA1 by reticulocyte ribosomes. Thus the synthesis of protein A in infected cells appears to be regulated at the level of translation.

The in vivo discrimination against RNA1 can also be demonstrated in vitro. Protein-synthesizing lysates prepared from BBV-infected *Drosophila* cells continue to translate the endogenous mRNAs, mimicking closely the in vivo production of BBV proteins (15). However, if BBV RNAs are extracted from infected cells and added to nuclease-treated *Drosophila* lysates, BBV protein A (as well as $\alpha$ and B) are synthesized in abundant amounts. This is also the case if the nuclease-treated lysate is derived from BBV-infected cells, suggesting the absence in infected cells of proteins or RNAs of viral origin capable of regulating the translation of RNA1.

The factor having the largest influence on the relative proportion of proteins A and $\alpha$ is the overall concentration of exogenous RNA added to the cell-free system. As increasing amounts of a 1 : 1 mixture of RNAs 1 and 2 are added to either rabbit reticulocyte or *Drosophila* cell lysates beyond their saturation levels, the ratio of A to $\alpha$ drops linearly (15). This phenomenon is diagnostic of translational competition between mRNAs and suggests that because of a lower intrinsic affinity, RNA1 competes less efficiently for an essential, as-yet-unidentified translational factor or factors (15). The effect is that late in infection, when the synthesis of messenger-active viral RNAs increases to very high levels (reaching 800 $\mu$g/ml in infected cells), competition for a factor or factors causes the synthesis of protein A to decline, whereas that of protein $\alpha$ continues to increase. It is thus possible to accommodate differential synthesis of those proteins required to accomplish both early RNA replication functions (protein A or B or both) and late assembly functions (coat precursor $\alpha$), while simultaneously synthesizing large equal amounts of the two genomic RNAs required for packaging.

This regulation was initially thought possibly to be due to RNA3, or its gene product (protein B), regulating the translational efficiency of RNA1 (14). This possibility seemed to be dispelled with the demonstration (discussed earlier) that the discrimination against RNA1 can be mimicked in protein-synthesizing cell lysates programmed only with BBV RNAs 1 and 2 (15). However, cells transfected with highly purified RNA1 alone make only small amounts of protein A, whereas RNA3 reaches abnormally high concentrations (14). A second mechanism for the control of protein A synthesis involving RNA3 is thus possible.

## The Roles of the Viral RNAs

### Requirement of Both Virion RNAs for Infectivity

The requirement of both virion RNAs 1 and 2 for infectivity was first demonstrated for NV (6). If the two RNAs of NV are separated by centrifugation on sucrose density gradients, each RNA alone has a very low infectivity when inoculated intracerebrally into neonatal mice. If the two RNAs are recombined, an infectivity level comparable to that of the original unfractionated RNA is regained. Dilutions of the 22 and 15 S RNAs which are noninfective kill mice when recombined before inoculation.

The situation regarding BBV RNAs was examined by transfecting *Drosophila* cells, pretreated with DEAE-dextran, with the separated virion RNAs (14). Although only about 5% of the cells become infected by this procedure, both virion RNAs are clearly required for the formation of infective virions. RNA1, however, is capable of independent replication.

## RNA Replication

### *RNA Species Synthesized in Infected Cells*

Four major species of RNA are detected in nodavirus-infected cells, whether NV-infected BHK cells (8) or BBV-infected *Drosophila* cells (53). These include the 22 and 15 S virion RNAs 1 and 2 (Figure 13), as well as two double-stranded RNA (dsRNA) species that migrate electrophoretically at the positions expected for ds forms of the two virion RNAs. This indicates that the two RNAs replicate separately. All the virus-induced RNAs are synthesized in the presence of actinomycin D, implying that the viral genome encodes a viral replicase (13).

A third single-stranded (ss) RNA arises in infected cells (Figure 13), most noticeably in cells infected with RNA1 alone (14). This is a 9 S subgenomic RNA (RNA3) derived from RNA1. A ds form of RNA3 can be isolated from BBV-infected *Drosophila* cells (17), suggesting that this RNA undergoes independent replication. In addition, a number of minor actinomycin-resistant ssRNAs of unknown origin arise in BBV-infected *Drosophila* cells (13).

Virus-specific RNA can be detected in infected cells as early as 2 hr p.i. (53) and accounts for more than 50% of total RNA synthesis within 8 hr. p.i. (13). Viral RNA synthesis reaches a maximum about 36 hr p.i., then declines gradually till at least 60 hr p.i. (13). These RNAs are actively synthesized even though virus particles are being produced, with cellular RNA synthesis being only partially inhibited (53). This inhibition could, however, be artifactual (13). The ratios of the major RNA species remains constant, with RNAs 1 and 2 being synthesized in nearly equimolar amounts and the level of viral RNAs remaining about five times that of the replicative species. The RNA species do not appear to be under any temporal control within the infected cell (53), although the production of RNA3 is inhibited by the presence of RNA2 (14).

### *Origin of the Subgenomic RNA*

The ability of RNA1 to replicate independently of RNA2 can be demonstrated by transfecting *Drosophila* cells, sensitized by pretreatment with DEAE-dextran,

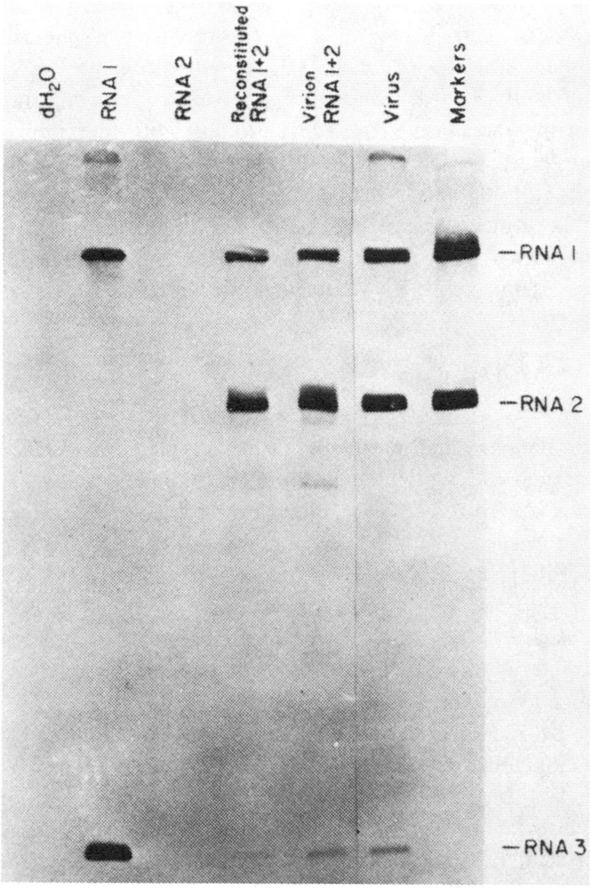


**Figure 13** Electrophoretic analysis of the actinomycin D-resistant RNAs produced in *Drosophila* cells infected with isolated BBV RNAs 1 and 2 (lanes 2–5) and with BBV virions (lane 6). Lane 7 contains marker RNAs purified from virions. (From Ref. 14.)

with highly purified preparations of separated BBV RNAs (14). Purified RNA1 stimulates the incorporation of [5-$^3$H]uridine into acid-insoluble material as actively as a mixture of RNAs 1 and 2, whereas RNA2 has no detectable activity. RNA1 is thus capable of independent replication in infected cells, indicating that RNA1 carries the viral gene(s) required for synthesis of the viral RNA polymerase. Cells infected with RNA1 also contain RNA3 (Figure 13), in a proportion much larger than is normally observed in an infection with both RNAs. Also, RNA1-infected cells make correspondingly large proportions of protein B. Thus RNA1 is the parent of subgenomic RNA3.

The amount of RNA3 synthesized in infected cells is regulated by the presence of RNA2 (14). Densitometer scans of fluorograms of the electrophoretically separated intracellular viral RNAs show that the proportion of RNA3 synthesized in RNA1-infected cells decreases progressively with increasing proportions of RNA2 in the inoculum. Thus, in addition to its function as messenger for the coat protein precursor $\alpha$, RNA2 also regulates the production of RNA3 in infected cells, a possible mechanism for which is discussed later. This is the second regulatory mechanism observed in BBV-infected cells, the other being the early shutdown of protein A synthesis due to its messenger (RNA1) competing unsuccessfully with RNA2 for host translational factors, as noted earlier.

### *Demonstration of a Viral Polymerase*

The presence of an RNA-dependent RNA polymerase activity not found in uninfected cells has been demonstrated in *Drosophila* cells infected with BBV (61). This activity is associated with particulate material that sediments at 20,000 $\times$ *g*, from which it can be solubilized by sonication in the presence of EDTA. Polymerase activity, as measured by the incorporation of [5-$^{3}$H]UTP into trichloroacetic acid–insoluble products, was not stimulated by exogenous BBV RNA and continued for 180 min at 29°C with an optimum of activity at pH 8.2. The polymerase exists as a 120 S complex with its template, synthesizing predominantly minus-strand BBV RNA. Protein A, the putative replicase, copurifies with this replication complex.

This polymerase can be rendered template-dependent by solubilizing the particulate material with detergent and destroying the endogenous template RNA with micrococcal nuclease (62). The addition of exogenous BBV RNA1 or RNA2 induces the synthesis of full-length negative-strand RNA isolated as a double-stranded complex with the added RNA. Secondary structure of the BBV RNAs is thought to be involved in replicase recognition, as there are no significant sequence homologies between these two RNAs (18). Of other RNAs tested, only brome mosaic virus (BMV) RNA catalyzes the synthesis of products similar to dsBMV RNAs. No RNAs are synthesized after the addition of RNA from southern bean mosaic virus, cowpea mosaic virus, or *Drosophila*.

## Structural and Functional Features of the Viral RNAs

The nodavirus BBV is the first insect virus of which the complete nucleotide sequence of the genome has been elucidated, using a combination of direct RNA sequencing of the 5' and 3' termini along with the chemical method of Maxam and Gilbert (67) to sequence different cDNA inserts subcloned into pBR322 (16–18). This information permits a detailed exploration of nodaviral genome organization, and the main structural and functional features are summarized in the following sections.

### *The Primary Structure of RNA1*

The only nodaviral RNA1 that has been sequenced is that of BBV (18) using strain W17 (28), a strain now recognized to be more closely related to FHV than to BBV (30). This RNA consists of 3106 bases (Figure 14) and not 3105 as originally published (18), the original sequence requiring an additional adenine following base 2688 (54). This implies a molecular weight of $1.04 \times 10^6$ for RNA1 (sodium salt).

The 5' end has the cap structure $m^7$GpppGp, while the 3' end has an unidentified moiety [probably a protein (18)] that prevents polyadenylation or ligation. The longest open reading frame in this RNA1 extends from the first AUG codon (bases 39–41) to a UGA codon at bases 3033–3035 (54). This sequence codes for a protein of 998 amino acids [and not 897, as reported (18)], with a calculated molecular weight of 112,290, which is in good agreement with previous estimates for the molecular weight of protein A. A second 5' proximal initiation codon (bases 564–566) would correspond to a protein too small to be protein A. The other two reading frames are tightly closed in the region 1 to 2700.

At position 2718 a sequence starts which corresponds to the subgenomic RNA3 (discussed later). One open reading frame exists in the RNA3 region, extending 318 bases from an AUG at bases 2737–2739. This open reading frame encodes a protein of molecular weight 11,633 corresponding to protein B that is found in infected cells (13).

### *The Primary Structure of RNA2*

The first nodaviral RNA to be sequenced was the RNA2 of BBV (16). This RNA contains 1399 bases (Figure 15), giving it a molecular weight of 0.47 $\times 10^6$.

The 5' end of RNA2 also has the cap structure $m^7$GpppGp, while the 3' end has a structure resisting polyadenylation. The first AUG codon in BBV RNA2 is situated at positions 23–25. The open reading frame following this AUG, which is terminated by a UAG at positions 1244–1246, encodes a protein of 407 amino acid residues. The calculated molecular weight of 44,000 for this protein agrees well with the estimates of 44, 46, 47, and about 40 kD obtained for the BBV coat protein precursor $\alpha$ by gel electrophoresis (10,63,12,60). A second open reading frame extends from base 1110 to 1325, coding for a putative protein of 72 amino acids. The negative strand also contains a long open reading frame corresponding to a putative protein of 267 amino acids. Computer predictions suggest that RNA2 probably has an exceptionally stable and distinctive secondary structure in solution, the 3' end folding into a single long imperfect hairpin and the 5' end folding into an intricate multiply bifurcated arrangement of stems and loops (16).

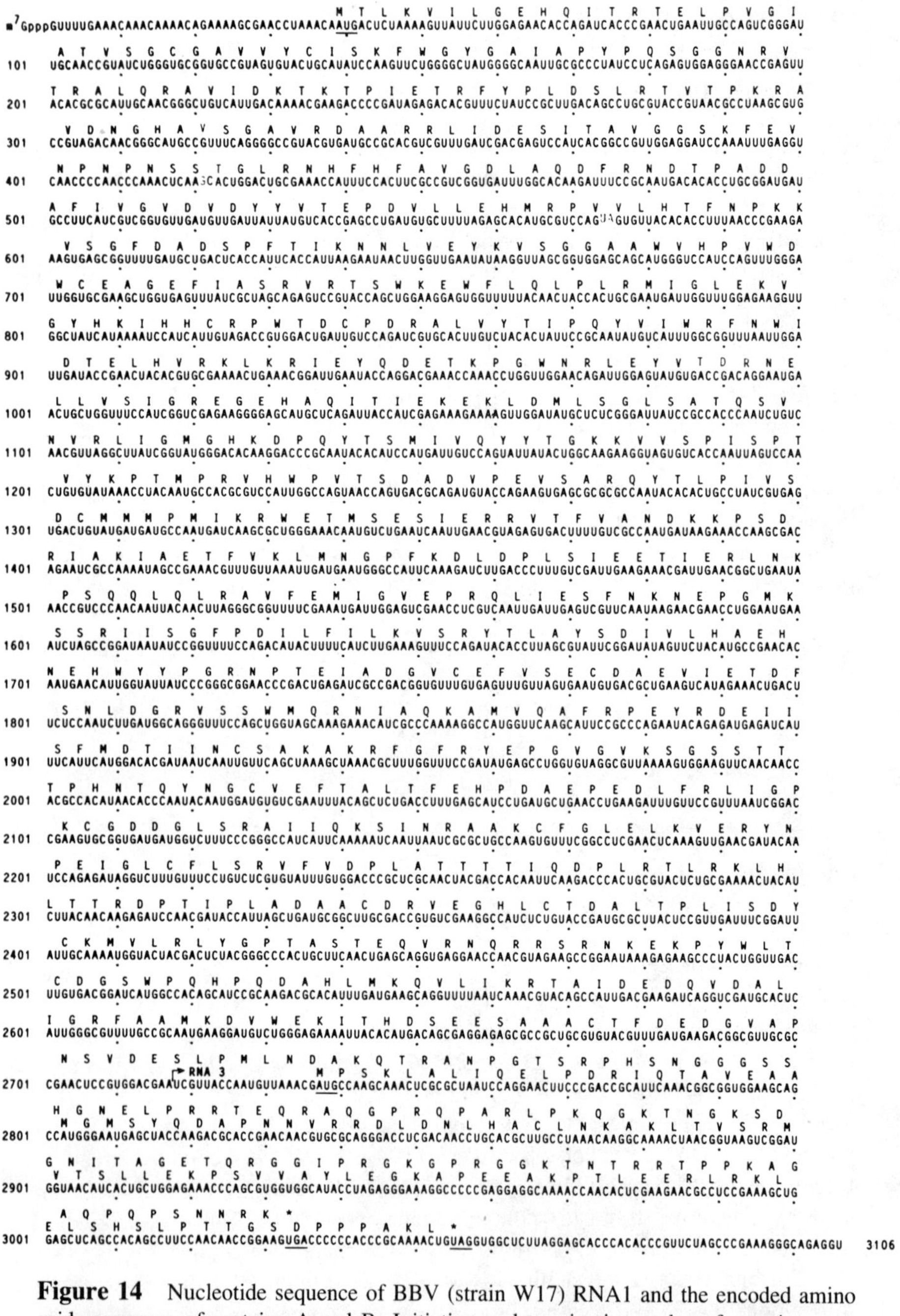

**Figure 14** Nucleotide sequence of BBV (strain W17) RNA1 and the encoded amino acid sequences of proteins A and B. Initiation and termination codons for major open reading frames are underlined, asterisks denoting termination codons. (From Ref. 18.)

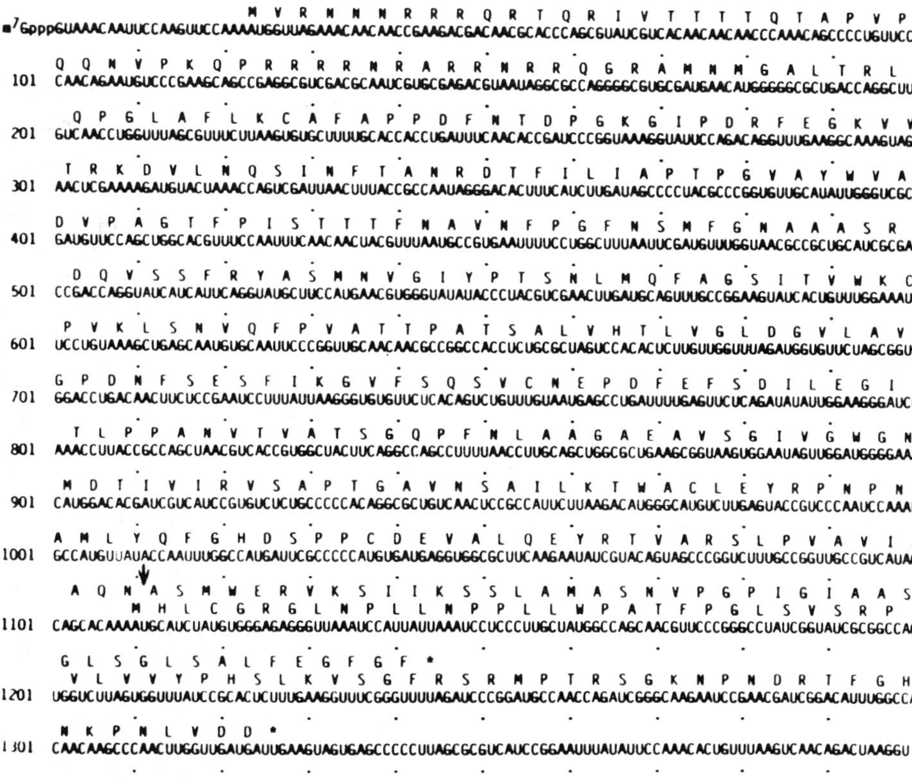

**Figure 15** Nucleotide sequence of BBV RNA2 and the encoded amino acid sequence of the viral coat protein precursor. The arrow indicates the site of the cleavage that generates the two capsid proteins. Asterisks show the termination codons for the two open reading frames. (From Ref. 16.)

### *The Primary Structure of RNA3*

The sequence of the subgenomic RNA3 consists of 389 bases (Figure 16) and is homologous to that of the 3'-terminal region of RNA1 extending from position 2718 (17). The molecular weight of RNA3 (sodium salt) is $0.13 \times 10^6$.

As found with RNAs 1 and 2, the 5' terminus of RNA3 is capped. One open reading frame is present, starting at an AUG at positions 20–22 and extending to a UAG at positions 338–340. The computer-generated secondary structure of RNA3 consists of nine stem and loop regions connected by either ss or ds segments (17).

```
Prot B                          M  P  S  K  L  A  L  I  Q  E  L  P  D  R
RNA      GpppUCGUUACCAAUGUUAAACGAUGCCAAGCAAACUCGCGCUAAUCCAGGAACUUCCCGACCG   60

Prot B    I  Q  T  A  V  E  A  A  M  G  M  S  Y  Q  D  A  P  N  N  V
RNA      CAUUCAAACGGCGGUGGAAGCAGCCAUGGGAAUGAGCUACCAAGACGCACCGAACAACGU   120

Prot B    R  R  D  L  D  N  L  H  A  C  L  N  K  A  K  L  T  V  S  R
RNA      GCGCAGGGACCUCGACAACCUGCACGCUUGCCUAAACAAGGCAAAACUAACGGUAAGUCG   180

Prot B    M  V  T  S  L  L  E  K  P  S  V  V  A  Y  L  E  G  K  A  P
RNA      GAUGGUAACAUCACUGCUGGAGAAACCCAGCGUGGUGGCAUACCUAGAGGGAAAGGCCCC   240

Prot B    E  E  A  K  P  T  L  E  E  R  L  R  K  L  E  L  S  H  S  L
RNA      CGAGGAGGCAAAACCAACACUCGAAGAACGCCUCCGAAAGCUGGAGCUCAGCCACAGCCU   300

Prot B    P  T  T  G  S  D  P  P  P  A  K  L  *
RNA      UCCAACAACCGGAAGUGACCCCCCACCCGCAAAACUGUAGGUGGCUCUUAGGAGCACCCA   360

RNA      CACCCGUUCUAGCCCGAAAGGGCAGAGGU   389
```

**Figure 16** Nucleotide sequence of BBV RNA3 and the amino acid sequence of protein B. (From Ref. 17.)

### *Mechanisms Regulating RNA and Protein Production*

Apart from revealing the open reading frames that specify the BBV proteins detected in infected cells, the genomic primary structure also provides possible explanations for some of the regulatory phenomena observed in vivo.

In infected cells, the total synthesis of BBV protein A is far less than that of protein $\alpha$, even though RNAs 1 and 2 exist in roughly equimolar amounts. Although BBV protein production may be regulated at various levels, regulation at the level of initiation of translation plays a major role (15). Both RNA1 and RNA2 have short leader sequences (38 and 22 bases, respectively) containing about 50% A, each with an A in the –3 position; only RNA2 has a G in the +4 position as well. The translational discrimination could thus be due to RNA2 more closely meeting the requirements favoring initiation of translation of eukaryotic mRNAs, namely, an A three nucleotides before and a G immediately after the AUG initiation codon (68). Also, the leader sequence of RNA1 can be folded into a stem and loop structure, which could impair initiation of translation (18).

The synthesis of subgenomic RNA3 from RNA1 in infected cells is drastically

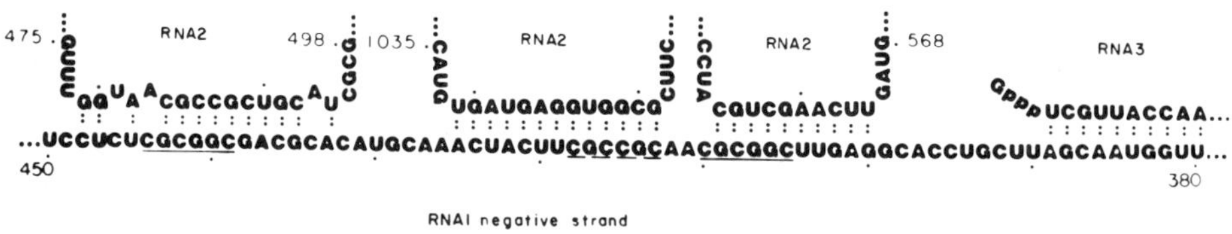

**Figure 17** Complementary regions between the RNA2 positive strand and the RNA1 negative strand. In RNA1 sequences, those above the solid lines are direct repeats, and that above the broken line is complementary to a direct repeat. (From Ref. 18.)

curtailed by the presence of RNA2. This inhibitory effect is insensitive to cycloheximide, suggesting a direct RNA–RNA interaction rather than one involving a translation product. Three regions on RNA2 are able to base pair with three complementary regions situated adjacent to each other on the negative-strand RNA1 just prior to the start of the RNA3 sequence (Figure 17). The three adjacent regions on negative-strand RNA1 also contain direct repeats able to form a stable stem and loop structure, possibly acting as a recognition site for the replicase (16). Assuming that RNA3 is derived by partial transcription of the genomic negative strand, the base pairing with RNA2 could impair replicase recognition and inhibit RNA3 production.

An unexpected finding is the limited sequence homology between the 3' ends of RNAs 1 and 2, as well as the apparent absence of 3' secondary structure homologies (18). A possible recognition site for replicase has thus not been identified.

## RNA Encapsidation and Virion Maturation

A puzzling feature of nodaviral replication is the mechanism whereby an RNA1 and an RNA2 molecule are selected for encapsidation within the same virion. One proposal is that the two RNAs could be linked by a protein molecule (8).

The foregoing sequence studies indicate that the two RNAs do not possess similar sequence regions serving as sites for initiation of encapsidation on each molecule. However, RNA1 and RNA2 are able to interact with each other via complementary regions (Figure 18) involving 16 out of 17 base pairs (18). Thus encapsidation of this RNA1–RNA2 complex would require only a single recognition site for coat protein.

Experiments employing FHV show that assembly starts with the rapid incorporation of coat protein precursor α into detergent-labile 140 S provirions (30). Analysis of FHV-infected *Drosophila* cell extracts, sedimented on sucrose gradients immediately after a 1 hr labeling period, indicates that about half of the newly synthesized protein cosediments with mature virions. This protein is

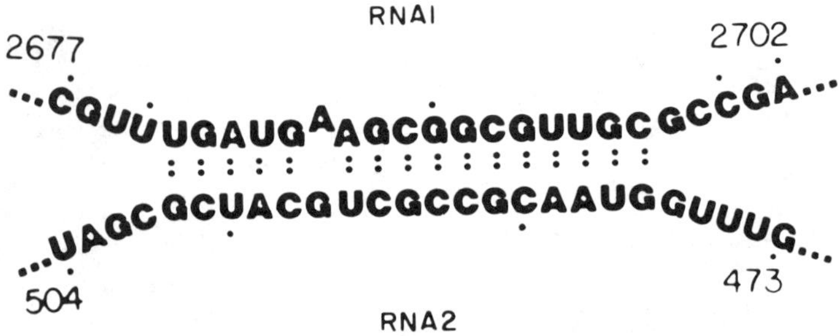


**Figure 18** Possible complementarity between RNA1 and RNA2 sequences. (From Ref. 18.)

predominantly in the uncleaved form but upon incubation is cleaved in the virions to $\beta$ and $\gamma$ chains. By contrast, no cleavage occurs after incubation of the free unassembled $\alpha$ in the gradient. Thus cleavage of $\alpha$ into $\beta$ and $\gamma$ requires assembly into provirions. Provirion formation is rapid, the availability of protein $\alpha$ and not of the virion RNAs being the limiting factor. In contrast, the maturation cleavage of $\alpha$ is slow, having a half-life of 3.6 hr. After this maturation cleavage, the virion acquires a stability to detergents and urea not possessed by the provirion.

Protease inhibitors, as well as antibody binding to the virus surface, do not inhibit cleavage (30). In addition, purified virions cleave as rapidly as those in cytoplasmic extracts. This is consistent with an unaccessible cleavage site buried deep within virions. It is also consistent with the deduction from sequencing and structural studies (20) that the site of cleavage, known to occur between an asparagine residue at position 363 and an alanine residue at position 364 of the $\alpha$ chain, is in a region of the $\alpha$ chain that lies near the RNA core of the virion. The cleavage of protein $\alpha$ thus appears to be autocatalytic, possibly requiring the involvement of a juxtaposed virion RNA base capable of cleaving peptide bonds by a proton-abstracting mechanism, as has been proposed for the maturation cleavage of picornaviruses (69). This postulated ability of RNA to mediate peptide bond cleavage could possibly explain the RNA-dependent cleavage observed in cell-free lysates, although cleavage of $\alpha$ in reassembled RNA-free capsids has been observed (72).

A possible clue to the mechanism of virus uncoating is the observation (30) that SDS-treated FH virions lose protein $\gamma$ to generate a slower sedimenting, possibly swollen, particle. The loss of protein $\gamma$ might thus be a stage in virus uncoating, analogous to the loss of VP-4 from picornaviruses (70). The cleavage of protein $\alpha$ may therefore confer stability while at the same time providing a mechanism for concerted virion disassembly (30).

Strategy of the Nodaviral Genome

The most important features of nodaviral gene function and morphogenesis are summarized in Figure 19 (30). Not illustrated are the regulatory effects that RNA2 has on RNA1, influencing the production of both protein A and RNA3.

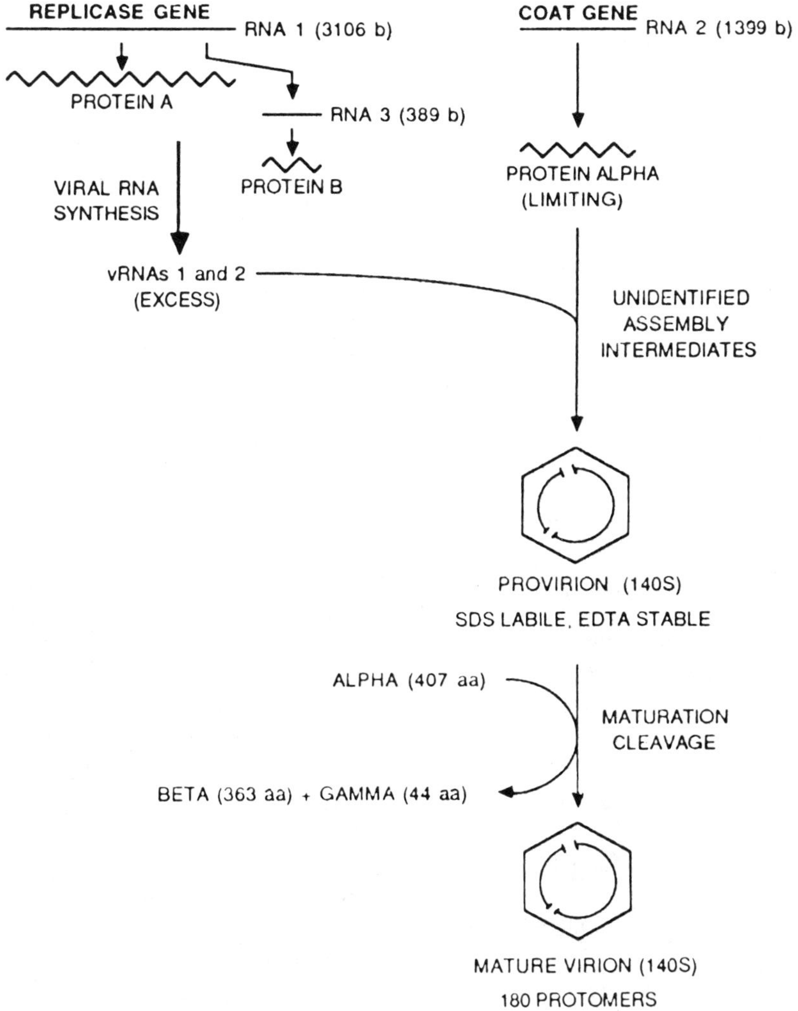


**Figure 19** Diagram illustrating our current understanding of nodaviral gene function and morphogenesis. (From Ref. 30.)

## Genetics

### Genomic Reassortment

The first investigations of nodaviral genetics were attempts to make hybrid viruses by transfecting *Drosophila* cells with mixtures of RNAs obtained from different nodaviruses (14). Although NV virions do not infect *Drosophila* cells, transfection with NV RNAs results in the production of three RNAs similar to those produced by BBV infection (suggesting that *Drosophila* cells lack exposed Nodamura viral receptors). Transfecting cells with BBV RNA1 and NV RNA2 results in the synthesis of only BBV RNAs 1 and 3, suggesting that the BBV polymerase is unable to recognize the heterologous RNA2.

The ability to plaque-purify BBV has made experiments with reassortant viruses feasible. Infectious virions can be made by transfecting *Drosophila* cells with RNA1 from BBV and RNA2 from FHV, and vice versa (11). Similarly, infectious reassortants can be made using RNAs isolated from the relatively noncytolytic wild-type BBV and the adapted W17 strain. The plaque technique now makes possible the use of assortants to pinpoint which of the two genomic RNAs carries traits such as plaque morphology and sensitivity to drugs and temperature. In this regard, it is of interest to note that the addition of 20 nonviral bases to the 5' terminus of genetically manipulated RNA1, but not of RNA2, causes a reduction in plaque size (19).

### Manipulation of the Nodaviral Genome

A new dimension was added to the study of nodaviral genetics by the demonstration that RNA transcripts, derived from cloned cDNA copies of the genomic RNAs of BBV-W17 (=FHV), are infectious to cultured *Drosophila* cells (19). This is the first demonstration of infectious transcript RNA for an insect or animal virus, and it opens the nodaviral genome to modification and analysis by recombinant DNA techniques.

Complementary DNAs (cDNAs), prepared by reverse transcription of the virion RNAs and converted to the ds form, are inserted into pSP64 or pPM1 transcription vectors, cloned, and transcribed with SP6 RNA polymerase (19). This procedure produces complete copies of FHV genomic RNAs. Individual transcripts have about 10% of the infectivity of the corresponding virion RNA, probably due to the presence of additional nonviral nucleotides (the same four at all their 3' ends and the same 20 at all the 5' ends). Progeny virus resulting from transcript infection is phenotypically indistinguishable from the original progenitor virus used to generate the cDNAs. Although the 5' nonviral transcript sequences are not reproduced in progeny viral RNA, evidently their presence does not fatally flaw either of the transcript RNAs. The absence of a 5' capping group on RNA1 demonstrably reduces infectivity, however, as do the additional 3' nonviral bases on RNA2.

Plasmids have been constructed with the 5' noncoding region of the BBV genome linked to a segment coding for the 3C protease of Coxsackie virus (71). Transcripts derived from these plasmids contain the ribosome binding sequences and initiator AUG of BBV as well as direct the efficient synthesis of biologically active protease in cell-free systems. This indicates that the BBV translation initiation signals can be used for the expression of non-BBV proteins.

This ability to manipulate nodaviral genes, as well as the availability of both a well-characterized host system (10) and a reliable plaque assay (11), provides a powerful means of investigating nodaviral functions. Also, as RNA1 is capable of independent replication (14), replicative functions can now be examined in the absence of other RNAs or in the presence of either RNA2 suitably modified to elucidate coat protein functions or of foreign RNA engineered to express new proteins in the host cells.

## CONCLUSION

The Nodaviridae comprise a small family of simple, unique viruses, possibly representing a primitive stage in ribovirus evolution from cellular genes (20). This simplicity has been instrumental in the rapid advances made in our understanding of the structure and molecular biology of certain members of this group. A number of questions still have to be answered, however, and these should prove amenable to molecular analysis, particularly now that the genome can be manipulated (19). Thus the role of protein B remains to be elucidated: could it be a replicase subunit, or perhaps be involved in some way with cell-to-cell movement of the virus? Similarly, although protein A is almost certainly the replicase (or a part thereof), replicase activity has not been conclusively demonstrated for this protein. Also, identification of both the trigger and the mechanism for the cleavage of coat protein precursor $\alpha$ into $\beta$ and $\gamma$ is within our grasp, now that the cleavage has been demonstrated to occur in purified FH provirions (30).

A start has also been made toward dissecting the control mechanisms employed by these bipartite viruses. The regulation of protein A synthesis is only partially understood, and identifying the translational factor(s) involved in this regulation will undoubtedly be facilitated by the demonstration that it (they) is abundant in rabbit reticulocyte lysate (15). The involvement, if any, of RNA3 or its product in this regulation needs to be clarified. Likewise, the mechanism proposed (18) for the inhibitory effect of RNA2 on RNA3 synthesis needs to be confirmed experimentally, for example, by site-directed mutagenesis. A similar approach could feasibly be used to identify the internal signal on minus-strand RNA1 that serves for the initiation of RNA3 synthesis or to locate the region on the NV genome governing the ability of this virus to recognize receptors on mammalian cells.

The relative sophistication of our understanding of nodavirus molecular biology contrasts rather starkly with our dearth of knowledge concerning the basic biology, ecology, and distribution of these viruses. This state of affairs is due mainly to nodaviruses having no discernible economic importance. Their distribution is apparently limited to Australasia, but we know very little of their seasonal occurrence, natural reservoirs, or modes of transmission. The advances in our understanding of nodaviral molecular biology may well stimulate, and even provide the means for, further investigations in basic nodaviral biology.

## ACKNOWLEDGMENTS

I would like to thank Professors John Newman, Roland Rueckert, and Paul Kaesberg for critically reading the manuscript and for helpful comments. Professors Jack Johnson and Edouard Kurstak and Drs. Tom Gallagher, John Bashiruddin, and Paul Scotti kindly provided photographs, unpublished manuscripts, and data. I am grateful to Shirley Pinchuk of the Rhodes University Electron Microscopy Unit for preparing some of the figures.

## REFERENCES

1. Matthews, R. E. F. Classification and nomenclature of viruses. Third report of the International Committee on Taxonomy of Viruses. *Intervirology* 17, 1, 1982.
2. Bruening, G. Plant covirus systems: Two component systems, in *Comprehensive Virology*, vol. 11, Fraenkel-Conrat, H. and Wagner, R. R., Eds., Plenum Press, New York, 1977, p. 2.
3. Scherer, W. F. and Hurlbut, H. S. Nodamura virus from Japan: A new and unusual arbovirus resistant to diethyl ether and chloroform. *Am. J. Epidemiol.* 86, 271, 1967.
4. Scherer, W. F., Verna, J. E., and Richter, G. W. Nodamura virus, an ether- and chloroform-resistant arbovirus from Japan. *Am. J. Tropical Med. Hyg.* 17, 120, 1968.
5. Murphy, F. A., Scherer, W. F., Harrison, A. K., Dunne, H. W., and Gary, G. W. Characterization of Nodamura virus, an arthropod transmissible picornavirus. *Virology* 40, 1008, 1970.
6. Newman, J. F. E. and Brown, F. Evidence for a divided genome in Nodamura virus, an arthropod-borne picornavirus. *J. Gen. Virol.* 21, 371, 1973.
7. Newman, J. F. E. and Brown, F. Absence of poly(A) from the infective RNA of Nodamura virus. *J. Gen. Virol.* 30, 137, 1976.
8. Newman, J. F. E. and Brown, F. Further physicochemical characterization of Nodamura virus. Evidence that the divided genome occurs in a single component. *J. Gen. Virol.* 38, 83, 1977.
9. Newman, J. F. E., Matthews, T., Omilianowski, D. R., Salerno, T., Kaesberg, P., and Rueckert, R. In vitro translation of the two RNAs of Nodamura virus, a novel mammalian virus with a divided genome. *J. Virol.* 25, 78, 1978.

10. Friesen, P., Scotti, P., Longworth, J., and Rueckert, R. Black beetle virus: Propagation in *Drosophila* line 1 cells and an infection-resistant subline carrying endogenous black beetle virus-related particles. *J. Virol.* 35, 741, 1980.
11. Selling, B. H. and Rueckert, R. R. Plaque assay for black beetle virus. *J. Virol.* 51, 251, 1984.
12. Friesen, P. D. and Rueckert, R. R. Synthesis of black beetle virus proteins in cultured *Drosophila* cells: Differential expression of RNAs 1 and 2. *J. Virol.* 37, 876, 1981.
13. Friesen, P. D. and Rueckert, R. R. Black beetle virus: Messenger for protein B is a subgenomic viral RNA. *J. Virol.* 42, 986, 1982.
14. Gallagher, T. M., Friesen, P. D., and Rueckert, R. R. Autonomous replication and expression of RNA 1 from black beetle virus. *J. Virol.* 46, 481, 1983.
15. Friesen, P. D. and Rueckert, R. R. Early and late functions in a bipartite RNA virus: Evidence for translational control by competition between viral mRNAs. *J. Virol.* 49, 116, 1984.
16. Dasgupta, R., Ghosh, A., Dasmahapatra, B., Guarino, L. A., and Kaesberg, P. Primary and secondary structure of black beetle virus RNA2, the genomic messenger for BBV coat protein precursor. *Nucl. Acids Res.* 12, 7215, 1984.
17. Guarino, L. A., Ghosh, A., Dasmahapatra, B., Dasgupta, R., and Kaesberg, P. Sequence of the black beetle virus subgenomic RNA and its location in the viral genome. *Virology* 139, 199, 1984.
18. Dasmahapatra, B., Dasgupta, R., Ghosh, A., and Kaesberg, P. Structure of the black beetle virus genome and its functional implications. *J. Mol. Biol.* 182, 183, 1985.
19. Dasmahapatra, B., Dasgupta, R., Saunders, K., Selling, B., Gallagher, T., and Kaesberg, P. Infectious RNA derived by transcription from cloned cDNA copies of the genomic RNA of an insect virus. *Proc. Natl. Acad. Sci. U.S.A.* 83, 63, 1986.
20. Hosur, M. V., Schmidt, T., Tucker, R. C., Johnson, J. E., Gallagher, T. M., Selling, B. H., and Rueckert, R. R. Structure of an insect virus at 3.0 Å resolution. *Proteins: Structure, Function and Genetics* 2, 167, 1987.
21. Longworth, J. F. Small isometric viruses of invertebrates. *Adv. Virus Res.* 23, 103, 1978.
22. Moore, N. F. and Tinsley, T. W. The small RNA viruses of insects: Brief review. *Arch. Virol.* 72, 229, 1982.
23. Reavy, B. and Moore, N. F. Replication of small RNA containing viruses of insects. *Microbiologica* 5, 63, 1982.
24. Moore, N. F., Reavy, B., and King, L. A. General characteristics, gene organization and expression of small RNA viruses of insects. *J. Gen. Virol.* 66, 647, 1985.
25. Reavy, B., Crump, W. A. L., Greenwood, L. K., and Moore, N. F. Characterization of a bipartite genome ribovirus isolated from *Lymantria ninayi*. *Microbiologica* 5, 305, 1982.
26. Scotti, P. D., Dearing, S., and Mossop, D. W. Flock House virus: A nodavirus isolated from *Costelytra zealandica* (White) (Coleoptera: Scarabaeidae).. *Arch. Virol.* 75, 181, 1983.
27. Reinganum, C., Bashiruddin, J. B., and Cross, G. F. Boolarra virus: A member of the Nodaviridae isolated from *Oncopera intricoides* (Lepidoptera: Hepialidae). *Intervirology* 24, 10, 1985.

28. Scotti, P. and Fredericksen, S. Manawatu virus: A nodavirus isolated from *Costelytra zealandica* (White) (Coleoptera: Scarabaeidae). *Arch. Virol.* 97, 85, 1987.
29. Rueckert, R. R. Personal communication, 1987.
30. Gallagher, T. M. and Rueckert, R. R. Assembly-dependent maturation cleavage in provirions of a small icosahedral insect ribovirus. *J. Virol.* 62, 3399, 1988.
31. Selling, B. H. Personal communication, 1987.

31.a. Wigley, P. J. and Miln, A. J. The distribution and ecology of Flock House virus in *Costelytra zealandica* (Coleoptera: Scarabaeidae), *Proc. 3rd Australasian Conf. Grassl. Invert. Ecol.*, Adelaide, November 30–December 4, 1981. p. 155.

32. Miura, Y., Inaba, Y., Hayashi, S., Takahashi, E., and Matumoto, M. A survey of antibodies to arthropod-borne viruses in Japanese cattle. *Vet. Microbiol.* 5, 277, 1980.
33. Longworth, J. F. and Archibald, R. D. A virus of black beetle, *Heteronychus arator* (F.) (Coleoptera: Scarabaeidae). *N. Z. J. Zool.* 2, 233, 1975.
34. Dearing, S. C., Scotti, P. D., Wigley, P. J., Dhana, S. D. A small RNA virus isolated from the grass grub, *Costelytra zealandica* (Coleoptera: Scarabaeidae), *N. Z. J. Zool.* 7, 267, 1980.
35. Greenwood, L. K. and Moore, N. F. The purification and partial characterization of a small RNA-virus from *Lymantria*: The identification of a Nodamura-like virus. *Microbiologica* 5, 49, 1982.
36. Tesh, R. B. Infectivity and pathogenicity of Nodamura virus for mosquitoes. *J. Gen. Virol.* 48, 177, 1980.
37. Bailey, L. and Scott, H. A. The pathogenicity of Nodamura virus for insects. *Nature (London)* 241, 545, 1973.
38. Garzon, S., Charpentier, G., and Kurstak, E. Morphogenesis of the Nodamura virus in the larvae of the lepidopteran *Galleria mellonella* (L.). *Arch. Virol.* 56, 61, 1978.
39. Longworth, J. F. and Carey, G. P. A small RNA virus with a divided genome from *Heteronychus arator* (F.) (Coleoptera: Scarabaeidae). *J. Gen. Virol.* 33, 31, 1976.
40. Mattern, C. F. T. and Daniel, W. A. Replication of poliovirus in HeLa cells: Electron microscopic observations. *Virology* 26, 646, 1965.
41. Dales, S., Eggers, H. J., Tamm, I., and Palade, G. E. Electron microscopic study of the formation of poliovirus. *Virology* 26, 379, 1965.
42. Bashiruddin, J. B. and Cross, G. F. Boolarra virus: Ultrastructure of intracytoplasmic virus formation in cultured *Drosophila* cells. *J. Invertebr. Pathol.* 49, 303, 1987.
43. Godman, G. C. The cytopathology of enteroviral infection. *Intern. Rev. Exp. Pathol.* 5, 67, 1966.
44. Crawford, A. M., Scotti, P., Sheehan, C., and Fredericksen, S. Replication of two coleopteran nodaviruses in the coleopteran cell line DSIR-HA-1179 from *Heteronychus arator* and the dipteran cell line from *Drosophila melanogaster* line 1. *N. Z. J. Zool.* 11, 93, 1984.
45. Bailey, L., Newman, J. F. E., and Porterfield, J. S. The multiplication of Nodamura virus in insect and mammalian cell cultures. *J. Gen. Virol.* 26, 15, 1975.
46. Schneider, I. Cell lines derived from late embryonic stages of *Drosophila melanogaster*. *J. Embryol. Exp. Morphol.* 27, 353, 1972.

47. Crawford, A. M. A coleopteran cell line derived from *Heteronychus arator* (Coleoptera: Scarabaeidae). *In Vitro* 18, 813, 1982.
48. Crump, W. A. L. and Moore, N. F. The proteins induced in black beetle virus-infected *Drosophila* cells at elevated temperature. *Ann. Virol.* 132 E, 495, 1981.
49. Crump, W. A. L. and Moore, N. F. The polypeptides induced in *Drosophila* cells by a virus of *Heteronychus arator*. *J. Gen. Virol.* 52, 173, 1981.
50. Hosur, M. V., Schmidt, T., Tucker, R. C., Johnson, J. E., Selling, B. H., and Ruckert, R. R. Black beetle virus: Crystallization and particle symmetry. *Virology* 133, 119, 1984.
51. Scherer, W. F. Variable results of sodium deoxycholate tests of Nodamura virus, an ether and chloroform resistant arbovirus. *Proc. Soc. Exp. Biol. Med.* 129, 194, 1968.
52. Clewley, J. P., Crump, W. A. L., Avery, R. G., and Moore, N. F. Two unique RNA species of the nodavirus black beetle virus. *J. Virol.* 44, 767, 1982.
53. Crump, W. A. L., Reavy, B., and Moore, N. F. Intracellular RNA expressed in black beetle virus-infected *Drosophila* cells. *J. Gen. Virol.* 64, 717, 1983.
54. Kaesberg, P. Organization of bipartite insect virus genomes: The genome of black beetle virus, in *The Molecular Biology of the Positive Strand RNA Viruses*, Rowlands, D. J., Mayo, M. A., and Mahy, B. W. J., Eds., Academic Press, London, 1987, p. 207.
55. Caspar, D. L. D. and Klug, A. Physical principles in the construction of regular viruses. *Cold Spring Harbor Symp. Quant. Biol.* 27, 1, 1962.
56. Rossmann, M. G., Arnold, E., Erickson, J. W., Frankenberger, E. A., Griffith, J. P., Hecht, H. J., Johnson, J. E., Kamer, G., Luo, M., Mosser, A. G., Rueckert, R. R., Sherry, B., and Vriend, G. Structure of a human common cold virus and functional relationship to other picornaviruses. *Nature* (*London*) 317, 145, 1985.
57. Abad-Zapatero, C., Abdel-Meguid, S. S., Johnson, J. E., Leslie, A. G. W., Rayment, I., Rossmann, M. G., Suck, D., and Tsukihara, T., Structure of southern bean mosaic virus at 2.8 A resolution. *Nature (London)* 286, 33, 1980.
58. Harrison, S., Olson, A. J., Schutt, C. E., Winkler, F. K., and Bricogne, G., Tomato bushy stunt virus at 2.9 A resolution. *Nature (London)* 276, 368, 1978.
59. Hogle, J., Chow, M., and Filman, J., Three-dimensional structure of poliovirus at 2.9 A resolution. *Science* 229, 1358, 1985.
60. Crump, W. A. L. and Moore, N. F., In vivo and in vitro synthesis of the proteins expressed by the RNA of black beetle virus. *Arch. Virol.* 69, 131, 1981.
61. Guarino, L. A. and Kaesberg, P., Isolation and characterization of an RNA-dependent RNA polymerase from black beetle virus-infected *Drosophila melanogaster* cells. *J. Virol.* 40, 379, 1981.
62. Saunders, K. and Kaesberg, P., Template-dependent RNA polymerase from black beetle virus-infected *Drosophila melanogaster* cells. *Virology*, 147, 373, 1985.
63. Guarino, L. A., Hruby, D. E., Ball, A., and Kaesberg, P., Translation of black beetle virus RNA and heterologous viral RNAs in cell-free lysates derived from *Drosophila melanogaster*. *J. Virol.* 37, 500, 1981.
64. Hendry, D. A., Blackhurst, D., and Rueckert, R., Cell-free translational behaviour of nodaviral RNAs, in *Fundamental and Applied Aspects of Invertebrate Pathology*, Samson, R. A., Vlak, J. M., and Peters, D., Eds., Foundation Fourth International Colloquium Invertebrate Pathology, Wageningen, 1986, p. 57.

65. Blackhurst, D. and Hendry, D. A., Cell-free processing of black beetle virus coat protein, in *Proc. VII International Congress of Virology,* Edmonton, August 9–14, 1987, p. 326.
66. Bashiruddin, J. B. and Martin, J. L. Proteins induced by Boolarra virus in *Drosophila melanogaster* line 1 cells. *Intervirology* 28, 122, 1987.
67. Maxam, A. M. and Gilbert, W. A new method of sequencing DNA. *Proc. Natl. Acad. Sci. U.S.A.* 74, 560, 1977.
68. Kozak, M. Possible role of flanking nucleotides in recognition of the AUG initiator codon by eukaryotic ribosomes. *Nucl. Acids Res.* 9, 5233, 1981.
69. Arnold, E., Luo, M., Vriend, G., Rossmann, M. G., Palmenberg, A. C., Parks, G. D., Nicklin, M. J., and Wimmer, E. Implications of the picornavirus capsid structure for polyprotein processing. *Proc. Natl. Acad. Sci. U.S.A.* 84, 21, 1987.
70. Koch, F. and Koch, G. *The Molecular Biology of Poliovirus*, Springer-Verlag, New York, 1985.
71. Dasmahapatra, B., Rozhon, E., and Schwartz, J. Cell-free expression of coxsackievirus 3C protease from a cDNA clone which contains the translational initiation signal of an insect virus, *Proc. VIIth Int. Congr. Virol.*, Edmonton, Alberta, August 9–14, 1987, p. 135.
72. Gallagher, T. Personal Communication, 1988.
73. Selling, B., Allison, R., and Kaesberg, P. Synthesis of insect viruses in plants, in Kaesberg, P., Keynote Lecture, *2nd Int. Symp. Positive Strand RNA Viruses*, Vienna, June 26–30, 1989.

# 9

# The *Nudaurelia* β Family of Insect Viruses

NORMAN F. MOORE
*Ministry of Defence, London, England*

## INTRODUCTION

The *Nudaurelia* β family of viruses is a small group all isolated from pest species of Lepidoptera, principally from Saturniid, Limacodid, and Noctuid moths. Although viruses in this group are important in that they naturally regulate a number of pest insects in the field, relatively little work has been performed on them because they are exotic to the majority of insect virus laboratories and there are no continuous tissue culture systems available to support the replication of these viruses. Consequently, work has been restricted to examination of physicochemical properties, in vivo pathogenicity, and replicative events utilizing in vitro translation systems.

## *NUDAURELIA* β VIRUS

Tooke and Hubbard (1) first reported a disease of the pine tree emperor moth (*Nudaurelia cytherea capensis* Stoll) in which they described flaccid, dark, diseased larvae. The larval stages of this moth are a major pest of pine and other trees in South Africa. On occasions the disease caused as much as 90% mortality in the larvae. Characterization of the causative agent of this disease demonstrated the presence of at least five distinct viruses, $\gamma$, $\delta$, $\alpha$, $\epsilon$, and $\beta$.

The *Nudaurelia* β virus had a diameter of 35 nm, contained a single piece of RNA with a molecular weight of $1.8 \times 10^6$, constituting about 11% of the virion weight, and a single protein of approximate molecular weight 60,000 (4,5). Finch et al. (6) found that the capsid contained 240 copies of the single protein arranged on a T = 4 icosahedral surface lattice. The virus had a low particle density of 1.298 g/ml in cesium chloride and a sedimentation coefficient of 210 S. The icosahedral capsid diameter is probably a little larger (38 nm) than originally estimated when accurately determined using catalase crystals as a size standard, and the structural protein has a molecular weight of 65,000 (7–9).

Prior to the identification of the *Nudaurelia* β virus, infected larvae of the emperor gum moth *Antheraea eucalypti* Scott (Saturniidae) were shown to contain an RNA virus by Grace and Mercer (10). Juckes et al. (11) demonstrated that the *Antheraea* and *Nudaurelia* β viruses were serologically related. However, while the first virus identified was the *Antheraea* virus, the majority of the work has been performed on the *Nudaurelia* β virus because it is a recurrent field isolate in South Africa. As stated earlier, the major problem with working with this family of viruses is the absence of a suitable tissue culture system to examine the replicative events; virus has to be isolated from field-collected insects in the case of *Nudaurelia* β. Tripconey (12) propagated the virus in primary cultures of ovarian cells of *Bombyx mori* and *Nudaurelia cytherea capensis* and demonstrated replication of the virus in the cytoplasm of cells by fluorescent antibody staining. However, repeated attempts by several workers have failed to produce an infection in continuous culture cells.

## OTHER MEMBERS OF THE FAMILY

The availability of hyperimmune rabbit antibodies against the *Nudaurelia* β virus has made it possible to readily identify other potential members of this group of viruses. Reinganum et al. (13) isolated viruses from the Limacodids *Darna trima* (Figure 1) and *Thosea asigna* and the Saturniid *Philosamia cynthia* x *ricini*, which reacted with *Nudaurelia* β virus antiserum. *Darna trima* Moore is a

**Table 1** Biophysical Properties of *Nudaurelia* β Family Viruses

| Virus | RNA × $10^{-6}$ | Capsid × $10^{-3}$ | Buoyant density in CsCl (g/ml) | (S) | Diameter (nm) |
|---|---|---|---|---|---|
| *Nudaurelia* | 1.8 | 66 | 1.3 | 210 | 38 |
| *Darna trima* | + | 66 | 1.3 | 199 | 38 |
| *Thosa asigna* | + | 60.8 | 1.275 | 194 | 35 |
| *Philosamia X* | + | 62.4 | 1.275 | 206 | 35 |
| *Trichoplusia ni* | 1.9 | 68 | 1.3 | 200 | 38 |
| *Dasychira pudibunda* | 1.8 | 66 | 1.31 | ND | 38 |
| *Nudaurelia* ω | ND | 65 | 1.285 | ND | 40 |

*Key*: ND, not determined; +, contains RNA.

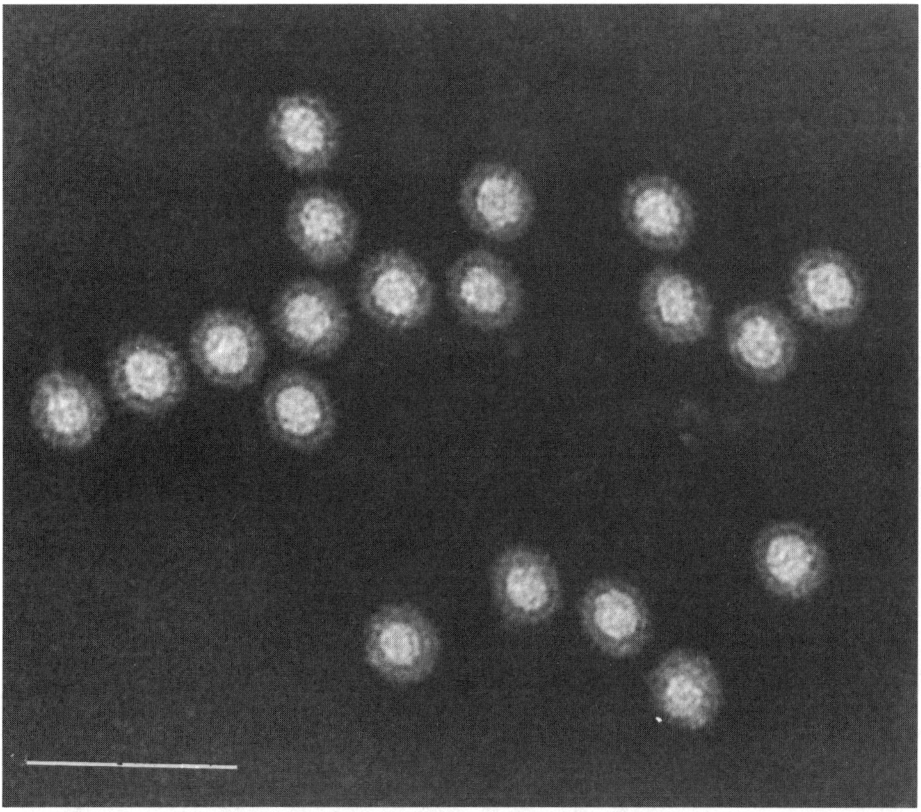

**Figure 1** Electron micrograph of purified *Trichoplusia ni* virus. Note partial entry of stain showing a core structure. Bar = 100 nm.

pest of several crops, particularly coconut and oil palms in Southeast Asia, and *Thosea asigna* is a pest of oil palms in parts of Malaysia. *Philosamia cynthia* x *ricini* is a hybrid maintained in the laboratory and the virus which was isolated from this insect is of unknown origin. The three viruses isolated had similar properties to *Nudaurelia* β virus with some slight variations that could be attributable to laboratory errors. The viruses listed in Table 1 show the more thoroughly characterized members of the family; other potential members have been serologically identified in several lepidoptera species (14–17).

One of the many problems with the production of baculoviruses for field control purposes is the presence of nonoccluded viruses, either as inapparent infections of the insects or as contaminants of the original baculovirus stock.

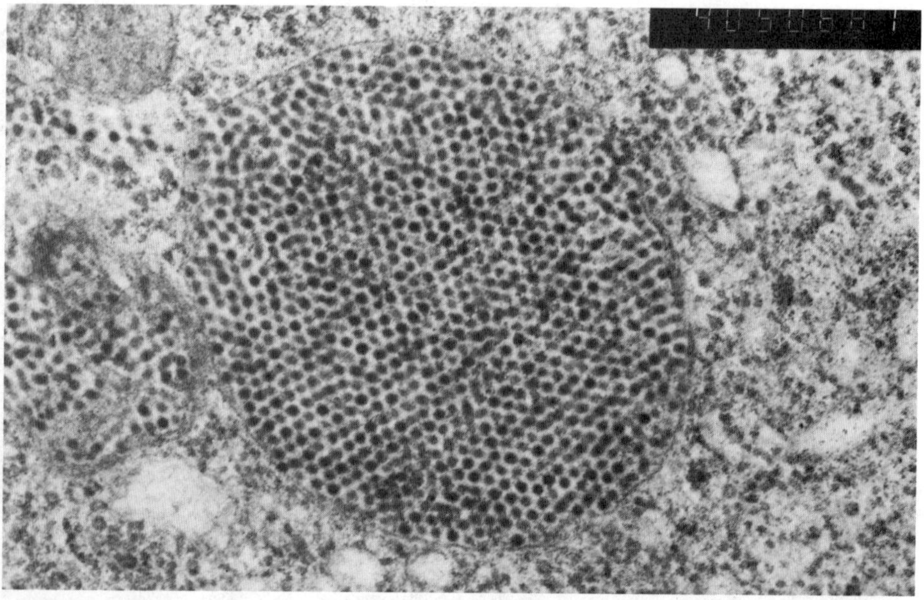
(A)

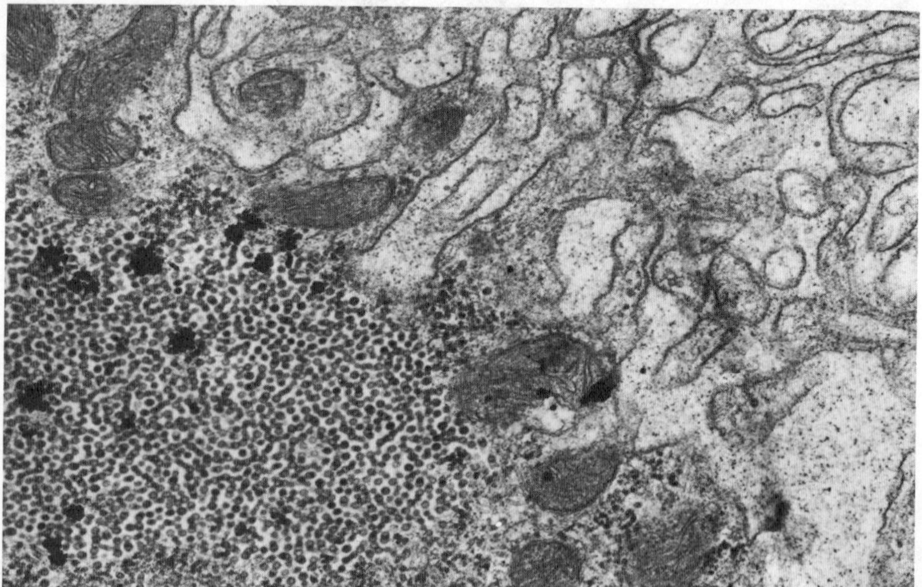
(B)

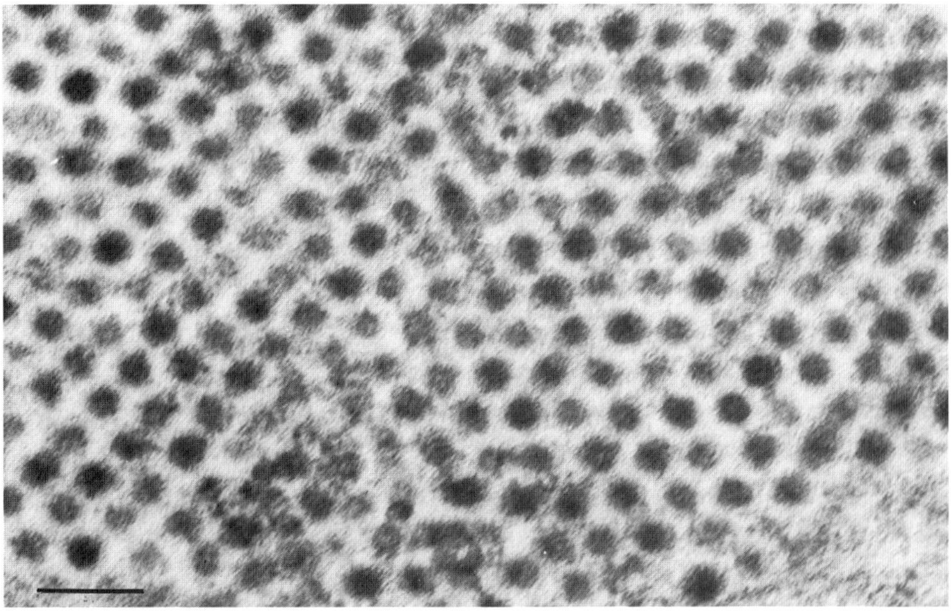

(C)

**Figure 2** (A) Electron micrograph of a section of the foregut of *Trichoplusia ni* infected with *Dasychira pudibunda* virus. A vesicle containing many particles is apparent. (B) Electron micrograph of the midgut of *Trichoplusia* infected with *Dasychira pudibunda* virus. (C) Enlargement of the virus particles contained within a vesicle. Bar = 100 nm.

Hess et al. (18,19) reported the presence of several types of virus particles when investigating the replicative cycle of *Autographa californica* NPV in the midgut cells of two insects, *A. californica* and *Trichoplusia ni*. The electron micrograph evidence appeared to indicate that the smaller nonoccluded virus, one of which was later demonstrated to be a member of the *Nudaurelia* β family (20), could be incorporated into the polyhedral structure of the baculovirus, that is, it became occluded. If this is the case, it will obviously be extremely difficult to produce a preparation of baculovirus free of contamination with other viruses.

The most recent member of this family of viruses was also isolated from the pine emperor moth. Naturally infected larvae contained a second serologically unrelated virus when infected with *Nudaurelia* β virus (21). This virus, called *Nudaurelia* ω virus, had a buoyant density of 1.285 g/ml in cesium chloride and a single major capsid polypeptide of 65,000 daltons. Tryptic peptide analysis demonstrated that the proteins of the two viruses were different. The second virus was distinguishable in having a protein of slightly higher molecular weight than *Nudaurelia* β virus and a slightly different morphology. This virus also failed to react with antisera to either *Antherea* virus or *Dasychira pudibunda* virus and yielded a reaction of nonidentity with *Nudaurelia* $\beta$ virus (21).

## REPLICATION AND PATHOLOGY

Most of the replicative studies have been performed on *Trichoplusia ni* and *Dasychira pudibunda* viruses and have been limited by the absence of susceptible tissue culture systems. The availability of an artificial diet for the maintenance of *T. ni* larvae has made it possible to study the effects of *T. ni* virus on this insect under controlled conditions.

The replicative studies of *T. ni* and *D. pudibunda* viruses implicate the cytoplasm of foregut and midgut cells (18,20,22) (Figure 2). When using electron microscopy and enzyme-linked immunosorbent assays (ELISAs) to examine various organs of the larvae, the viruses appear to be restricted to the gut until very late in the infection cycle (23) (Figure 3). Vail et al. (24) studied the pathology of *T. ni* virus on its homologous insects and used ELISA and weight depression as criteria of infection state. They found that infected larvae were severely retarded in growth and could weigh as little as 13 times less than uninfected larvae. However, when larvae were infected with very low concentrations of virus, they did not demonstrate a reduction in weight gain, although the infection was established as demonstrated by ELISA.

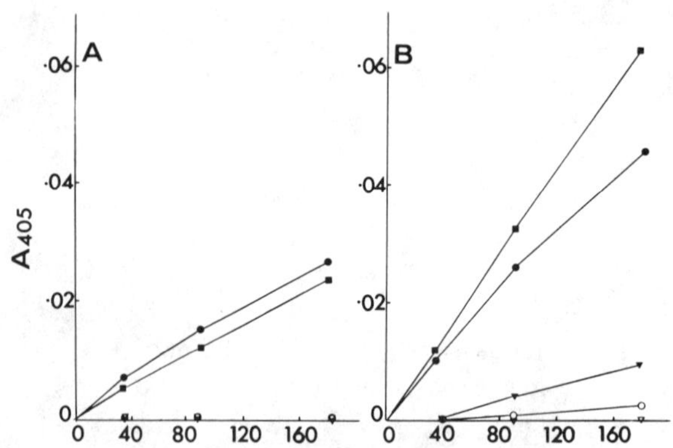


**Figure 3** Determination of the location of infection in *Trichoplusia ni* larvae by *Trichoplusia ni* virus using ELISA (23). Absorbance values for *Trichoplusia ni* virus in various tissues at 6 days (A) and 10 days (B) postinfection. The tissues examined were midgut (●), hindgut (■), epidermal tissue (▲), rudimentary gonads (○), and fat body (▽). The head, Malpighian tubules, and salivary glands contained no detectable virus particles and all values were on the baseline.

Studies on the mechanism of replication of these viruses have been restricted to examination of the in vitro translation products from the viral RNA. King and Moore (25) were unable to bind the purified viral RNA to oligo (dT), demonstrating the absence of a poly-A tract. The efficiency of translation of the RNAs of *T. ni* and *D. pudibunda* viruses is extremely low when compared with insect picornaviruses, and numerous products are found in the rabbit reticulocyte lysate system. Only small quantities of capsid protein are produced, and translation products are found with molecular weights in excess of 100,000. Immunoprecipitation indicates high molecular weight potential polymerase and capsid protein precursors in addition to early termination products of these two proteins (26,27).

The viruses from *Nudaurelia cytherea capensis* and *Darna trima* are potent control agents for the larval stages of their respective hosts, and the latter virus has been utilized as an applied control agent. However, there may be some reluctance to use these viruses in the field because of serological reaction between these viruses and the serum of various animals, including man. Longworth et al. (28) reported a serological reaction between pig sera and *Darna trima* virus. MacCallum et al. (29) reported the presence of precipitating antibodies against *Darna trima* virus collected in East Malaysia in the sera of several species of domestic animals in the United Kingdom and some wild animals from East Africa. More recently MacCallum et al. (30) reported the presence of precipitating antibodies to the same virus in a small percentage of human sera from individuals in West Malaysia and the United Kingdom. These authors suggested that antigenically related viruses may be present in the environment and associated with asymptomatic or inapparent infections in man.

## REFERENCES

1. Tooke, F. G. C. and Hubbard, C. S. Department of Agriculture and Forestry, Union of South Africa, Bulletin 210 (Entemology Series No. 7), 1941.
2. Juckes, I. R. M. Viruses of the pine emperor moth. *Bull. S. Afr. Soc. Plant Pathol. Microbiol.* 4, 18, 1970.
3. Juckes, I. R. M. Purification and characterization of the viruses of the pine emperor moth *Nudaurelia cytherea capensis* (Fabr.), Ph.D. (diss.), University of Cape Town, 1974.
4. Struthers, J. K. Physico-chemical and substructural studies on *Nudaurelia capensis* beta virus, M.Sc. thesis, Rhodes University, South Africa, 1973.
5. Struthers, J. K. and Hendry, D. A. Studies on the protein and nucleic acid components of *Nudaurelia capensis* beta virus. *J. Gen. Virol.* 22, 355, 1974.
6. Finch, J. T., Crowther, R. A., Hendry, D. A., and Struthers, J. K. The structure of *Nudaurelia capensis* beta virus: The first example of a capsid with icosahedral surface symmetry T = 4. *J. Gen. Virol.* 24, 191, 1974.

7. Matthews, R. E. F. Classification and nomenclature of viruses. Fourth Report of the International Committee on Taxonomy of Viruses. *Intervirology* 17, 1, 1982.
8. Moore, N. F. and Tinsley, T. W. The small RNA-viruses of insects. *Arch. Virol.* 72, 229, 1982.
9. Hendry, D. A., Moore, N. F., Morris, T. J., Reinganum, C., and Scotti, P. An overview of the *Nudaurelia* β virus group. *Abstract, 6th Int. Congr. Virol.*, Sendai, Japan, 1984, p. 341.
10. Grace, T. D. C. and Mercer, E. H. A new virus of the saturniid *Antherea eucalypti* (Scott). *J. Invertebr. Pathol.* 7, 241, 1965.
11. Juckes, I. R. M., Longworth, J. F., and Reinganum, C. A serological comparison of some non-occuluded insect viruses. *J. Invertebr. Pathol.* 21, 119, 1973.
12. Tripconey, D. Studies on a non-occuluded virus of the pine tree emperor moth. *J. Invertebr. Pathol.* 15, 268, 1970.
13. Reinganum, C., Robertson, J. S., and Tinsley, T. W. A new group of RNA viruses from insects. *J. Gen. Virol.* 40, 195, 1978.
14. Greenwood, L. K. and Moore, N. F. A single protein *Nudaurelia* β-like virus of the pale tussock moth, *Dasychira pudibunda. J. Invertbr. Pathol.* 38, 305, 1981.
15. Greenwood, L. K. and Moore, N. F. The *Nudaurelia*-β group of small RNA viruses of insects: Serological comparison of four members. *Microbiologica* 4, 271, 1981.
16. Greenwood, L. K. and Moore, N. F. The *Nudaurelia*-β group of small RNA-containing viruses of insects: Serological identification of several new isolates. *J. Invertbr. Pathol.* 39, 407, 1982.
17. Moore, N. F., Greenwood, L. K., and Rixon, K. R. Studies on the capsid protein of several members of the *Nudaurelia* β group of small RNA viruses of insects. *Microbiologica* 4, 59, 1981.
18. Hess, R. T., Summers, M. D., Falcon, L. A., and Stoltz, D. B. A new icosahedral insect virus: Apparent mixed nuclear infection with the baculovirus of *Autographa californica. IRCS Med. Sci.* 5, 562, 1977.
19. Hess, R. T., Summers, M. D., and Falcon, L. A. A mixed virus infection in midgut cells of *Autographa californica* and *Trichoplusia ni* larvae. *J. Ultrastr. Res.* 65, 253, 1978.
20. Morris, T. J., Hess, R. T., and Pinnock, D. E. Physico-chemical characterization of a small RNA virus associated with baculovirus infection in *Trichoplusia ni. Intervirology* 11, 238, 1979.
21. Hendry, D., Hodgson, V., Clark, R., and Newman, J. Small RNA viruses co-infecting the pine emperor moth *Nudaurelia cytherea capensis. J. Gen. Virol.* 66, 627, 1985.
22. Reavy, B. and Moore, N. F. The replication of small RNA-containing viruses of insects. *Microbiologica* 5, 63, 1982.
23. Greenwood, L. K., and Moore, N. F. Determination of the location of an infection in *Trichoplusia ni* larvae by a small RNA-containing virus using enzyme-linked immunosorbent assay electron microscopy. *Microbiologica* 7, 97, 1984.
24. Vail, P. V., Morris, T. J., Collier, S. S., and Mackey, B. An RNA virus in *Autographa californica* nuclear polyhedrosis virus preparation: Incidence and influence on baculovirus activity. *J. Invertbr. Pathol.* 41, 171, 1983.

25. King, L. A. and Moore, N. F. The RNAs of two viruses of the *Nudaurelia* β family share little homology and have no terminal poly A tracts. *FEMS Microbiol. Lett.* 26, 41, 1985.
26. Reavy, B. and Moore, N. F. In vitro translation of the genome of a small RNA virus from *Trichoplusia ni*. *J. Invertbr. Pathol.* 44, 244, 1984.
27. King, L. A., Merryweather, A. T., and Moore, N. F. Proteins expressed in vitro by two members of the *Nudaurelia* β family of viruses, *Trichoplusia ni* and *Dasychira pudibunda* viruses. *Ann. Virol. (Institut Pasteur)* 135, 335, 1984.
28. Longworth, J. F., Robertson, J. S., Tinsley, T. W., Rowlands, D. J., and Brown, F. Reactions between an insect picornavirus and naturally occurring IgM antibodies in several mammalian species. *Nature (London)* 242, 314, 1973.
29. MacCallum, F. O., Robertson, J. S., and Tinsley, T. W. The significance of serological reactions between certain pathogenic viruses and mammalian sera. *Proc. Int. Conf. Comp. Virol.*, Quebec, Canada, 1977.
30. MacCallum, F., Brown, G., and Tinsley, T. W. Antibodies in human sera reacting with an insect pathogenic virus. *Intervirology* 11, 234, 1979.

# 10

# Identification, Pathology, Structure, and Replication of Insect Picornaviruses

NORMAN F. MOORE
*Ministry of Defence, London, England*

## IDENTIFICATION AND CHARACTERISTICS OF INSECT PICORNAVIRUSES

More than 30 small RNA viruses of insects have been called picornaviruses or enteroviruses by analogy with mammalian viruses. The mammalian picornaviruses have buoyant densities in cesium chloride of 1.33–1.45 g/cm$^3$, $S_{20W}$ of 140–165, diameters of 22–30 nm, contain four major polypeptides (60 molecules of each per virion), three of molecular weight 24–41 $\times$ 10$^3$ (VP1–3) and one of 5.5–13.5 $\times$ 10$^3$ (VP4), and have one molecule of infectious, positive-sense, single-stranded RNA with a molecular weight of 2.5 $\times$ 10$^6$. The RNA is polyadenylated at the 3' end and there is a protein (with a molecular weight of about 2500) covalently attached at the 5' terminus. There are relatively few instances in which the insect picornaviruses have been examined in sufficient detail to fulfill these criteria. However, on the basis of physicochemical characteristics of size, sedimentation coefficient, buoyant density, and the presence of the correct complement of proteins and the correct size of RNA, three insect viruses have been classified as being members of the family Picornaviridae, though not yet assigned to genera (1). These are Cricket paralysis virus, *Drosophila* C virus, and *Gonometa* virus (see Table 1).

Enteroviruses (a genus of the Picornaviridae) are often found associated with the gastrointestinal tract. A number of morphologically similar insect viruses have been seen as arrays in the cytoplasm of infected cells in proximity to the gut, hence the desire to place them in this genus of the Picornaviridae. The enteroviruses are stable to acid pH to protect them against the conditions of the gut. Several workers have examined the acid stability of the insect "enterovi-

**Table 1** Physicochemical Properties of Insect Picornaviruses

| Virus | Molecular weight of RNA ($\times 10^{-6}$) | Molecular weight of capsid proteins ($\times 10^{-3}$) | Buoyant density in CsCl (g/ml) | Sedimentation coefficient | Diameter (nm) |
|---|---|---|---|---|---|
| Cricket paralysis | 2.5–2.8 | (43)[c], 35, 34, 30 | 1.34 | 167 | 27 |
| *Drosophila* C | 2.5–3.0 | (37), 31, 30, 28, 8.5 | 1.34 | 153 | 27 |
| *Gonometa* | —[b] | (47.5), 36.5, 32, 29, 12 | 1.35 | 180 | 32 |
| Poliovirus[a] | 2.4 | (41), 35, 28, 24, 5.5 | 1.34 | 155 | 28 |

[a]Example of mammalian picornavirus.
[b]Not done.
[c]Brackets indicate minor amounts of a protein corresponding to VPO (see table).

ruses" and found them stable at low pH. However, the significance and relevance of these findings are difficult to assess as insect gut contents tend to be of much higher pH than those of mammalian species.

Cricket paralysis virus (CrPV) was identified during a mass breeding program of the Australian field cricket *Teleogryllus oceanicus*. Early instar nymphs suffered paralysis of their rear limbs and eventually died. *T. commodus* colonies in the same laboratory showed similar effects (2,3). When the virus particles were injected into crickets mortality was caused within three days. Field populations of the crickets were found to contain the same virus (2). Jousset et al. (4) found a virus in extracts of laboratory-bred fruit flies *(Drosophila melanogaster)* which had died, and this virus was found to be very pathogenic for the flies. This virus was called *Drosophila* C virus (DCV) to distinguish it from other "types" of virus isolates from *D. melanogaster*. A small RNA virus, first isolated from *Gonometa podocarpi* by Harrap et al. (5), was characterized in detail by Longworth et al. (6). This small virus stimulated much interest, because, when applied to larval populations of *G. podocarpi* that were defoliating exotic pines (primarily *Pinus patula*) in Uganda, it killed the insect and proved to be an extremely effective controlling agent.

The RNA of *Gonometa* virus was not examined and there was no evidence for a VP4 with CrPV. However, in all other characteristics these viruses appear to be similar to the mammalian viruses. A problem arises, however, with *Gonometa* virus in that no laboratory stocks of live virus are in existence. The virus was collected from field-infected insects and no system was developed in the laboratory for propagation. Hence, as previously indicated (7), it is probably unacceptable to include this virus in any listing of known insect picornaviruses. Undoubtedly the virus still exists in field populations of insects.

Although this virus should not be listed, several recently described viruses fulfill the characteristics of the Picornaviridae. Pudney et al. (8) found an enterolike virus in a pool of 100 *Mansonia uniformis* mosquitoes collected near the village of Kawino in the Kano Plain district of Kenya. The virus (which they named Kawino virus) multiplied in *Aedes pseudoscutellarius* AP-61, *A. malaysensis* AM-60, and *A. albopictus* cell lines at 28°C. The virus particles were shown to be very similar to those of a classical mammalian picornavirus. However, while the viral RNA was approximately the correct size at a molecular weight of $2.6 \times 10^6$, only a very small proportion bound to Sepharose poly-U. Hence the Kawino virus genome is unlike those of vertebrate picornaviruses in that it does not contain a poly-A tract. Kawase et al. (9) described a virus of the silkworm *Bombyx mori* which was a contributing factor to Flacherie disease, which results in a major loss of cocoons in Japan. The virus contained four structural proteins and a polyadenylated RNA with a covalently attached VpG of the correct size to allow the virus to be classified as a picornavirus (10–14). A virus from the oat bird cherry aphid *(Rhopalosiphum padi)* also had the structural protein and RNA content associated with picornaviruses (14). Bailey and his colleagues also identified several small viruses of the European honeybee, *Apis mellifera*, and the Eastern hive bee, *Apis cerana*, which fulfilled at least part of the criteria necessary for inclusion in the Picornaviridae (15–18).

## PATHOLOGY, TRANSMISSION, AND HOST RANGE

Relatively little is known about the pathology of insect picornaviruses or how they are transmitted under natural conditions. The most studied viruses are CrPV and DCV. If the eggs from crickets or *Drosophila* are surface-sterilized, the virions are not transmitted (2,19). However, if the infected insects survive to maturity, there is a strong possibility that the eggs could be surface-contaminated with these viruses, due to (for example) fecal contamination. Plus and Jousset (19) injected *Drosophila* flies with DCV and introduced them into colonies of uninfected flies. A reduced proportion of the colony became infected if the inoculated flies were removed as a function of time. This would imply some removal of the inoculum from the environment, by virus breakdown or other means. Scotti et al. (20) hypothesized that the soil might act as a reservoir for CrPV, replenished each year by more cricket cadavers, and that bird and animal movement in conjunction with wind and rain may aid virus spread. However, the transmission modes are not known and it must be conjectured that they involve surface contamination of eggs and contamination of the environment by the rotting cadavers of infected insects. Fecal contamination by infected but actively feeding insects must be presumed to be a major route of transmission to uninfected insects. Reinganum (21) suggested cannibalism as a possibility for virus transmission. CrPV was detected in the lepidopteran *Antheraea eucalypti* and Reinganum suggested that crickets might have become infected with the virus by eating the infected cadavers of *A. eucalypti* larvae dropping from branches to the ground.

Natural epizootics have not been found in field populations of *Drosophila* or crickets and so laboratory mortalities could be attributable to stress factors not found in the wild. Jousset (22) examined the effect of DCV on many *Drosophila* species and found that in *D. ananasae* 89% mortality was obtained but in *D. immigrans* the virus multiplied without obvious disease. Unfortunately, a number of the studies of susceptibility and host range have been performed by injecting the test insect, which is an unrealistic method to mirror the condition in the wild, unless the insect is being parasitized. Plus and Golubovsky (23) found a complex relationship between genetically dissimilar *Drosophila* flies and resistance to infection by DCV. While their results were difficult to explain, the line merits further work as the insect model might be a suitable system for more general studies of resistance to disease.

The host ranges of CrPV and DCV have been studied by the experimental introduction of virions into other hosts and serological identification of these viruses in other insects. The first procedure suffers in that several workers have used injection rather than the more natural route of infection feeding and also have not determined that the virus replicating in the insect was the same as that introduced. The second, serological, procedure is inadequate on its own as it can be difficult to discriminate the degree of similarity between the test virus and the original antigen. CrPV has been serologically "identified" in crickets in the United States and New Zealand and in several other insects (see Ref. 20). DCV has been detected in six natural populations of *D. melanogaster* in France, the French Antilles, and North Africa (24,25). Reinganum (21) experimentally infected a range of insects with CrPV and found that 3 Orthoptera (in addition to the two species of *Teleogryllus*) and 9 out of 10 species of Lepidoptera tested were susceptible, in terms of lethality. A few species of Hymenoptera, Coleoptera, and Dictyoptera failed to support virus replication (21). Jousset (22) found that two isolates of DCV replicated in the Dipteran *Ceratitis capitata* but not in *Calliphora erythrocephala*, *Aedes aegypti*, *Culex pipiens*, or in the Lepidopteran *Arctia caja*.

Plus et al. (24) found that CrPV replicated in the Orthopteran *Gryllus bimaculatus* but two strains of DCV failed to grow. Plus and Scotti (26) presented evidence that isolates of CrPV and DCV could be separated into three groups on the basis of host range tests. However, with all these studies the mode of infection may be an important factor, and this has yet to be fully investigated. For example, when *Galleria mellonella* are infected with CrPV per os mortality is greatly decreased compared to those infected with injections.

Electron microscopic examination of infected insects reveals, with each of the insect picornaviruses, that the virions are found in crystalline arrays in the cytoplasm of infected cells (2,4,22,27). In crickets infected with CrPV virus particles were present in the alimentary canal cells, nerve ganglion cells, and epidermal cells; DCV is found mainly in tracheal cells and cells surrounding the cerebral cortex in infected *D. melanogaster*. Jousset (22) found that DCV appeared to multiply in various sites in other insect hosts.

## SEROLOGICAL INTERACTIONS OF INSECT PICORNAVIRUSES

Several studies have been performed to examine the serological relationships within the insect picornaviruses and to detect whether there is evidence of multiplication in mammals or relationship between insect and mammalian picornaviruses. One study also utilized CrPV to investigate the basis of the "antiviral system" in insects by inoculating *G. mellonella* larvae with inactive virus and subsequently challenging with live virus (28). However, the *G. mellonella* were not protected by "vaccination." Whereas there is some evidence that insects have a responsive defense system against bacterial surface proteins, there is no known responsive defense system against viruses which corresponds to the mammalian immune system. Screening of mammalian sera by reaction with purified CrPV has revealed a number of serological interactions (29,30; MacCallum, F., unpublished). Scotti and Longworth (29) demonstrated that a significant proportion of cattle sera, collected from an area in which CrPV was known to occur, produced immunoprecipitation with the purified virus. The antibodies were of the IgM type and at a low level, which suggested a low-grade exposure to viral antigens rather than an active viremia. Moore et al. (30) used hyperimmune rabbit sera and randomly collected cattle sera, both of which precipitated radiolabeled viral proteins from CrPV-infected *Drosophila* cells. They failed to precipitate low molecular weight nonstructural proteins with either sera, which they took as evidence that the animals were exposed to intact, nonreplicating virus. However, this assumes that the low molecular weight proteins would not have been further degraded and would elicit an immunological response.

Tinsley et al. (31) took the examination of the possible interactions between CrPV and mammalian sera further by identifying viruses that appeared to be serologically related. When anti-encephalomyocarditis virus (EMCV) antibodies were diffused in agar against EMCV and CrPV in adjacent cells, there was a confluence of precipitant bands, which demonstrated a strong serological interaction. On the reciprocal converse reaction of anti-CrPV sera with CrPV and EMCV antigens, spur formation showed that the antigens were not identical. Unlike the previous study of Scotti and Longworth (29) on naturally occurring reactions of CrPV with cattle sera, where they demonstrated that the reactive antibodies were of the IgM class, Tinsley et al. (31) demonstrated the active species in their hyperimmune sera to be IgE. These authors further examined the serological interaction between CrPV and EMCV by investigating the enhancement of the antibody response to one virus by subsequent inoculation of the other virus into guinea pigs. They found that injections of either virus sensitized the animal so that injection with the second virus increased the neutralizing antibody titer against EMCV. However, the increases in titer were small and the authors' conclusion "that the virions of CrPV and EMCV are antigenically related, and are strains of one and the same virus" appears an overinterpretation.

## TISSUE CULTURE STUDIES AND REPLICATIVE EVENTS

Two of the insect picornaviruses (in addition to Kawino virus; 8) have been adapted to growth in insect cell lines. Reingamun (3) found an increase in titer of CrPV when he infected *Aedes aegyptii* cells with CrPV. This virus also multiplied in *Lymantria dispar*, *Trichoplusia ni*, and *Drosophila* cell line GM1, and Scotti (32,33) found that CrPV multiplied in Schnieder's DL1 cells with a reproducible and distinctive cytopathic effect, which could be used to titrate the virus (20). DCV was demonstrated to establish persistent infections in *D. melanogaster* cell lines (34–36) and the authors considered that the contamination originated from the bovine sera used as a necessary component of the growth medium for the cells. DCV and CrPV have been titrated in virus-free, laboratory-reared *D. melanogaster* larvae and cricket nymphs (or *G. mellonella* larvae) (2,24,32). However, in view of the susceptibility of *Drosphila* cells to both viruses, it was possible to develop more accurate plaque assay systems (33,37). Both CrPV and DCV replicate to high titers in *Drosophila* cells (37,38), the infection caused by CrPV being more lytic and that with DCV generally being more cell associated when production of intracellular and extracellular virus is compared.

The intracellular proteins expressed by CrPV and DCV in infected *Drosophila* cells have been examined in detail by Moore and his co-workers. Using a combination of different one-dimensional gel systems and isoelectric focusing SDS-polyacrylamide gels it is possible to identify about 20 virus-specified proteins with CrPV (39). This constitutes a much greater potential coding capacity than would be anticipated of the single-stranded RNA of molecular weight approximately $2.5 \times 10^6$. The virions appear to have a similar pattern of production of their structural proteins from high molecular weight precursors to that which has been characterized with mammalian picornaviruses. Earlier studies demonstrated the synthesis of large amounts of two of the structural proteins, equivalent to VP1 and VP3 of mammalian picornaviruses, even with relatively short pulses with radioactive amino acid precursors (38,40). This preemptive appearance of VP1 and VP3 suggested that the polycistronic protein found with mammalian picornaviruses to give rise to the four structural proteins, VpG, polymerase, and protease was not formed. There were three possibilities to explain the fast appearance of the structural proteins: internal initiation, synthesis of the proteins from a subgenomic messenger, and/or fast processing of the viral structural proteins from a polyprotein. Examination of the radiolabeled intracellular RNAs in infected cells demonstrated the presence of both viral RNAs and high molecular weight dsRNAs, which were partially resistant to ribonuclease and corresponded to replicative forms (41,42). No low molecular weight subgenomic mRNAs were detected. When DCV- or CrPV-infected *D. melanogaster* cells were treated with protease inhibitors and amino acid analogues prior to pulsing with radioactive amino acid precursors it was possible to stop some of the processing events resulting in the appearance of high molecular weight proteins (38,43,44). By using heat shock (i.e., elevating the temperature of infected cells from 28 to 37°C) it was also possible to produce high molecular weight precursors of the DCV structural proteins which could be "chased" into viral proteins when the temperature was reduced (45–47). Translation of DCV and CrPV in rabbit reticulocyte lysates also produced a series of high molecular weight precursor proteins (48,49). Although these differed from those found in the infected cell, the proteins produced in vitro acted as precursors of two of the viral capsid proteins (VP1 and VP3). The structural proteins that were obtained in the cell-free system were produced by the action of a viral coded protease, and the DCV protease could be shown to act on the CrPV precursor proteins, and vice versa (50). Hence it appears that the replicative events are similar to those found with mammalian picornaviruses with the appearance of a polyprotein that is subsequently cleaved, at least in part, by the action of a virus-coded protease to give rise to viral structural proteins. A putative VPO (the precursor protein of VP2 and VP4 in mamalian picornaviruses) was identified in DCV- and CrPV-infected cells, which in the case of DCV could be readily chased into VP2 and VP4, though there was no evidence of VP4 with CrPV (40,51). Pactamycin mapping also demonstrated that VP0, VP3, and VP1 of CrPV are encoded at the 5' end of the genome, as is found with mammalian picornaviruses (52).

## GENOMIC STRUCTURE

King and Moore (53) demonstrated that CrPV and DCV has a genome-linked protein (VpG) of an approximate molecular weight of $3.9 \times 10^3$ at the 5' end of the viral RNA and that this RNA was polyadenylated at the 3' end. King et al. (56) cloned DNA complementary to the RNA of CrPV into the plasmid pBR322 by a hybrid RNA–cDNA cloning strategy and obtained overlapping clones spanning 7.5 kb of the estimated 8.5 kb genome. Comparison of the restriction enzyme map with those of mammalian picornaviruses demonstrated no similarities and comparison of the sequence of the 3'-terminal 1600 nucleotides revealed no significant homology. Analysis of the stop codons in the CrPV sequence demonstrated the presence of a large open reading frame extending from the 5'-terminal nucleotide to the stop-codon 242 bases from the 3' poly (A) tract. The 1600 3'-terminal nucleotides of the CrPV cDNA showed four possible hairpin loop structures, a similarity with the mammalian picornaviruses, which all show varying degrees of secondary structure. Furthermore, King et al. demonstrated the presence of a conserved amino acid region indicative of the presence of a polymerase. This was a Gly-Asp-Asp (GDD) triplet surrounded by hydrophobic amino acids and a second 11 amino acid stretch located 22–37 amino acids upstream of the GDD sequence. This has recently been identified in the viral polymerase of several plant and animal RNA viruses.

## COMPARISON OF INSECT PICORNAVIRUSES

A serological relationship between CrPV and DCV was demonstrated by Reinganum and Scotti (57). Plus et al. (24) used several isolates of DCV ($DCV_o$, $DCV_z$, $DCV_c$, $DCV_v$, and $DCV_{AN}$) in osmoelectrophoretic cross-reactions with CrPV and DCV antisera and demonstrated the presence of spurs between CrPV antiserum and the purified DCV, showing nonidentity. CrPV isolated in Berkeley, California, and CrPV from New Zealand crickets have also been demonstrated to be nonidentical (20). The DCV isolates were obtained from the French Antilles ($DCV_{AN}$), France [Charolles ($DCV_C$) and Vigier ($DCV_V$)], and Morocco [Quarzazate ($DCV_O$), Zagora ($DCV_Z$), and Taroudant ($DCV_T$)]; though they have different pathogenicities in *Drosophila* flies serological techniques used were not sufficiently sensitive to detect antigenic differences. Both the structural proteins and the proteins expressed in infected *D. melanogaster* cells by the different isolates were compared, and relatively small differences in the molecular weight of virus-induced proteins were found (58). Clewley et al. (59) demonstrated differences between the various DCV isolates using oligonucleotide fingerprinting and showed the French isolates to be more similar to each other than the Moroccan viruses. DCV and CrPV were demonstrated to be very different by olignucleotide mapping, and no common oligonucleotides were detected (55). King et al. (60) used complementary DNA hybridization to compare the DCV isolates both with each other and with CrPV. While homologies were detected between the DCV isolates, no sequence homology was found between CrPV and DCV (60).

## FIELD CONTROL OF PEST INSECTS

Efforts to control pest species of insects have largely been confined to work with baculoviruses and cytoplasmic polyhedrosis viruses, although some representatives of the other families of insect viruses have been used on various occasions. Relatively few attempts have been made using small RNA viruses. A virus from the *Nudaurelia* $\beta$ family, *Darna trima* virus, was used to control its homologous pest on oil and coconut palms in Malaysia and, as previously mentioned, *Gonometa podocarpi* was controlled in Uganda by a possible picornavirus. Manousis and Moore (61) tested representatives of several families of insect viruses on the olive fruit fly (*Dacus oleae* Gmelin), which is the most serious pest of olives. Adult flies fed a solution containing CrPV demonstrated a high mortality. The authors suggested future approaches to test the possible use of CrPV as a control agent for this insect and described methods of application.

## REFERENCES

1. Matthews, R. E. F. Classification and nomenclature of viruses. *Intervirology* 17, 1, 1982.
2. Reinganum, C., O'Loughlin, G. T., and Hogan, T. W. A non-occluded virus of the field crickets *Teleogryllus oceanicus* and *T. commodus* (Orthoptera: Gryllidae). *J. Invertebr. Pathol.* 16, 214, 1970.
3. Reinganum, C. Studies on a non-occluded virus of the field crickets *Teleogryllus oceanicus* and *T. commodus*, M. Sc. Thesis, Monash University, Melbourne, Australia, 1973.

4. Jousset, F.-X., Plus, N., Croizier, G., and Thomas, M. Existence chez *Drosophila* de deux groupes de picornavirus de propriétés sérologiques et biologiques différentes. *C. R. Hbebd. Séances*, ser. 275, 3043, 1972.
5. Harrap, K. A., Longworth, J. F., Tinsley, T. W., and Brown, K. W. A noninclusion virus of *Gonometa podocarpi* (Lepidoptera: Lasiocampidae). *J. Invert. Pathol.* 8, 270, 1966.
6. Longworth, J. F., Payne, C. C., and Macleod, R. Studies on a virus isolated from *Gonometa podocarpi* (Lepidoptera: Lasiocampidae). *J. Gen. Virol.* 18, 199, 1973.
7. Moore, N. F., Reavy, B., and King, L. A. General characteristics, gene organisation and expression of small RNA viruses of insects. *J. Gen. Virol.* 66, 647, 1985.
8. Pudney, M., Newman, J. F. E., and Brown, F. Characterization of Kawino virus: An entero-like virus isolated from the mosquito *Mansonia uniformis* (Diptera: Culicidae). *J. Gen. Virol.* 40, 433, 1978.
9. Kawase, S., Hasimoto, Y., and Nakagaki, M. Characterisation of flacherie virus of the silkworm *Bombyx mori*. *J. Sericult. Sci. Jpn.* 49, 477, 1980.
10. Hashimoto, Y. and Kawase, S. Characteristics of structural proteins of infectious flacherie virus from the silkworm, *Bombyx mori*. *J. Invertebr. Pathol.* 41, 68, 1983.
11. Hashimoto, Y., Watanabe, A., and Kawase, S. In vitro translation of infectious flacherie virus RNA in a wheat germ and a rabbit reticulocyte system. *Biochim. Biophys. Acta* 781, 76, 1984.
12. Hashimoto, Y., Watanabe, A., and Kawase, S. Preliminary studies of terminal structures of infectious flacherie virus RNA. *Microbiologica* 7, 91, 1984.
13. Hashimoto, Y., Watanabe, A., and Kawase, S. Evidence for the presence of a genome-linked protein in infectious flacherie virus. *Arch. Virol.* 90, 301, 1986.
14. D'Arcy, C. J., Burnett, P. A., Hewings, A. D., and Goodman, R. M. Purification and characterization of a virus from the aphid *Rhopalosiphum padi*. *Virology* 112, 346, 1981.
15. Bailey, L. Viruses attacking the honey bee. *Adv. Virus Res.* 20, 271, 1976.
16. Bailey, L. and Woods, R. D. Bee viruses, in *The Atlas of Insect and Plant viruses*, Maramorosch, K., Ed., Academic Press, New York, 1977, p. 141.
17. Bailey, L., Carpenter, J. M., and Woods, R. D. Egypt bee virus and Australian isolates of Kashmir bee virus. *J. Gen. Virol.* 43, 641, 1979.
18. Bailey, L., Carpenter, J. M., and Woods, R. D. A strain of sacbrood virus from *Apis cerana*. *J. Invertebr. Pathol.* 39, 264, 1982.
19. Jousset, F.-X. and Plus, N. Étude de la transmission horizontale et de la transmission verticale des picornavirus de *Drosophila melanogaster* et de *Drosophila immigrans*. *Ann. Microbiol.* 126B, 231, 1975.
20. Scotti, P. D., Longworth, J. F., Plus, N., Croizier, G., and Reinganum, C. The biology and ecology of strains of an insect small RNA virus complex. *Adv. Virus Res.* 26, 117, 1981.
21. Reinganum, C. The isolation of cricket paralysis virus from the emperor gum moth, *Antheraea eucalypti* Scott, and its infectivity towards a range of insect species. *Intervirology* 5, 97, 1975.

22. Jousset, F.-X. Étude experimentale du spectra d'hôtes du virus C de *Drosophila melanogaster* chez quelques diptèrese et lépidopteres. *Ann. Microbiol. Inst. Pasteur,* 127A, 529, 1976.
23. Plus, N. and Golubovsky, M. D. Resistance to *Drosophila* C virus of fifteen 1 (2) GL/CY stocks carrying 1 (2) GL lethals from different geographical regions. *Genetika* 12, 227, 1980.
24. Plus, N., Croizier, G., Reinganum, C., and Scotti, P. D. Cricket paralysis virus and *Drosophila C* virus: Serological analysis and comparison of capsid polypeptides and host range. *J. Invertebr. Pathol.* 31, 296, 1978.
25. Plus, N., Croizier, G., Duthoit, J. L., David, J., Anxolabhere, D., and Periquet, G. Découverte, chez la *Drosophile*, de virus appartenant à trois nouveaux groupes. *C. R. Acad. Sci.* 280, 1501, 1975.
26. Plus, N. and Scotti, P. D. The biological properties of eight different isolates of Cricket paralysis virus. *Ann. Virol. (Inst. Pasteur)* 135E, 257, 1984.
27. Moore, N. F. Pathology associated with small RNA viruses of insects, in *Viral Insecticides for Biological Control*, Maramorosch, K. and Sherman, K. E., Eds., Academic Press, New York, 1985, p. 233.
28. Rohel, D. Z., Chadwick, J., and Faulkner, P. Tests with inactivated cricket paralysis virus as a possible immunogen against a virus infection of *Galleria mellonella* larvae. *Intervirology* 14, 61, 1980.
29. Scotti, P. D. and Longworth, J. F. Naturally occurring IgM antibodies to a small RNA insect virus in some mammalian sera in New Zealand. *Intervirology* 13, 186, 1980.
30. Moore, N. F., McKnight, L., and Tinsley, T. W. Occurrence of antibodies against insect virus proteins in mammals: Simple model to differentiate between passive exposure and active virus growth. *Infect. Immun.* 31, 825, 1981.
31. Tinsley, T. W., MacCallum, F. O., Robertson, J. S., and Brown, F. Relationship of encephalomyocarditis virus to cricket paralysis virus of insects. *Intervirology* 21, 181, 1984.
32. Scotti, P. D. Cricket paralysis virus replicates in cultured *Drosophila* cells. *Intervirology* 6, 333, 1976.
33. Scotti, P. D. End-point dilution and plaque-assay methods for Cricket paralysis virus in cultured *Drosophila* cells. *J. Gen. Virol.* 35, 393, 1977.
34. Plus, N. Endogenous viruses of *Drosophila melanogaster* cell lines: Their frequency, identification and origin. *In Vitro* 12, 1015, 1978.
35. Gateff, E., Gissmann, L., Shrestha, R., Plus, N., Pfister, H., Shröder, J., and Zurhausen, H. Characterization of two tumorous blood cell lines of *Drosophila melanogaster* and the viruses they contain, in *Invertebrate Systems In Vitro*, Kurstak, E., Maramorosch, K., and Dübendorfer, A., Eds., Elsevier/North-Holland Biomedical Press, Amsterdam, 1980, p. 517.
36. Plus, N. Further studies on the origin of the endogenous viruses of *Drosophila melanogaster* cell lines. In *Invertebrate Systems In Vitro*, Kurstak, E., Maramorosch, K., Dübendorfer, A., Eds., Elsevier/North-Holland Biomedical Press, Amsterdam, 1980, p. 435.

37. Moore, N. F. and Pullin, J. S. K. Plaque purification of Cricket paralysis virus using an agar overlay on *Drosophila* cells. *J. Invertebr. Pathol.* 39, 10, 1982.
38. Moore, N. F., Pullin, J. S. K., Crump, W. A. L., Reavy, B., and Greenwood, L. K. Comparison of two insect picornaviruses, *Drosophila C* and cricket paralysis viruses. *Microbiologica* 4, 359, 1981.
39. Moore, N. F., Hunt, R. C., Pullin, J. S. K., and Marshall, L. M. Cricket paralysis virus specific polypeptides in infected *Drosophila* cells analysed by isoelectric focusing and 2-D analysis. *Microbiologica* 5, 215, 1982.
40. Moore, N. F., Kearns, A., and Pullin, J. S. K. Characterization of Cricket paralysis virus-induced polypeptides in *Drosophila* cells. *J. Virol.* 33, 1, 1980.
41. Reavy, B. Replication of Small RNA-Viruses of Insects, D. Phil. diss. Oxford University, Oxford, England.
42. Reavy, B., Crump, W. A. L., and Moore, N. F. Characterization of Cricket paralysis virus and *Drosophila* C virus induced RNA species synthesized in infected *Drosophila melanogaster* cells. *J. Invertebr. Pathol.* 41, 397, 1983.
43. Moore, N. F., Pullin, J. S. K., and Reavy, B. The intracellular proteins induced by Cricket paralysis virus in *Drosophila* cells: The effect of protease inhibitors and amino acid analogues. *Arch. Virol.* 70, 1, 1981.
44. Moore, N. F., Reavy, B., and Pullin, J. S. K. Processing of Cricket paralysis virus-induced polypeptides in *Drosophila* cells: Production of high molecular weight polypeptides by treatment with iodoacetamide. *Arch. Virol.* 68, 1, 1981.
45. Moore, N. F., Pullin, J. S. K., and Reavy, B. Inhibition of the induction of heat-shock proteins in *Drosophila melanogaster* cells infected with insect picornaviruses. *FEBS Lett.* 128, 93, 1981.
46. Moore, N. F. and Pullin, J. S. K. Heat shock used in combination with amino acid analogues and protease inhibitors to demonstrate the processing of the proteins of an insect picornavirus (*Drosophila* C virus) in *Drosophila melanogaster* cells. *Ann. Virol. Inst. Pasteur* 134E, 285, 1983.
47. Reavy, B., Pullin, J. S. K., and Moore, N. F. Translational inhibition of heat-shock induced gene expression in picornavirus-infected *Drosophila melanogaster* cells. *Microbios* 38, 91, 1083.
48. Reavy, B. and Moore, N. F. Cell-free translation of Cricket paralysis virus RNA: Analysis of the synthesis and processing of virus-specific proteins. *J. Gen. Virol.* 55, 429, 1981.
49. Reavy, B. and Moore, N. F. *In Vitro* translation of Cricket paralysis virus RNA. *Arch. Virol.* 67, 175, 1981.
50. Reavy, B. and Moore, N. F. Cell-free translation of *Drosophila C* virus RNA: Identification of a virus protease activity involved in capsid protein synthesis and further studies on *in vitro* processing of cricket paralysis virus specified proteins. *Arch. Virol.* 76, 101, 1983.
51. Moore, N. F., Reavy, B., Pullin, J. S. K., and Plus, N. The polypeptides induced in Drosophila cells by *Drosophila C* virus (strain Ouarzazate). *Virology* 112, 411, 1981.
52. Reavy, B. and Moore, N. F. The gene organisation of a small RNA-containing insect virus: Comparison with that of mammalian picornaviruses. *Virology* 131, 551, 1983.
53. King, L. A. and Moore, N. F. Evidence for the presence of a genome-linked protein

in two insect picornaviruses, Cricket paralysis and *Drosophila* C virus. *FEMS Lett.* 50, 41–44, 1988.
54. Eaton, B. T. and Steacie, A. D. Cricket paralysis virus RNA has a 3' terminal poly (A). *J. Gen. Virol.* 50, 167, 1980.
55. Pullin, J. S. K., Moore, N. F., Clewley, J. P., and Avery, R. J. Comparison of the genomes of two insect picornaviruses, cricket paralysis virus and *Drosophila* C virus, by ribonuclease T1 oligonucleotide fingerprinting. *FEMS Microbiol. Lett.* 15, 215, 1982.
56. King, L. A., Pullin, J. S. K., Stanway, G., Almond, J. W., and Moore, N. F. Cloning of the genome of cricket paralysis virus: Sequence of the 3' end. *Virus Res.* 6, 331, 1987.
57. Reinganum, C. and Scotti, P. D. Serological relations between twelve small RNA viruses of insects. *J. Gen. Virol.* 31, 131, 1976.
58. Moore, N. F., Pullin, J. S. K., Crump, W. A. L., and Plus, N. The proteins expressed by different isolates of *Drosophila* C virus. *Arch. Virol.* 74, 21, 1982.
59. Clewley, J. P., Pullin, J. S. K., Avery, R. J., and Moore, N. F. Oligonucleotide fingerprinting of the RNA species obtained from six *Drosophila* C virus isolates. *J. Gen. Virol.* 64, 503, 1983.
60. King, L. A., Massalski, P. R., Cooper, J. I., and Moore, N. F. Comparison of the genome RNA sequence homology between cricket paralysis virus and *Drosophila* C virus by complementary DNA hybridization analysis. *J. Gen. Virol.* 65, 1193, 1984.
61. Manousis, T. and Moore, N. F. Cricket paralysis virus, a potential control agent for the olive fruit fly, *Dacus oleae* Gmel. *Appl. Environ. Microbiol.* 53, 142, 1987.

# 11

# Birnaviridae

PETER DOBOS
*College of Biological Science, University of Guelph, Guelph, Ontario, Canada*
ÉVA NAGY
*Ontario Veterinary College, University of Guelph, Guelph, Ontario, Canada*
ROY DUNCAN
*University of Calgary Health Sciences Centre, Calgary, Alberta, Canada*

## INTRODUCTION

Birnaviruses are medium-sized unenveloped icosahedrons containing a bisegmented dsRNA genome. The suffix bi– signifies double-strandedness as well as the bisegmented nature of the virus genome, whereas "rna" indicates the nature of the viral nucleic acid. The family Birnaviridae was officially established by the International Committee on Taxonomy of Viruses in 1984 (1). The prototype of birnaviruses is infectious pancreatic necrosis virus (IPNV), which can cause high mortality epizootics in young salmonid fish (2). Another member is infectious bursal disease virus (IBDV) of chickens, which infects the lymphoid cells of the bursa of Fabricius of domestic fowl, resulting in humoral immunosuppression (3). The only insect birnavirus is *Drosophila* X virus (DXV), which causes anoxia sensitivity and death of infected fruit flies (4). It appeared in France in 1979 as a contaminant in *Drosophila melanogaster* in a laboratory where the replication of sigma virus (an insect rhabdovirus) was being investigated. DXV-infected flies died after a 15 min exposure to pure $CO_2$ or $N_2$, whereas uninfected flies did not exhibit similar sensitivity. Virus-infected flies died in 5–15 days, depending on the titer of the injected virus. Anoxia sensitivity developed 3–4 days before death. Flies could also be infected from contact transmission where approximately a third of the flies died in 30–40 days. Electron microscopic observations showed cytoplasmic viroplasms in brain, trachea, and ovary cells.

Most of our knowledge about the molecular biology of birnaviruses comes from research with the prototype IPNV; however, the strategy of the virus genome is applicable to both IBDV and DXV.

## GROWTH AND MOLECULAR STRUCTURE OF BIRNAVIRUSES

### Virus Growth

Growth of IPNV in tissue culture can be achieved in a variety of continuous cell lines from the teleost fishes at temperatures below 24°C, resulting in a characteristic cytopathic effect (CPE). Virus strains that do not readily develop defective interfering (DI) particles (IPNV Jasper) and cells that do not produce interferon (e.g., CHSE 214 cells) give higher yields (approximately $10^8$ pfu/ml) (2).

IBDV can be grown at 37°C in primary chick embryo fibroblasts or in HeLa or Vero cells. The yield is about one order of magnitude lower than that of IPNV, probably because of interferon production (birnaviruses do not cause rapid shutoff of host macromolecular synthesis) (5).

DXV may be grown in fruit flies or tissue culture. Fly colonies can be infected through contaminated feed and dead flies can be harvested periodically and stored at –20°C until a desired amount is accumulated for virus purification. This is an inexpensive, albeit not very rapid, method to produce milligrams of purified virus; however, it has two drawbacks: (1) radiolabeling of the virus components is not feasible; and (2) endogenous virus (such as *Drosophila* picornavirus or reovirus) may become activated, which can interfere with the growth of DXV (4).

The virus can also be grown to high titers ($10^9$ PFU/ml) in established *Drosophila* cell lines either in suspension or in monolayers, resulting in a characteristic CPE (cells appear fragmented) in 24–48 hr (6).

All three viruses can be plaqued, but in the case of fish and *Drosophila* cells, low gelling temperature (Seaplaque) agarose is used in the overlay since these cells are sensitive to higher temperatures (e.g., 42°C), where ordinary agarose solidifies. Birnaviruses do not exhibit hemagglutinating activity (2,6).

### Biophysical and Biochemical Characteristics

Viruses grown in tissue culture can be concentrated by polyethylene glycol precipitation, clarified by Freon 113 extraction, and purified by sucrose and cesium chloride gradient centrifugation (7). The buoyant density of infectious particles is 1.33–1.34 g/cm$^3$ in cesium chloride and that of empty capsid (top component) 1.29 g/cm$^3$. The sedimentation constants of DXV, IBDV, and IPNV are 503, 460, and 435 S, respectively (8). Electron micrographs of purified virus reveal hexagonal-shaped icosahedrons with a side-to-side diameter of 60 nm (8) (Figure 1). Four capsomers can be seen per edge ($T = 9$), indicating that the virus capsid is made up of 92 capsomers. Birnaviruses look identical under the

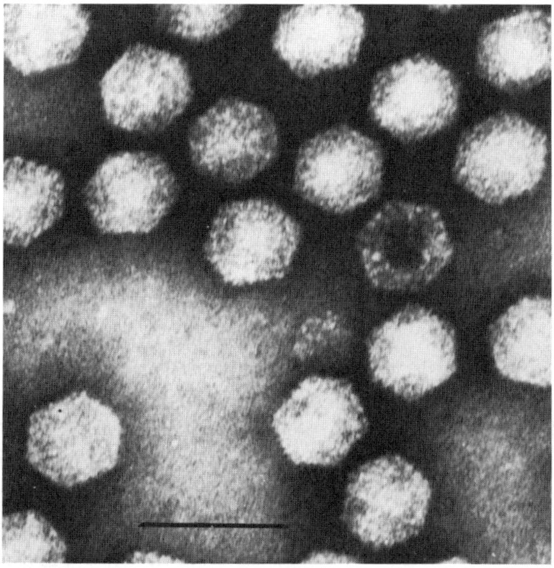

**Figure 1** Electron micrograph of DXV. Bar = 100 nm. (From Ref. 2.)

electron microscope but can be easily distinguished from reoviruses since they lack the characteristic double capsid of the latter (8).

## The Virus Genome

Two large-size class dsRNA segments make up the genomes of birnaviruses (8). Linear dose–response curve of virus infectivity indicates that the divided genome resides in a single particle (6). The RNase-resistant virus genome sediments at 14–16 S in sucrose gradients. After denaturation it sediments at 24 S and becomes RNase sensitive (9). The size difference between the larger (segment A) and smaller (segment B) genome is the least in the case of DXV (see Figure 2). Segment A is 3800 bp (MW $2.3 \times 10^6$) and segment B 3600 bp (MW $2.2 \times 10^6$). The sizes of IBDV genome segments A and B have been reported to be 3400 and 2900 bp, respectively; those of IPNV, 3097 and 2855 bp (10). The genomes of all three birnaviruses have been cloned, and segment A of IBDV and IPNV have been sequenced (11,12). The viral RNA contains no 3' poly-A tails, and in the case of IPNV, the two strands of each genome segment are completely base paired with some conserved terminal sequences in both segments, which presumably serve for RNA polymerase recognition. IPNV and DXV have been

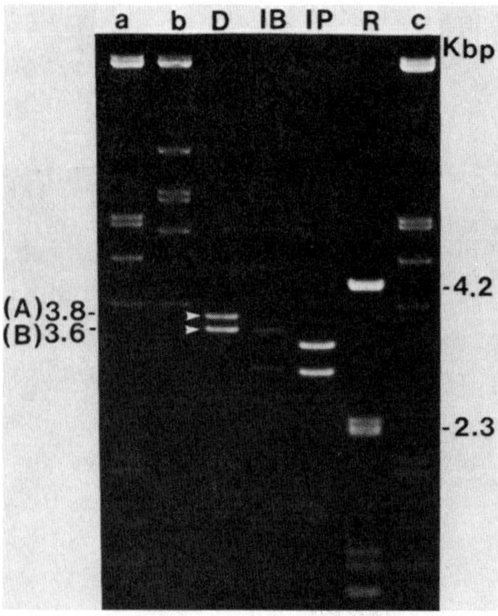


**Figure 2** Agarose gel electrophoresis of genomes of birnaviruses stained with ethidium bromide. Lane a, lambda phage DNA digested with EcoRI and HindIII; lane b, lambda phage DNA digested with EcoRI; lane D, DXV dsRNA; lane IB, IBDV dsRNA; lane IP, IPNV dsRNA; lane R, reovirus dsRNA; lane c, lambda phage DNA digested with EcoRI and HindIII. The numbers along the side indicate kilobase pairs.

reported to contain a genome-linked protein (VPg; 13,14), presumably linked to the 5' end since the 3' ends of both strands of each genome segment can be poly-A tailed in vitro. It is not known if the VPg is present at the 5' end of both plus and minus strands. No 5' cap structure has been reported in any of the birnaviruses. The virus genome constitutes approximately 9% of the mass of the virion; the remaining 91% is protein (8).

## Structural Proteins

The structural proteins of birnaviruses fall into three size classes (large, medium, and small) which are present in different relative proportions (Figure 3). The designations, molecular weights, and percent distribution of the different birnavirus polypeptides are listed in Table 1. In all cases the minor high molecular weight VP1 polypeptide is encoded by genome segment B (10,15,16), it does not undergo posttranslational cleavage, and it is thought to be the virion-associated RNA polymerase (15). The major outer capsid polypeptide VP2 is a late-maturation cleavage product of pVP2 (preVP2). This cleavage may not be complete; thus variable amounts of pVP2 are usually found in purified virus preparations (6). Polypeptides VP3 and VP4 are thought to be internal com-

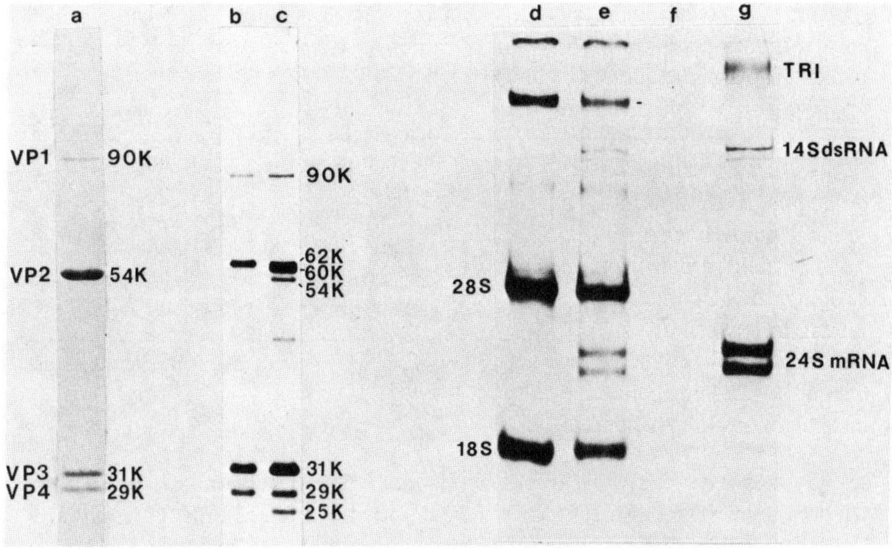


**Figure 3** Polyacrylamide gel patterns of IPNV-specific macromolecules. Lane a, polypeptides of purified virus (VP); lanes b and c, autoradiograms of labeled infected cell-specific polypeptides. The cellular background was reduced by UV irradiation and infected cell monolayers were labeled with [$^{35}$S]methionine 6 hr after infection for 1 hr (lane b) and 4 hr (lane c). Lanes d–g, fluorographs of [$^{3}$H]uridine-labeled RNA of uninfected and IPNV-infected cells analyzed in composite agarose–acrylamide gels. RNA from uninfected, nonirradiated cells showing 28 and 18 S rRNA (lane d), IPNV-infected, nonirradiated cells showing cellular and virus-specific RNAs (14 S dsRNA and 24 S ssRNA) (lane e), UV-irradiated, uninfected cells (lane f), and UV-irradiated, infected cells (lane g). TRI indicates the position of transcription intermediate. (From Ref. 2.)

ponents of virions since they are not found in top components (empty capsids; 17). They are qualitatively different from each other and from the other polypeptides except in the case of some IPNV serotypes (e.g., Jasper), where VP4 is a cleavage product of VP3. In this case both precursor and product are found in the virion (18). Polypeptides pVP2, VP3, and VP4 are encoded by genome segment A (10,15,16).

## VIRUS-SPECIFIC RNA AND PROTEIN SYNTHESIS

### RNA Synthesis in Infected Cells

Virus-specific RNA synthesis has been investigated only in the case of IPNV (19). Actinomycin D-treated infected cells were labeled with [$^{3}$H] uridine and the

**Table 1** Molecular Weights and Percent Distribution of Birnavirus Polypeptides

| Designation | Molecular weight (percent distribution) | | |
|---|---|---|---|
| | IPNV | IBDV | DXV |
| VP1 | 90,000 (3) | 90,000 (3) | 110,000 (3) |
| pVP2 | 62,000 | 50,000 | 49,000 |
| | (64)[a] | (51) | (70) |
| VP2 | 54,000 | 41,000 | 45,000 |
| VP3 | 31,000 (17) | 32,000 (40) | 34,000 (18) |
| VP4 | 29,000 (15) | 28,000 (6) | 27,000 (9) |

[a]Since the amount of pVP2 varies in different preparations, the percent contributions of pVP2 and VP2 are combined.

cell lysates were analyzed by polyacrylamide gel electrophoresis (PAGE) and fluorography. Three forms of RNA were detected in infected cells: (1) a partially RNase resistant, heterogeneous 14–16 S transcription intermediate (TR1); (2) two species of 24 S RNase-sensitive, polysome-associated RNA that could be hybridized to the two genome segments; and (3) two segments of progeny dsRNA (see Figure 3). No subgenomic RNA species were found. It is not known if RNA synthesis proceeds by a conservative or semiconservative mechanism. Cores or subviral particles could not be found in infected cells and the 24S mRNAs were found not to contain a 3' poly-A tail. RNA synthesis was very sensitive to cycloheximide at any time during the viral growth cycle, indicating that some polypeptide has to be synthesized continuously for RNA synthesis to take place.

## In Vitro RNA Synthesis

Virion-associated RNA polymerase activity has been detected in both purified IPNV and DXV preparations (20,21). The enzyme was found to be active without any pretreatment of the virus. Optimum activity was measured in the presence of 6 m*M* $Mg^{2+}$, pH 8.0 at 30°C. Between 50 and 75% of the reaction product remained associated with the dsRNA template and was resistant to RNase; the remainder was found as extravirion single-stranded RNA broken down to 5–7 S fragments by virus-associated RNase(s) (20). The use of RNase inhibitors permitted the synthesis of some genome-length 24 S ssRNA similar to that found in infected cells. Attempts to increase the yield of 24 S ssRNA by modifying the reaction met with little success. Neither the addition of spermidine, *S*-adenosyl-**L**-methionine, cytoplasmic extracts from CHSE cells, nor the addition of capped cellular or reovirus ssRNA (as possible cap donors) resulted in increased 24 S RNA synthesis (20).

When [$^3$H] uridine-labeled purified IPNV was used in the polymerase reaction (using unlabeled UTP), some labeled 24 S ssRNA was obtained, indicating that in vitro transcription may proceed by a semiconservative (strand-displacement) mechanism (20).

## Protein Synthesis in Infected Cells

The mechanism of virus-specific protein synthesis has been investigated in IPNV-infected CHSE cells (22), IBDV-infected CEFs (5), and DXV-infected *Drosophila* cell cultures (6). Since the overall pattern of protein synthesis is very similar in the different systems, only DXV-infected cells are described in detail.

DXV infection of cell cultures does not result in a rapid inhibition of host protein synthesis, but a reduction in cellular background could be achieved by heat shocking the cells at 37°C for one hour in the presence of actinomycin D. Replicate infected cultures were so treated, followed by pulse-labeling with [$^{35}$S]methionine at different times during the virus replication cycle. The cells were lysed and analyzed by PAGE and autoradiography. The data in Figure 4 demonstrate the time course of protein synthesis in DXV-infected *Drosophila* cells. All five structural polypeptides listed in Table 1 could be detected in infected cells from about 4 hr postinfection onward. One additional polypeptide (67 kD) not present in the virions was also detected. Pulse-chase experiments and peptide map comparisons revealed that the 110, 67, 34, and 27 kD polypeptides are different from one another. Posttranslational cleavage of the 67 kD polypeptide yields the 49 kD protein, which subsequently undergoes slow (incomplete) maturation-associated cleavage to generate the 45 kD VP2 major capsid protein (6).

A polypeptide comparable to the 67 kD proteins is not found in IPNV-infected cells (22). In the case of the latter, VP3 is partially cleaved during virus maturation to VP4. In addition, a 29 kD nonstructural protein is found in infected cells, which, during pulse-chase experiments, were found to be processed to a 25 kD polypeptide (18; see Figure 3). Recent experiments indicated that this truncated (25 kD) NS polypeptide is also present in minute amounts in purified virions (Dobos, unpublished data).

Since both IPNV genome segments generate genome length 24 S mRNAs (one of which codes for VP1, the other for VP2, VP3, and VP4), it was of interest to find the mechanism by which VP2, VP3, and VP4 are produced. Experiments using amino acid analogues, protease inhibitors, $ZnCl_2$, and supraoptimal temperatures showed that either the medium and small class proteins of IPNV do not arise from a high molecular weight precursor polypeptide by proteolytic cleavage, or that processing is rapid (cotranslational) and the protease–substrate reaction is insensitive to these manipulations, or that the mRNA coding for these polypeptides is polycistronic (22). Experiments using in vitro protein synthesis were thought to resolve these questions.

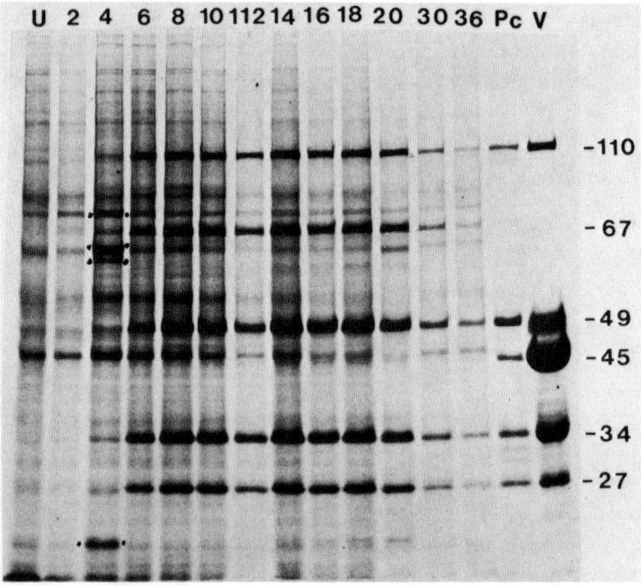


**Figure 4** Time course of protein synthesis in DXV-infected *Drosophila* cells. Heat shocked, actinomycin D-treated cells were pulse-labeled for 1 hr at 26°C using [$^{35}$S]methionine and processed for polyacrylamide gel electrophoresis and autoradiography. The number above each lane indicates the time, after infection, when the radiolabel was added. U, uninfected cell; Pc, pulse-chase, after a 1 hr pulse at 14 hr postinfection the cells were incubated for an additional 10 hr in medium that contained an excess of unlabeled methionine. V, labeled, purified virus. Cells represented in lane 4 were heat shocked without actinomycin D treatment (the dots mark the position of heat-shock proteins). The numbers along the side indicate the molecular sizes of polypeptides in kilodaltons. (From Ref. 6.)

## In Vitro Protein Synthesis

The translation of genome segment B in a reticulocyte protein-synthesizing system yielded VP1 in all birnaviruses tested (10,16,23). The translation of genome segment A yielded three polypeptides that had identical electrophoretic mobilities and peptide maps with pVP2, VP3, and VP4 (or in the case of IPNV, NS29). Interestingly, in the case of DXV, the 67 kD polypeptide was completely processed to its 49 kD product within 20–30 min, indicating that the protease is virus coded (16).

Analysis of the in vitro time course of protein synthesis revealed no high molecular weight precursor polypeptide for these products. Furthermore, in the case of IPNV, all polypeptides became labeled when synthesized in the presence of [$^{35}$S]fMet-tRNA$_f$Met, which indicated that the RNA is polycistronic, and internal initiation of translation took place in vitro (23).

## MAPPING AND SEQUENCING OF THE VIRUS GENOME

To further investigate the mechanism of gene expression of birnaviruses, genome segment A of both IPNV and IBDV were cloned and sequenced (11,12; DXV is also cloned but sequencing is not complete; Dobos, unpublished data). Both IPNV and IBDV genome segment A contain a single, large, open reading frame (ORF) capable of encoding for 100–105 kD polyprotein. Partial amino acid sequencing and hybrid-arrested translation experiments established that the order of coding regions on RNA segment A is 5'pVP2-VP4-VP3-3' for IBDV and 5' pVP2-NS-VP3-3' for IPNV (11,12,24). Experiments involving in vitro synthesis of 3' truncated IPNV mRNAs from viral cDNA and the subsequent translation of these mRNAs in reticulocyte translation systems revealed that the NS polypeptide is the virus-coded protease and the active site of this enzyme is located at the carboxy terminal sequence (25). There is considerable sequence homology between the pVP2 and VP3 regions of IBDV and IPNV with no homology whatsoever in the middle (VP4, NS) protease encoding sequences (P.J. Hudson, personal communication).

When full-length, in vitro-produced, segment A-specific mRNA was translated in the reticulocyte protein-synthesizing system, a 106 kD polyprotein was detected (product of the large ORF) in addition to the pVP2, NS, and VP3 polypeptides. The combined peptide maps of the three polypeptides overlapped with that of 106 kD protein, indicating a precursor–product relationship (25). The 106 kD precursor polyprotein has also been immune-precipitated recently from infected cell lysates (É. Nagy, unpublished results). These data indicate that genome segment A of birnaviruses generate three polypeptides by rapid cotranslational cleavage, which is not sensitive to those inhibitors that can prevent the cleavage of picornavirus-specific polypeptides. A similar situation is found in flaviviruses, which also code for a large polyprotein that undergoes rapid cotranslational processing (26); in fact, the virus genome was thought to be polycistronic until the establishment of its nucleotide sequence, which revealed a single large ORF (27).

Hybrid-arrested translation experiments, with denatured IPNV segment A RNA, revealed that internal initiation of translation can also occur at in-phase methionine codons to produce VP3 (or VP3 plus NS) without producing pVP2 (24). This mechanism may also take place in vivo since the three polypeptides are not produced in equimolar amounts in infected cells (22). Further experiments are needed to resolve this point.

Although the DXV genome has not been sequenced, the following data indicate that the order of polypeptides on genome segment A is also 5'-pVP2-VP4-VP3-3', that is, (5'67 kD-27 kD-34 kD-3'). During in vitro translation experiments, it was found that undenatured DXV dsRNA could be faithfully translated in the reticulocyte system (16). Treating the dsRNA with S1 nuclease abolished its messenger activity, but if the nuclease was removed by phenol extraction and the dsRNA denatured, it could again be faithfully be translated. This characteristic was most likely due to a large (160 bp) AU-rich region of the RNA which is situated about 30 bp from the 5' end and appeared as a terminal "bubble" under the electron microscope when the virus genome was partially denatured by 70% formamide (14). Most likely, this region is at least partially single-stranded under physiologic conditions, allowing ribosomal binding to the 5' end of the plus strand in vitro. Upon further incubation, ribosomal movement would "unzipper" the dsRNA, thereby allowing the downstream cistrons to be translated. In vitro pulse-chase experiments revealed that the temporal order of

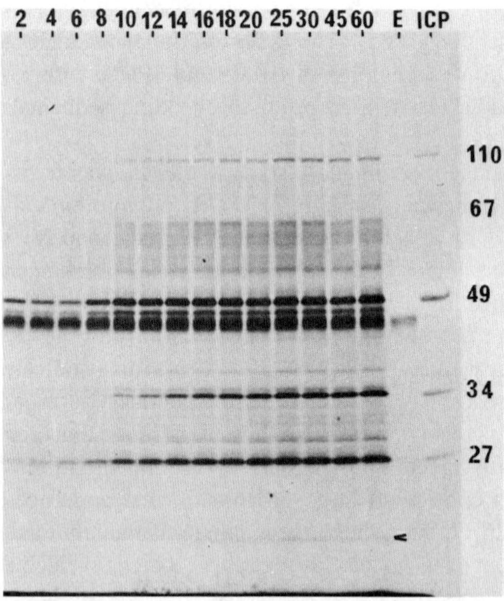


**Figure 5** Pulse-chase experiment in rabbit reticulocyte lysates primed with native (undenatured) DXV dsRNA. Cell-free protein synthesis was initiated using native dsRNA in the presence of [$^{35}$S]methionine. At intervals, aliquots of the reaction mixture were removed and mixed with a 1000-fold excess of unlabeled methionine and incubation continued at 37°C for a total of 60 min. The labeled products were analyzed by PAGE and autoradiography. The numbers above the lanes indicate the time (in minutes) when the "chase" began. E, endogenous product; ICP, infected-cell polypeptide. Arrow indicates possible additional cleavage product. (From Ref. 6.)

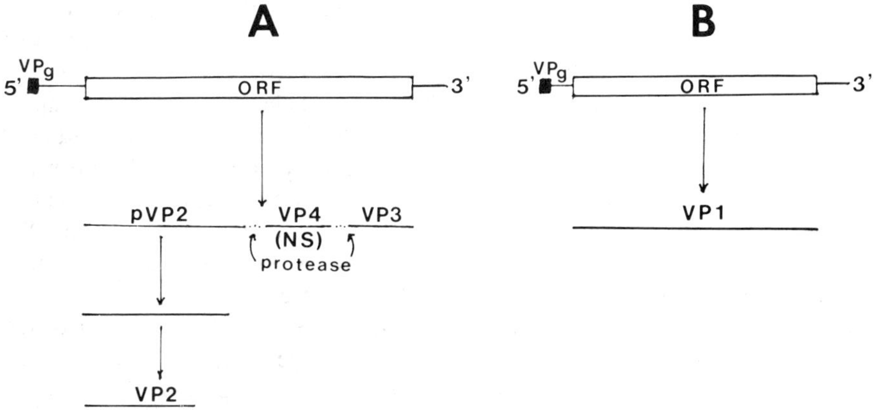


**Figure 6** Genomic organization and expression strategy of dsRNA segments A and B of birnaviruses.

incorporation of [$^{35}$S]methionine into segment A-specific polypeptides is 49, 27, 34; thus an order of 5'67 kD-27 kD-34 kD-3' was tentatively established (16) (Figure 5).

It appears that S1 nuclease treatment of native DXV RNA removes at least part of the single-stranded "bubble" sequence, where ribosomes would bind since the nuclease-treated dsRNA could not serve as messenger in vitro. Subsequent denaturation restored messenger activity, indicating that the nuclease-resistant sequences contained all the coding regions, as well as the essential sequences needed for polypeptide chain initiation.

Figure 6 shows a generalized birnavirus map which is consistent with the available data. In all cases, the large size class VP1 polypeptide is encoded by genome segment B. Sequence A contains a large ORF encoding a large polyprotein, which is cleaved cotranslationally to produce 5'-VP2-VP4-VP3-3'. VP4 (or in the case of IPNV, the NS polypeptide) is the virus-coded protease. In all cases the major outer capsid polypeptide precursor is mapped to the 5' end of the segment A plus strand. In all cases, pVP2 undergoes a number of cleavages (most likely involving its carboxy terminal) to generate VP2. It is not known where the putative VPg is encoded.

## VIRAL PERSISTENCE AND DEFECTIVE INTERFERING PARTICLES

The reduction of virus yield in IPNV-infected cultures following serial passage was reported by a number of workers. CHSE cells that survived a lytic infection grew into persistently infected cultures which continued to produce low levels of infectious virus without any CPE. These cells did not produce interferon and were resistant to infection by homologous virus but were susceptible to heterologous viruses (28). Thus the establishment of carrier cultures was thought to be due to DI particles. Although IPNV DI particles cannot be physically separated from the standard virus, by correlating cell survival with DI particle input, they could be quantitated (29). A 20-fold increase in DI/PFU ratio was obtained as a result of multiple passages; however, this was due to a reduction in the number of plaque-forming units rather than an increase in the number of DI particles. Thus virus samples from diluted passage contained as many DI particles as samples from undiluted passage. Single hit cell-survival kinetics was ascribed to IPNV DI particles which did not replicate at the expense of standard virus; rather, the yield of both particle types was reduced (30). Even when given months to replicate during persistent infection, DI particles of IPNV did not become enriched in CHSE cells. It appears that these cells are poor hosts for DI replication, although they rapidly generate interfering activity. Furthermore, the homologous interference is group specific because it extends to all IPNV serotypes. Persistent cultures can be cured by repeated subculturing in the presence of anti-IPNV antibody (30). Clearly, then, IPNV DI-mediated interference differs from the classic vesicular stomatitis virus system.

The study of viral persistence of DXV is rather limited. Tenninges et al. (4) found that when *Drosophila* cell cultures were infected with DXV at high multiplicities of infection (MOI), the cells lysed in 26 hr; however, some cells survived and were able to colonize the tissue culture flask within 10 days. Only 2% of these cells were positive in immunofluorescence tests, and this percentage did not change after 13 subculturings. At this time, $4 \times 10^7$ infectious units per milliliter (or 90 IU per fluorescent cell per day) were released. The cells were resistant to superinfection at high MOI up to 120 days. These cultures could not be cured by subculturing once a week for nine weeks in the presence of anti-DXV serum (4).

## CONCLUSION

Investigations conducted since the mid-1970s revealed that birnaviruses have the following novel features not found in other dsRNA-containing viruses: (1) only two segments of genomic dsRNA; (2) the presence of VPg; (3) transcription of both monocistronic (genome segment B) and polycistronic (segment A) mRNA, the latter giving rise to three polypeptides by rapid proteolytic cleavage; (4) the presence of a viral protease which is encoded in the central region of genome segment A, similar to the virus-coded protease in CPMV B-RNA (31); (5) possible semiconservative mechanism of RNA replication; (6) in vivo RNA synthesis that is highly sensitive to cycloheximide; (7) a virion-associated RNA polymerase which, unlike that of other dsRNA viruses, cannot be used to generate viral mRNAs in vitro; (8) absence of subviral particles in infected cells; and (9) unusual mechanism of DI particle-mediated homologous interference. These features ensure that further advances in our understanding of the molecular biology of birnaviruses will continue to be exciting.

## ACKNOWLEDGMENTS

The work in our laboratory was supported by the National Science and Engineering Research Council of Canada.

## REFERENCES

1. Brown, F. The classification and nomenclature of viruses: Summary of results of meetings of the International Committee on Taxonomy of Viruses in Sendai, September 1984. *Intervirology* 25, 141, 1986.
2. Dobos, P. and Roberts, T. E. The molecular biology of infectious pancreatic necrosis virus: A review. *Can. J. Microbiol.* 29, 377, 1983.
3. Becht, H. Infectious bursal disease virus. *Curr. Top. Microbiol. Immunol.* 29, 337, 1983.
4. Teninges, D., Ohanessian, A., Richard-Molard, C., and Contamine, D. Isolation and biological properties of *Drosophila* X virus. *J. Gen. Virol.* 42, 241, 1979.
5. Müller, H. and Becht, H. Biosynthesis of virus-specific proteins in cells infected with infectious bursal disease virus and their significance as structural elements for infectious virus and incomplete particles. *J. Virol.* 44, 384, 1982.
6. Nagy, E. and Dobos, P. Synthesis of *Drosophila* X virus proteins in cultured *Drosophila* cells. *Virology* 134, 358, 1984.
7. Chang, N., Macdonald, R. D., and Yamamoto, T. Purification of infectious pancreatic necrosis virus and comparison of polypeptide composition of different isolates. *Can. J. Microbiol.* 24, 19, 1978.
8. Dobos, P., Hill, B. J., Hallett, R., Kells, D. T., Becht, H., and Teninges, D. Biophysical and biochemical characterization of five animal viruses with bisegmented, double-stranded RNA genomes. *J. Virol.* 32, 593, 1979.
9. Dobos, P. Size and structure of the genome of infectious pancreatic necrosis virus RNA. Nucl. Acid Res. 3, 1903, 1976.
10. Azad, A. A., Barrett, S. A., and Fahey, K. J. The characterization and molecular cloning of the double-stranded RNA genome of an Australian strain of infectious bursal disease virus. *Virology* 143, 35, 1985.
11. Hudson, P. J., McKern, N. M., Power, B. E., and Azad, A. A. Genomic structure of the large segment of infectious bursal disease virus. *Nucl. Acid Res.* 14, 501, 1986.
12. Duncan, R. and Dobos, P. The nucleotide sequence of infectious pancreatic necrosis virus (IPNV) dsRNA segment A reveals one large ORF encoding a precursor polyprotein. *Nucl. Acid Res.* 14, 5934, 1986.
13. Persson, R. H. and Macdonald, R. D. Evidence that infectious pancreatic necrosis virus has a genome linked protein. *J. Virol.* 44, 437, 1982.

14. Revet, B. and Delain, E. The *Drosophila* X virus contains a 1 $\mu$M, double stranded RNA circularized by a 67-kDa terminal protein: High resolution denaturation mapping of its genome. *Virology* 123, 29, 1982.
15. Macdonald, R. D. and Dobos, P. Identification of the proteins encoded by each genome segment of infectious pancreatic necrosis virus. *Virology* 114, 414, 1981.
16. Nagy, E. and Dobos, P. Coding assignment of *Drosophila* X virus genome segments: *In vitro* translation of native and denatured virion dsRNA. *Virology* 137, 58, 1984.
17. Dobos, P., Hallett, R., Kells, D. T. C., Sorenson, O., and Rowe, D. Biophysical studies of infectious pancreatic necrosis virus. *J. Virol.* 22, 150, 1977.
18. Dobos, P. and Rowe, D. Peptide map comparison of infectious pancreatic necrosis virus-specific polypeptides. *J. Virol.* 24, 805, 1977.
19. Somogyi, P. and Dobos, P. Virus-specific RNA synthesis in cells infected by infectious pancreatic necrosis virus. *J. Virol.* 33, 129, 1980.
20. Mertens, P. P. C., Jamieson, P. B., and Dobos, P. *In vitro* RNA synthesis by infectious pancreatic necrosis virus-associated RNA polymerase. *J. Gen. Virol.* 59, 47, 1982.
21. Bernard, J. *Drosophila* X virus RNA polymerase: Tentative model for *in vitro* replication of double stranded virion RNA. *J. Virol.* 33, 717, 1980.
22. Dobos, P. Virus-specific protein synthesis in cells infected by infectious pancreatic necrosis virus. *J. Virol.* 21, 242, 1976.
23. Mertens, P. P. C. and Dobos, P. Messenger RNA of infectious pancreatic necrosis virus is polycistronic. *Nature (London)* 297, 234, 1982.
24. Nagy, E., Duncan, R., Krell, P., and Dobos, P. Mapping of the large RNA genome segment of infectious pancreatic necrosis virus by hybrid arrested translation. *Virology* 158, 211, 1987.
25. Duncan, R., Nagy, É., Krell, P.J., and Dobos P. Synthesis of the infectious pancreatic necrosis virus polyprotein, detection of a viral coded protease, and fine structure mapping of the genome segment A coding regions. *J. Virol.* 61, 3655, 1987.
26. Rice, C. M., Lenches, E. M., Eddy, S. R., Shin, S. J., Sheets, R. L., and Strauss, J. H. Nucleotide sequence of yellow fever virus: Implications for flavivirus gene expression and evolution. *Science* 229, 726, 1985.
27. Westway, E. G. Replication of flaviviruses, in *The Togaviruses,* Schlesinger, R. W., Ed., Academic Press, New York, 1980, p. 531.
28. Macdonald, R. D. and Kennedy, J. C. Infectious pancreatic necrosis virus persistently infects chinook salmon embryo cells independent of interferon. *Virology* 95, 260, 1979.
29. Macdonald, R. D. and Yamamoto, T. Quantitative analysis of defective interfering particles in infectious pancreatic necrosis virus preparations. *Arch. Virol.* 55, 77, 1978.
30. Kennedy, J. C. and Macdonald, R. D. Persistent infections with infectious pancreatic necrosis virus mediated by defective-interfering (DI) virus particles in a cell line showing strong interference but little DI replication. *J. Gen. Virol.* 58, 361, 1982.
31. Franssen, H., Goldbach, R., and van Kammen, A. Translation of bottom component RNA of cowpea mosaic virus in reticulocyte lysate: Faithful proteolytic processing of the primary translation product. *Virus Res.* 1, 39, 1984.

# 12

# Parvoviridae of Invertebrates: Densonucleosis Viruses

SHIGEMI KAWASE
*Nagoya University, Chikusa, Nagoya, Japan*
EDOUARD KURSTAK
*University of Montreal, Montreal, Quebec, Canada*

## INTRODUCTION

The family of Parvoviridae includes several small, single-stranded DNA-containing viruses isolated from invertebrate and vertebrate animals as well as from humans (1,2). This family of viruses consists of three genera: *Parvovirus*, *Dependovirus*, and *Densovirus* (1).

The parvoviruses isolated from invertebrates, mainly insects, form a unique group of viruses. All are linked to fatal diseases, well described in *Galleria mellonella* larvae infected by the densonucleosis virus (3–5). The common name of densovirus and densonucleosis virus (DNV) was first used by Kurstak and Vago (6). Two invertebrate parvoviruses, the densovirus isolated from the greater wax moth *G. mellonella* and from the silkworm *Bombyx mori*, have been investigated in detail (7–10). These densoviruses possess cubic symmetry with a diameter of 20–25 nm and a weight of $6 \times 10^6$ daltons.

Densoviruses possess a small amount of polyamines along with DNA. About 70% of the viral particle is taken by four proteins. Viruses do not contain lipids and therefore are resistant to organic solvents and have a high buoyant density in cesium chloride of 1.40–1.45 g/ml.

Densoviruses are autonomous viruses characterized by the separate encapsidation of complementary single-stranded DNA. However, these single complementary strands anneal upon extraction to form a double-stranded DNA. Parvovirus and dependovirus are vertebrate viruses. The former is able to replicate autonomously but the latter requires a coinfection with either a adenovirus or a herpesvirus. In this chapter emphasis is given to *B. mori* densovirus.

## VARIETY AND DISTRIBUTION

The first densonucleosis virus was isolated in 1964 from larvae of the greater wax moth, *G. mellonella* (11). Subsequently, similar viruses were isolated in order Lepidoptera, Diptera, Orthoptera, Blattariae, and Odonata. The host insect species infected with accepted and possible members of DNV so far are distributed all over the world (Table 1).

Of these DNVs, *Bombyx* DNV (BmDNV) is fairly complicated on classification. In 1968, a disease of the silkworm *B. mori* broke out in sericultural farms around the suburbs of Ina city, Nagano Prefecture, Japan, and caused serious economic damage. The disease was initially believed to be caused by the infectious flacherie virus (IFV), a picornavirus, since the symptoms in the moribund larvae were closely similar to those of IFV. Based on the histopathological studies of the infected larvae, however, Shimizu (12) suggested that the "Ina flacherie virus" (Ina isolate) was not an IFV but a heretofore undescribed

**Table 1** Densonucleosis Viruses (DNV)

| Host | Country of Isolation | Year of Isolation |
|---|---|---|
| Lepidoptera | | |
| *Aglais uricae* | United Kingdom | 1973 |
| *Agraulis vanillae* | United Kingdom | 1980 |
| *Bombyx mori* | Japan | 1973 |
| | China | 1982 |
| *Diatraea saccharalis* | Guadeloupe | 1977 |
| *Euxoa auxilliaris* | United States | 1970 |
| *Galleria mellonella* | France | 1964 |
| *Junonia coenia* | United Kingdom | 1973 |
| *Pieris rapae* | China | 1984 |
| *Sibine fusca* | France | 1977 |
| Diptera | | |
| *Aedes aegypti* | Soviet Union | 1972 |
| *Simulium vittatum* | United States | 1976 |
| Odonata | | |
| *Lencorrhinia dubia* | Sweden | 1979 |
| Orthoptera | | |
| *Periplaneta fuliginosa* | Japan | 1979 |

virus in the silkworm. This new disease and its virus have been studied by several Japanese investigators (10). On the basis of these data, they concluded that the virus differed from IFV and was similar in cytopathological and physicochemical characteristics to a DNV of *G. mellonella* (GmDNV) (4). They proposed that the virus be named *Bombyx* densonucleosis virus (BmDNV). Subsequently, similar DNVs were isolated in the silkworm independently by Matsui (13) and Furuta (14). By chemical and serological studies, the virus isolated by Furuta was found to be identical or closely similar to the Ina isolate (15).

Recently, another new type of DNV was discovered in diseased silkworms in mainland China (16). The histopathological features of the midgut cells infected with China isolate were clearly different from those of the Ina isolate. New types of DNV now have been isolated from the silkworm at Saku city (Saku isolate) (17) and Yamanashi Prefecture (Yamanashi isolate) (18), in Japan. They caused a fatal disease in silkworm strains, which are resistant to the Ina isolate.

The serological relationships among several isolates of BmDNVs have been investigated (19). The DNVs isolated by Matsui and by Furuta were closely related serologically to Ina isolate. No serological relationships occur between Ina and China, Ina and Yamanashi, and Ina and Saku isolates. It has also been determined that Saku isolate is related to Yamanashi isolate. Although China and Yamanashi isolates have a common antigen, according to Iwashita (20) these isolates are not identical. Therefore, the BmDNVs are at least classified into two or three groups. To avoid confusion in names, according to Siegl et al.'s proposal in 1985 (1), the authors wish to designate these DNVs as follows: Ina isolate and DNVs isolated by Matsui and by Furuta as the BmDNV type 1; Saku and Yamanashi isolates, type 2; China isolate, type 3.

BmDNV-1 is not related serologically to GmDNV (21). A virus that multiplies in the midgut of the mulberry pyralid, *Glyphodes pyloalis*, is not distinguishable serologically from BmDNV-1 (22,23). *Pseudoplusia* DNV is related to GmDNV, but they are not identical (24). *Casphalia* DNV is not related to *Gm* or *Junonia* DNV (25).

## STRUCTURE

### Structure of the Viral Capsid

#### Size and Buoyant Density of the Virion

Although sizes of all DNVs seem to have diameter of the order of 20–25 nm, some variations exist among different DNVs. On the BmDNV-1, Nakagaki and Kawase (26) obtained a value of 22 ± 0.5 nm based on tobacco mosaic virus particles as an internal standard. The size of BmDNV-2 and 3 are almost same or slightly larger than type 1, i.e., type 2, 21–23 nm (18); type 3, 23–24 nm (16).

The sizes of other DNVs are as follows: *Junonia coenia*, 20–22 nm (27); *Aedes aegypti*, 20 nm (28); *Acheta domestica*, 22 nm (29); *Diatraea saccharalis*, 22 nm (30); *Sibine fusca*, 20–22 nm (31); *Periplaneta fuliginosa*, 20–21 nm (32); *Agraulis vanillae*, 20 nm (33); *Pieris rapae*, 25 nm (34); *Casphalia externa*, 22 nm (25); *Pseudoplusia includens*, 20 ± 1 nm (24). According to a Chinese book published recently, *Catopsilia*, *Dendrolimus*, and *Odontosia* DNVs are 20 ± 1, 22–25, and 24 nm, respectively (35).

The occurrence of particles with two different densities has been observed in BmDNV-1 (1.40 and 1.45 g/ml) (26) and *Periplaneta* DNV (1.41 and 1.44 g/ml) (36). *Pseudoplusia* DNV has bouyant density of 1.40 g/ml for the full virus particles, but another component denser than the value was always observed (24). These dense particles are probably the "dense precursor particles."

### Structure of the Viral Capsid

Viral capsid is constructed by structural units, capsomers. In Parvoviridae including DNV, determination of the fine structure is difficult because of the small size of the virion. There are several reports on the number of capsomers in parvoviruses (37). For example, in Kilham rat virus (KRV), H-1, H-3, HT, and HB viruses, the number reported was 32; in adeno-associated virus (AAV) it was 12 or 32.

Structure of the BmDNV-1 capsid was investigated in detail with highly magnified micrographs of the virion using rotation technique (38). The external shape of the virion appeared to be polygonal, with a maximum linear dimension of 22 nm; some of the capsids were clearly hexagonal and others nearly circular of intermediate in outline. Observation of higher magnification revealed that the virion had twofold, threefold, and fivefold axes of symmetry, and the fine structure of the virus seemed to be in agreement with a model of an icosahedron with 12 capsomers. The structure of BmDNV particles is similar to GmDNV (Fig. 1) (5).

In some electron micrographs, globular structures with a diameter of about 11 nm were seen. To cover the surface of a virion of a diameter of 22 nm by means of 12 globular structures, the theoretical diameter of a capsomer would be about 10 nm. This could be interpreted as suggesting that the globular structures were possibly capsomers from disrupted virus particles. The presence of such capsomerlike globules supports the 12-capsomer model (38).

Protein analysis of BmDNV-1 revealed that the particle contains about 60 molecules of VP1 (major viral protein, *Mr* 50,000) which appear to be the repeating units for the capsid structure (39). On the basis of electron micrographs, it was concluded that each capsomer is a pentamer of VP1. The functions of remaining three proteins (VP2, VP3, VP4) for the construction of the capsid are not known. From these results, it is highly probable that the surface of BmDNV-1 particle has the form of an icosahedron constructed by 12 capsomers.

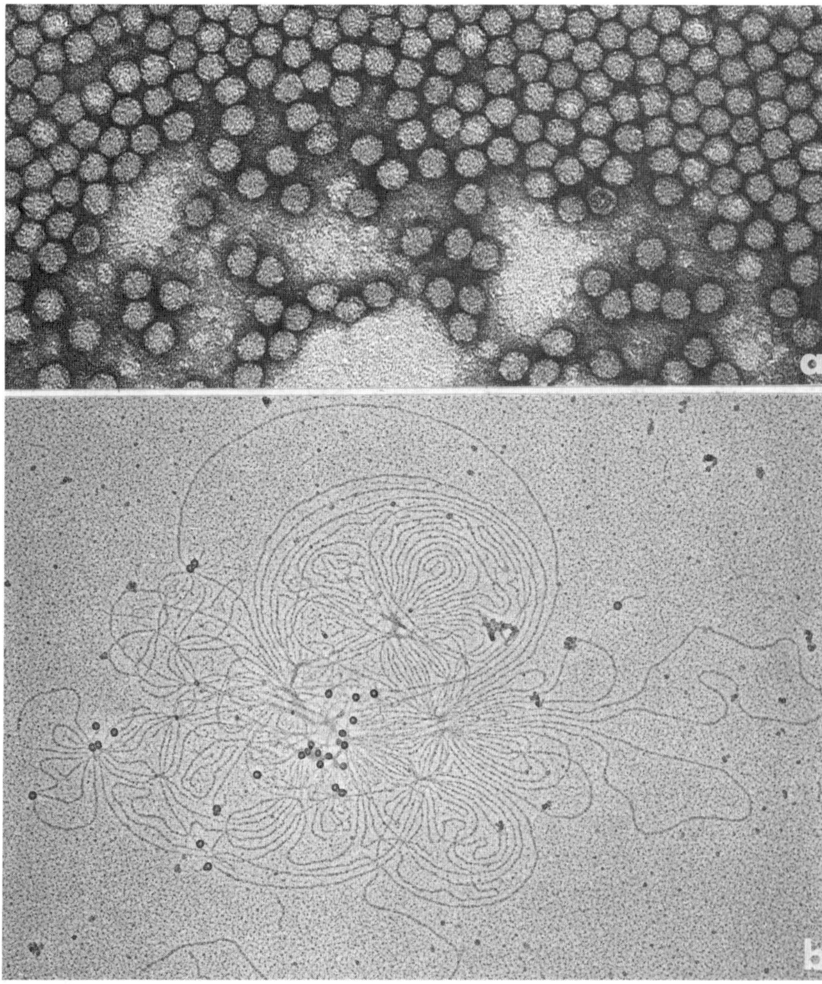


**Figure 1** *Galleria mellonella*. Densonucleosis virus particles (a) and DNA preparations (b) of GmDNV. (*From Professor Kurstak's collection.*)

## Structure of the Viral Genome

Structures of the viral genome of several DNVs have been investigated by electron microscopy (Figs. 2–4). Electron micrographs of the double-stranded (ds) DNA of BmDNV-1 revealed that the DNA was mainly composed of linear molecules with an average length of 1.7 μm, and less well-defined structures were often observed (26). The double-stranded molecule had a molecular weight of $3.4 \times 10^6$, which was determined by both electron microscopy and agarose gel electrophoresis (Fig. 5). Although no circular DNA molecule was observed in the preparations, 3.4 μm length molecules, which were assumed to be linear duplex dimers, were rarely observed. This fact indicates that a cohesive end is also present in the DNA molecule of BmDNV-1, through which two molecules are held together by short complementary single-stranded (ss) regions.

In the case of BmDNV-2, it has been known that the DNA preparation from BmDNV-2 is resolved in an agarose gel electropherogram (Fig. 5) into two closely migrating bands, DNA I and II (17,40). Since the size of BmDNV-2 is very small (24 nm in diameter), it is unlikely that a virion has two DNA species ($2.1 \times 10^6$ and $1.7 \times 10^6$) in its capsid. To explain this phenomenon, Kurihara et al. (17) suggested that the two DNA bands may have been derived from two different configurations of ssDNA during the course of DNA extraction from the virion. On the other hand, Kawase et al. (40) found that DNA I and II reextracted from agarose gels consist of different base sequences if these molecules are the same ds linear strands, since the electrophoretic profiles of fragments treated with restriction enzymes from the two DNA preparations differed from each other. They suggested that BmDNV-2 may be a mixture of two similar but different DNVs.

When DNA preparations from BmDNV-2 were observed by electron microscope, two groups of linear dsDNA strands of different length, mean lengths 1.7 and 2.1 μm, were observed (41). Infrequently, very long dsDNA molecules partially attached to ssDNA were observed. This phenomenon suggests that the DNA molecules have cohesive ends.

To determine whether the two different-length DNA molecules are derived from DNA I and II molecules, DNA I and II were reextracted from these bands separately and observed by electron microscope (41). Both DNA I and II were linear dsDNA molecules, and no difference in structure of their DNA strands was observed, but DNA I was slightly longer than DNA II. The molecular weights of DNA I and II were calculated by their length. The average lengths were 2.1 (DNA I) and 1.7 μm (DNA II), which were equivalent to $M_r$ $4.2 \times 10^6$ and $3.4 \times 10^6$ as dsDNA, respectively. These data almost coincide with the values described by Kawase et al. (40).

On *Pseudoplusia* DNV, inverted terminal repetition of the DNA was observed by electron microscopy (24). The terminal repetition comprised about 6–7% of

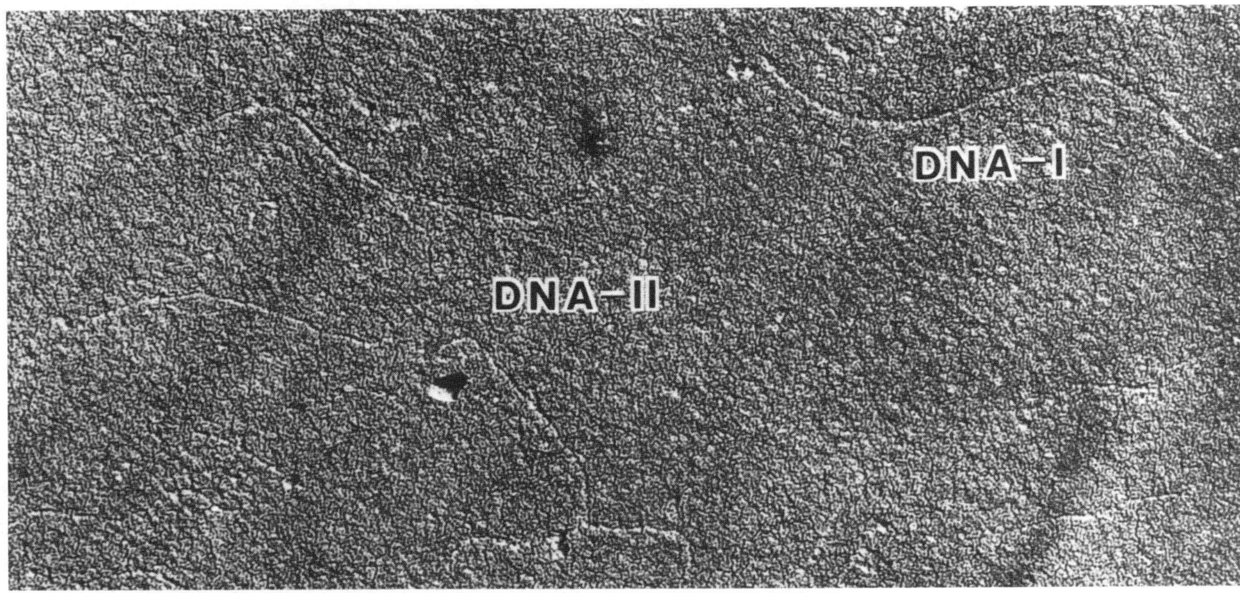


**Figure 2** Electron micrograph of DNA I and II molecules in the DNA preparation of BmDNV-2. (Courtesy of Dr. Y. Y. Hu.)

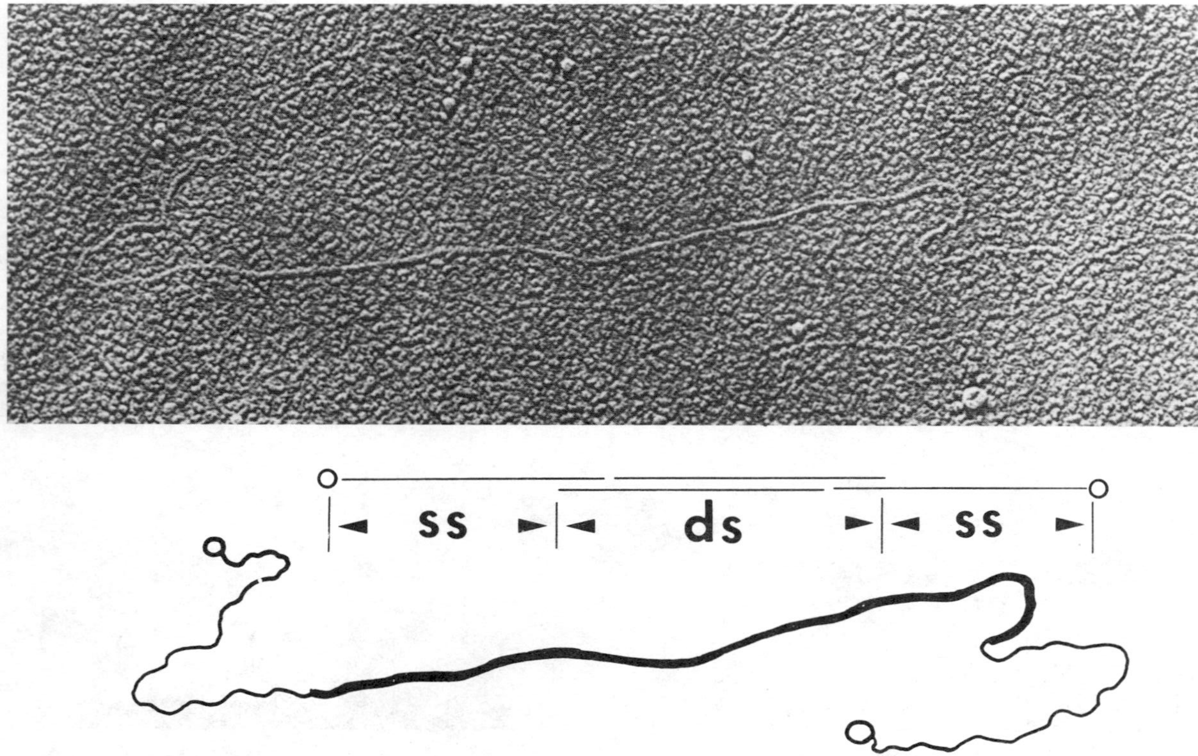


**Figure 3** Double-stranded DNA molecule attached to single-stranded DNAs and empty capsids at both ends (BmDNV-2). (Courtesy of Dr. Y. Y. Hu.)

**Figure 4** Electron micrographs of DNA I and II reextracted from agarose gel electropherograms (BmDNV-2). (Courtesy of Dr. Y. Y. Hu.)

the genome. Circular ssDNA molecules with a ds "panhandle"-like protrusion or dsDNA with two such protrusions were observed. These protrusions are most simply explained by the presence of inverted terminal repetition in the genome. The size of these panhandle structures was estimated to be 440 $\pm$ 50 base pairs. The size of the linear dsDNA as determined by electron microscopy was 6300 $\pm$ 90 base pairs, or $4.1 \times 10^6$ daltons.

In the case of *Casphalia* DNV, linear dsDNA was observed by electron microscopy, and the length was about 2 $\mu$m. The molecular weight of the ssDNA was $1.9 \times 10^6$ (25).

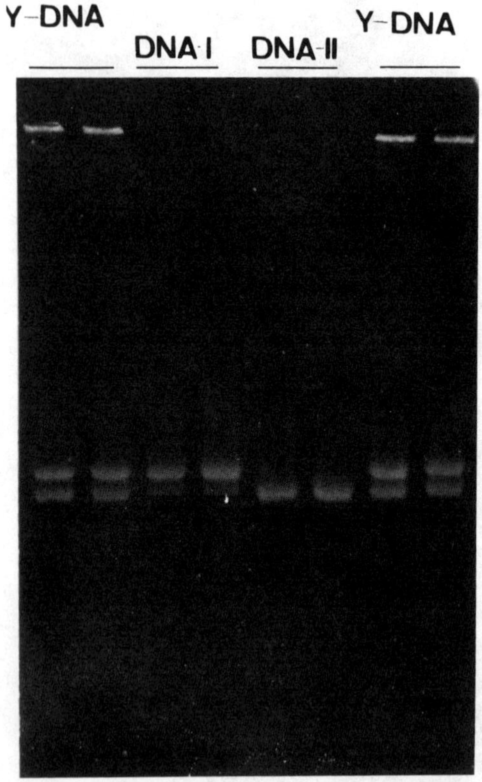


**Figure 5** Electropherograms of DNA from BmDNV-2 (Y-DNV) and reextracted DNA I and II. (Courtesy of Dr. Y. Y. Hu.)

# PHYSICOCHEMICAL PROPERTIES

## General Characteristics

All DNVs contain DNA and protein, but no essential lipid, since they are resistant to ether and chloroform treatments. In addition, small quantities of polyamines have been detected in *Galleria mellonella*, *Aglais urticae*, and *Bombyx mori* DNVs (42,43).

Generally, viruses are inactivated by certain physicochemical treatments, and they are one of the fundamental bases for the classification of viruses. Furthermore, these properties are important factors for protection of silkworm from this disease. However, only a few data are available on the effects of physicochemical treatments on DNVs.

It has been reported that BmDNV-1 is not inactivated at 65°C for 20 min but inactivated considerably at 80°C for 10 min (44). The virus was stable for ether treatment but inactivated by 3% formalin treatment for 2 h.

Recently, effects of physicochemical treatments and some characteristics on BmDNV-2 were investigated by Seki (45). The virus was not inactivated by ether or sodium desoxycholate treatments. Under 0.5 and 1% formalin treatments for 10 min, it was inactivated considerably; it was almost completely inactivated at 3%. On the effects of pH, it was not inactivated in the range of 3.0–12.0 but was inactivated at pH 2.0 for 1 h. By ultraviolet irradiation, the virus was noticeably inactivated for 20 min and completely for 60 min. It was not inactivated at 50°C for 60 min but was inactivated to some extent at 60°C and completely at 70°C for 20 min.

*Periplaneta* DNV was stable at 56°C for 20 min and was not inactivated by lipid solvent (36). *Catopsilia* and *Dendrolimus* DNVs were also heat stable, but they were inactivated at 60–70°C (35). These general characteristics are similar to those of other parvoviruses.

## Nucleic Acid

As stated earlier, DNVs possess an ssDNA in the capsid, since the addition of formaldehyde to the virus preparation produced an increase in absorption over a period of 24 h and a shift in absorption maximum. Furthermore, acridine orange staining of the virion provides the evidence of ssDNA in the capsid. It is confirmed by several workers that DNV contains two complementary ssDNA strands separately in different particles. When DNA was extracted under a solution of high ionic strength, the two complementary single strands were joined together by complementary base pairing and became one dsDNA (3,4).

It was reported that BmDNV-1 contains 28% DNA in the virion (26), and the base ratio (G:A:C:T) is 22.1:30.0:20.6:27.2 (46). Interestingly, in BmDNV-2, the viral DNA was resolved in agarose gel electrophoresis into closely migrating bands, 3.8 and $4.0 \times 10^6$ daltons, as dsDNA. In the case of *Pseudoplusia* DNV, DNA content in the virion was 38% (24).

## Protein

Generally, vertebrate parvoviruses consist of three structural proteins (37). In contrast, when examined on 10% polyacrylamide gels, all DNVs reported so far have four polypeptides with $M_r$ 43,000–120,000 except for BmDNV-2 (10,40). In most DNVs, the protein with the lowest molecular weight is the major one.

The structural proteins of BmDNV-1 were investigated in detail using SDS-PAGE (39). Four structural proteins were found as well as other DNVs (8,9). The molecular weight was estimated from the relative mobility in various gel concentrations and the retardation coefficient. The major protein (VP1), accounting for 65% of the total virion protein, had $M_r$ 50,000. The BmDNV-1 particle contained about 60 molecules of VP1, which is believed to be capsid protein, as previously stated. In GmDNV, the estimates of the molecular weights of viral proteins varied considerably with a change in gel concentrations (47). However, in BmDNV-1, the variations did not occur when the electrophoresis was conducted at different gel concentrations (39).

BmDNV-1 possesses four structural proteins with a total molecular weight of about $2.5 \times 10^5$ (39). If the total molecular weight represents the products of individual genes, this would exceed the coding capacity of the DNA.

The amino acid compositions of the four structural proteins were investigated (48). The data indicate some similarities among the four proteins. These proteins, in common, contained acidic rather than basic amino acids, and their relative concentrations of each amino acid were very similar. The peptide map of VP1 almost coincided with that of VP2. All of the fragments of VP3 obtained after the digestion with *Staphylococcus aureus* V8 protease were also found in VP4, whereas the digests of VP3 and VP4 contained at least three extra bands as compared to VP1 or VP2. The strikingly similar peptide patterns of the structural proteins suggested that VP3 and VP4 might have extensive sequence homology. Moreover, two smaller proteins, VP1 and VP2, might be derived, at least partially, from a common DNA sequence.

In the case of BmDNV-2, the results on structural proteins are fairly complicated. According to Kurihara et al. (17) the DNV (Saku isolate) consists of four structural proteins. Peptide mapping of these proteins revealed that the pattern of VP1 was similar to those of VP2, and that of VP3 to VP4. On the other hand, Kawase et al. (40) reported that BmDNV-2 (Yamanashi isolate) consisted of six proteins. They suggested that the DNV may be a mixture of two similar but different DNVs consisting of three or four polypeptides each, since the DNV preparation contained two different DNA species, as noted. According to Iwashita, BmDNV-3 seems to consist of six proteins (49).

In the case of *Periplaneta* DNV, four polypeptides of $M_r$ 48,000–76,000 were detected on 10% gel system (36). About 60% of the total viral protein was composed of the major peptide, which was not the lowest molecular weight. Both *Casphalia* (25) and *Pseudoplusia* DNVs (24) also consisted of four polypeptides.

## Polyamine

It has been reported that the naturally occurring polyamines spermine, spermidine, and putrescine and related polyamines occur in many viruses as well as in biological materials. Recently, functions of polyamines have been reported by several investigators (50–53).

Data from quantitative analyses of polyamine contents in BmDNV-1 indicate that the contents and ratio of spermine, spermidine, and putrescine clearly are different from those of Gm and *Aglais* DNVs (42,43). Furthermore, the ratio of the three polyamines found in BmDNV-1 differed considerably from that of silkworm body (50).

Despite many reports about polyamines contained in various virus particles (51–53), the function of polyamines in viruses is not fully understood. It is possible, however, that they could be an aid in packaging the genome inside the virus particles by neutralizing some of the negative charge (51). An in vitro study supports this hypothesis (52). It has also been reported that polyamines sufficient to neutralize 9–13% of the genome are incorporated into EMC virus particles preferentially from the host cell pool (53), and these results suggest that polyamines may serve an important function in the virus life cycle.

## MOLECULAR BIOLOGY

### Molecular Biology of Multiplication

Molecular biology study of DNV replication is limited since a suitable cell line for the virus replication has not yet been established.

According to Kobayashi (54), on analysis of DNV polypeptides translated in vitro through mRNA from the midgut in the silkworm, the amounts of polypeptides translated changed significantly in the uninfected midgut during larval development, and those of the midgut infected with BmDNV-2 followed a similar pattern to that in uninfected midgut. This result suggested that the DNV infection had little effect on the amount of translatable mRNA in the midgut at least until 10 days after infection. When the translation products were analyzed by SDS-PAGE and fluorography, it was found that more than 40 polypeptides were translated through mRNAs from the uninfected midgut. On the other hand, four polypeptides were additionally translated in the midgut infected with the DNV from 2 days after infection. Relative amounts of these polypeptides were different from one another but changed little throughout the experiment. Approximate molecular weights of the specific polypeptides were 115,000, 53,500, 49,000, and 46,500, respectively. This was similar to those of structural polypeptides of BmDNV-2. Detailed experiments are in progress to explore the possibility that these polypeptides translated in vitro are structural proteins of BmDNV-2 using the antibody.

### Molecular Evolution

To clarify the relationship between mammalian and insect parvoviruses from a viewpoint of molecular evolution, the genome sequence homology was compared, since over 85% of BmDNV-1 genome sequence, which is described in the next section, has been determined (55). By a dot matrix method, the genome sequence of rodent parvovirus, H-1 virus (H-1), minute virus of mice (MVM), a human dependovirus, adeno-associated virus 2 (AAV-2), and BmDNV-1 were compared.

The sequence comparison among these virus genomes indicates that there is a region showing more than 80% homology between H-1 and MVM. In the nucleotide sequence of 440 bases in that region, a subregion of more than 40% homology existed between H-1 and AAV-2. Furthermore, within that region the BmDNV-1 genome was also shown to have 35% homology with the H-1, MVM, and AAV-2 genomes in a 300 base sequence.

Therefore, the nucleotide sequence in this region has been conserved in each of the evolutionary diverged members of the family Parvoviridae. Interestingly, this conserved region is located not only within open reading frame 2 (ORF2) of the BmDNV-1 genome but also within sequences encoding the nonstructural proteins of the other viruses examined. Thus this domain may code an important active site in the gene product.

These regions of the four viruses were aligned with each other to estimate the number of nucleotide substitutions, and the proportion of different nucleotides in these homologous regions was computed among the different viruses. By using these proportions, the total number of nucleotide substitutions per site at all codon positions was estimated by the four-parameter method.

These results confirmed that the substitution number at the third position of a codon is much larger than those at the first two positions. Hydrophobicity in this region was also conserved during the evolution of these viruses. The observations indicate that this conserved region may be important in biological function. Based on the homology of the region, a phylogenetic tree for these four viruses was constructed (55).

Although the significance of the conserved region still remains obscure, it is tentatively concluded that the rodent parvoviruses (H-1 and MVM), a human dependovirus (AAV-2), and an insect densovirus (BmDNV-1) share a common ancestor.

The rate of evolution of these DNA viruses is of interest, because it is not clear whether these virus genomes evolve at the same rate as the host genomes. If we use the divergence time between the orders Rodentia and Primates (80 million years) as that of the rodent parvoviruses H-1 and MVM and the human dependovirus AAV-2, the nucleotide substitution rate of these viruses can be estimated to be about $3.5 \times 10^{-9}$ per year per site.

## GENETICS

### Physical Map of DNV Genome

A physical map of the BmDNV-1 genomic DNA was constructed by cleaving it with 16 restriction endonucleases, and the digests were analyzed by agarose gel electrophoresis (56).

The fragments obtained from the various single and double digestions were also investigated. From these data, it was concluded that XbaI, XhoI, BbeI, SmaI, EcoRV, KpnI, HpaI, and StuI did not cleave the viral DNA. BamHI, HindIII, PvuII, and SalI cut the DNA at one site; PstI and BglII at two sites; and DraI at three sites. BamHI and HindIII sites could be deduced from the analysis of double digests almost equal to that of uncut viral DNA, and no extra bands were seen.

The map shows that BglII cut the DNA to three fragments. The size of the largest fragment is $1.3 \times 10^6$ daltons, which covers more than 85% of the full length of viral DNA. This fragment was analyzed for the nucleotide sequence of BmDNV-1.

On the BmDNV-2 (Yamanashi isolate), as stated earlier, the viral DNA was resolved in agarose gel electrophoresis into two closely migrating bands, DNA I and II (40). When DNA II was treated with HindIII, three fragments were obtained. However, the physical map of DNA I and II is not yet constructed.

## Nucleotide Sequence of the Viral Genome

More than 85% (4273 nucleotides) of the complete nucleotide sequence of the BmDNV-1 has been determined (55) (Fig. 6). The computer analysis showed

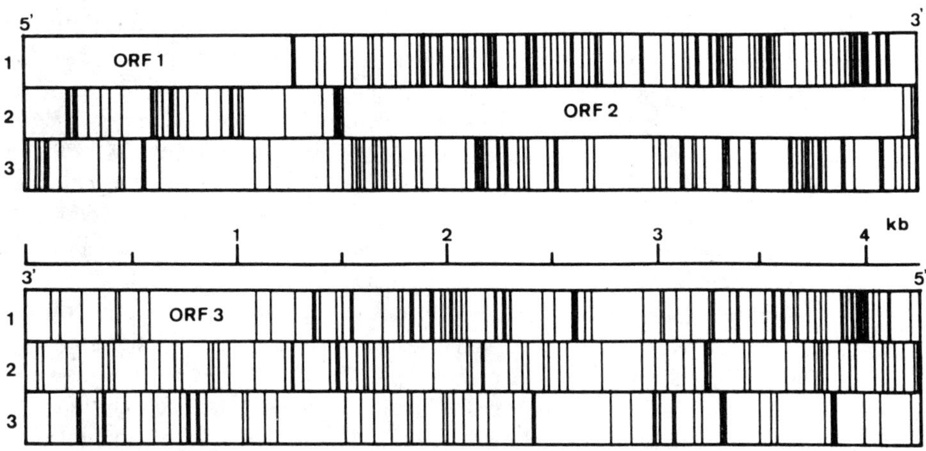


**Figure 6** ORFs in the BmDNV genome. The upper three lines correspond to the three ORFs in a genome strand and the lower three lines are those in the complementary strand. The vertical lines represent the position of the termination codon in each frame. kb = kilobases. (Courtesy of Dr. H. Bando.)

that there are no recognition sites for restriction enzymes HpaI, KpnI, SmaI, XbaI, XhoI; and there is only one site each for BamHI, HindIII, SalI, and PvuII; two sites for PstI and BglII; and four sites for DraI. These results were consistent with the physical map of the genomic DNA described previously. The presence of inverted repeats at both ends of mammalian parvovirus genomes have been reported, but that of BmDNV-1 is not known, since about 300 nucleotides at either end of the genome DNA remain to be sequenced.

Open reading frames in the DNA sequence were determined by computer analysis (55). The BmDNV-1 genome contains at least three major open reading frames, two of which (ORF1 and 2) lie in the same strand, and the third (ORF3) is in the complementary strand. ORF1 contains 1290 nucleotides from nucleotide 1 to 1290. If ORF1 is translated without interruption through RNA splicing to the terminator at nucleotide 1291, a protein of 43,000 daltons will be produced. However, this domain corresponds to the region starting from the upstream region undetermined, which indicates that there is a possibility of the frame coding a protein of $M_r$ ~50,000 at maximum.

Another larger open reading frame (ORF2) containing 867 codons is located between nucleotide 1529 and 4210 and will generate a protein of about 87,000 daltons. The rather small open reading frame (ORF3) in the complementary strand consists of 120 codons. To further define the genetic organization and coding potential of BmDNV-1 genome, the distributions of transcriptional and translational signals within the sequence are analyzed (55).

In eukaryotes, the control regions of translation by RNA polymerase II are often chracterized by a TATA box which occurs approximately 30 nucleotides upstream of the actual sites of initiation of the RNA chain (57). Although the precise function of the TATA sequence is not clear, it is generally thought to specify the starting site for transcription. Other sequences in the –70 to –80 region have also been implicated in RNA polymerase II transcription of both viral and nonviral genes, and the consensus sequence is GGPyCAATCT (CCAAT) (58). On the other hand, only AUG codon is used for the translation initiation signal in eukaryotes. It must, however, be noted that the AUG codons at the actual initiation sites usually have a purine at their –3 and +4 positions (59). In particular, more than 95% of the –3 positions are purine nucleotides.

A search of the determined BmDNV-1 genome sequence for such transcription and translation initiation signals indicates that there are three initiation signal sets for ORF2 (55). The first set preceding the frame contains CCAAT box, TATA box, and Pu**AUG sequences. Therefore, if these signal sets are functional, proteins of $M_r$ about 88,000, 66,000, and 54,000 will be produced, respectively. On the other hand, the promotor region from ORF1 will be located on the upstream region undetermined. Consequently, it is difficult to identify the actual coding potential at present.

As for ORF3 in the complementary strand, it seems that this frame also has signals for transcription (55). However, the –3 and +4 positions of all AUG codons within ORF3 are pyrimidine nucleotide. Therefore, it is doubtful whether ORF3 has codons for any protein.

Although the exact arrangements of the coding sequences into the viral proteins have yet to be determined, the analysis was extended to compute the frequence of usage of codons for the putative genes in BmDNV-1 genome (55). The strategies of codon usage for three frames were similar, but the frequency of usage of the codons for ORF2 was more biased than those for ORF1. Comparison of the amino acid composition of the structural proteins and those of the three frames derived from each nucleotide sequence revealed that the patterns obtained from structural proteins were similar to those for ORF2 but very different from those for ORF1 or 3 (55).

From these results, Bando et al. (55) concluded that all four structural proteins of BmDNV-1 are coded for on ORF2. This also indicates that nonstructural protein will be coded for on the left half of the genome (ORF1) as well as other parvoviruses from mammals.

## CYTOPATHOLOGY

### Histopathology

Histopathological studies on the midgut epithelium of the silkworm infected with BmDNV-1 showed that nuclei of the columnar cells hypertrophied and stained markedly with Feulgen reaction or methyl green (60). The degenerated columnar cells were liberated into midgut lumen.

In the case of BmDNV-2 and DNV-3, almost same features were observed under a light microscope; i.e., at the last stage of infection, the infected nuclei were more than 2.5 times as large as normal nuclei (16,18). The whole area of the nucleus changes into a dense and homogeneous structure which is strongly stained with methyl green and Feulgen reaction. With degradation of the infected nucleus, pyronine stainability of the damaged cytoplasm decreased as a result of vacuolation. Eventually the degenerating cytoplasm is pushed by the hypertrophied nuclei and begins to break down, and cells infected with the virus are liberated into gut lumen.

The midgut cells infected with BmDNV-3 have been studied by fluorescent microscope after staining with acridine orange (16). The nuclei of noninfected cells are green, but when infection progresses, a green-yellow fluorescent area appears in the nucleus. At the last stage of infection, the orange-yellow or red intranuclear inclusion occupies the whole nucleus. This phenomenon indicates the presence of virus particles with ssDNA.

The DNA content of the individual BmDNV-3–infected nuclei was examined by microspectrophotometry (16). The relative Feulgen-DNA content was measured for 10–20 nuclei from each slide by absorbance of one-wavelength and two-wavelength methods. For comparison, the DNA contents of 200 nuclei were measured and the average was taken. The result revealed that DNA content in the case of infected nuclei was 13–16 times as large as in the case of noninfected nuclei.

## Electron Microscopy

The first ultrastructural change that occurs in the silkworm infected with BmDNV-1 is the appearance of electron-dense bodies in the nucleus, suggesting that there is virogenic stroma. Virogenic stroma is not observed in the cytoplasm (60).

The ultrastructural changes of midgut cells infected with BmDNV-3 occur in both the cytoplasm and nucleus (16). In the nucleus at an early stage of infection, the chromatin disappears and many nucleoli appear. The nucleus hypertrophies greatly and is gradually filled with a network of electron-dense granules, which is probably the precursor of the virogenic stroma. In this replication, two different patterns occur in the infected nuclei: In one the virions replicate in linear array; in the other the virions disperse as patches, gradually condense, and finally form large masses. In the case of BmDNV-2, the pathological changes in thin sections were almost the same as those of BmDNV-3 (18).

In *Aedes* DNV, the first changes are observed in the cytoplasm of DNV-infected cells (61). They consist of the formation of paracrystalline structures containing particles of 18–20 nm in diameter. Virogenic stroma and paracrystalline virion arrays are found in the nuclei of the virus-infected cells. A similar crystalline array of virus particles was observed in the cytoplasm of the fat body, muscle, and pericardial cells of *P. fuliginosa* infected with *Periplaneta* DNV (36).

In the case of *Sibine* DNV, cell alterations, mainly at the nuclear level, have been found in thin sections of the thick-end intestinal wall (31). At a less advanced stage of infection, the structure of the nucleus is already altered, and the nuclear substance is divided into opaque masses. The virogenic mass continues to grow with small cavities around which virus particles are formed. The mitochondria hypertrophy and burst, and their contents are released leaving vacuoles in their place.

On the *C. extranea* infected with *Casphalia* DNV, accumulation of viral particles was observed in the periphery of the virogenic stroma (25).

The electron microscope observation of thin sections on the cells infected with *Pseudoplusia* DNV revealed that most cytopathic effects were found in the nuclei, in which the chromatin was greatly reduced in amount and confined to the vicinity of the nuclear envelope (24). In this case, crystallized virus particles in the nucleus were observed.

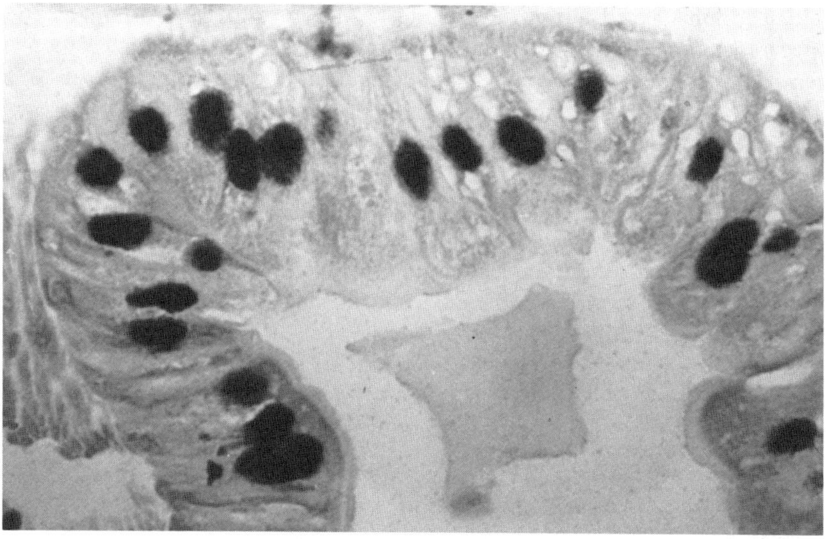

**Figure 7** Incorporation of [$^3$H]thymidine in the nuclei of the silkworm midgut infected with BmDNV-3. (Courtesy of Dr. Y. Iwashita.)

## Autoradiography

On both Gm and BmDNVs, the site of synthesis of the viral nucleic acid and protein has been studied by autoradiography.

In the case of BmDNV-1, the DNA synthesis occurs predominantly in the infected nuclei of columnar cells of the silkworm midgut, indicating that the virus multiplies in the nucleus (60,62). On the BmDNV-2 and DNV-3, a large accumulation of [$^3$H]thymidine was observed in the infected nuclei (Fig. 7).

The multiplication of BmDNV-1 was greatly reduced when the infected larvae reared at 25–28°C were transferred to an environment at the high temperature of 37°C (62). Autoradiographic results using [$^3$H]thymidine and [$^3$H]-tyrosine revealed that the synthesis of both viral DNA and protein was greatly reduced in the infected larvae maintained at 37°C. These results indicate that high temperature (37°C) may reduce the activity of enzymes concerned with viral DNA and protein syntheses (62).

## Immunofluorescence and Immunoperoxidase Techniques

Immunofluorescence technique has been applied with BmDNV-1 (62). Of the various tissues in the infected silkworm larvae, only the midgut epithelium shows specific fluorescence. The fluorescence is restricted to the nuclei of midgut columnar cells, showing the site of virus multiplication. Fluorescence appears sporadically in a few columnar cells 24 h after virus inoculation; it becomes more intense and the number of fluorescing cells increases with time. At 72 h, most of the columnar cells near the cardiac valve do not show any fluorescence even at the late stage of the disease (62,63). On BmDNV-2, specific fluorescence was observed in the hypertrophied nuclei of midgut infected with the DNV by fluorescent antibody technique (64).

As noted, BmDNV-1 multiplication is greatly reduced when the infected larvae are reared at high temperatures (62). Fluorescent antibody studies also confirmed that the synthesis of the DNV antigen in the larvae was inhibited under the high temperature.

Recently, on BmDNV-2 and DNV-3, immunoperoxidase technique was applied to detect the DNV antigen in an early infection state. This technique was very effective for detection of the DNV antigen, because the sensitivity was very high.

## PATHOBIOLOGY

### Multiplication

In the silkworm *Bombyx mori*, the multiplication process of BmDNV-1 was investigated by applying the single radial immunodiffusion method, and relative amounts of DNV antigen were determined by the size of precipitated ring (50). When BmDNV-1 was inoculated at the first day of the fourth instar, virus titer increased rapidly from 24 h after inoculation and reached maximum at 96 h after inoculation. At the late stage of infection, considerable decrease of virus titer was observed. The excretion of the virion into gut lumen from midgut epithelium is considered to be one cause of the matter, since in the last stage of infection a large quantity of the virion was recovered from the feces of DNV-infected larvae (65).

In the case of BmDNVs, the target organ is the midgut. The changes in protein, nucleic acid, and polyamine concentrations in the midgut epithelium during the course of the BmDNV-1 infection were studied to establish the dynamic changes in the metabolism of the midgut infected with DNV (50). No marked differences were observed between the infected and the healthy midgut epithelia with regard to the concentrations of DNV, RNA, and protein, except for the rather high concentration of DNA in the infected midgut during the last 4 days. Three kinds of polyamines, spermine, spermidine, and putrescine, were found in the DNV-infected and healthy midguts. The spermine/spermidine ratio in the infected midgut during the high viral multiplication period was lower than that of the healthy midgut.

In BmDNV-2–infected midgut, however, marked increase of DNA concentration was recognized during the progress of disease, and at the last stage of infection the DNA concentration in infected midgut was about 2.2 times that of healthy midgut (66). This phenomenon suggests that a large quantity of virion is produced in infected midgut nuclei.

Glycogen content in the DNV-infected midgut was found to be significantly lower than that in the uninfected midgut (66).

## Host Range and Cross-Infection

The host range of GmDNV is restricted to *G. mellonella*. In contrast, *Junonia* DNV infects *Mamestra brassicae*, *Bombyx mori*, and *Lymantria dispar*, but not *G. mellonella* (27). Interestingly, GmDNV can be transmitted from an infected *G. mellonella* larva to a healthy larva via the ovipositor of the parasite *Nemeritis canescens* (6).

*Aedes* DNV was infectious for different species of mosquitoes: *Aedes vexans*, *A. geniculatus*, *A. caspius dorsalis*, *A. cantans*, *A. albopictus*, *Culex pipiens pipiens*, *C. pipiens molestus*, and *Culiseta annulata* (11). The infectivity of the DNV of *Periplaneta fuliginosa* against other species of *Periplaneta* has been investigated (36). Adults of *P. austrasiae* and *P. brunnea* and the nymphs of *P. japonica* are susceptible, but the adults of *P. americana* are less susceptible.

In the case of BmDNV-1, there is no information concerning its cross-infection to other insects except between *B. mori* and mulberry pyralid, *Glyphodes pyoalis*; i.e., the DNV of *Glyphodes* infects BmDNV-1–susceptible strains of the silkworm (22). On the other hand, BmDNV-2 does not infect *G. pyoalis* but infects *B. mandarina* (67).

## Inheritance of Nonsusceptibility

Per os inoculation tests of BmDNV-1 to various silkworm strains revealed that most strains available in Japan were nonsusceptible, whereas some strains were highly susceptible to the DNV infection (68). On the BmDNV-2, it was infectious for strains resistant to BmDNV-1; in contrast, almost all susceptible strains for BmDNV-1 were resistant to BmDNV-2 (69). It is of interest to note that susceptible–nonsusceptible relations for several silkworm strains used between DNV-1 and DNV-2 showed almost negative correlation. Generally speaking, about half the Japanese strains and most Chinese strains were nonsusceptible to BmDNV-1 (69,70).

To determine the mode of inheritance of nonsusceptibility to DNV infection in silkworm strains, the infectivity tests on BmDNV-1 have been conducted with susceptible (S) and nonsusceptible (N) parent strains, their reciprocal $F_1$ hybrids, and the backcrossed hybrids to either of the parents (68). The $-\log IC_{50}$ values of reciprocal $F_1$ hybrids and those of their backcrossed hybrid to the S strain were nearly the same as that of the S strain, while the $-\log IC_{50}$ value of $F_2$ hybrid was a little less, and the value of the backcrossed hybrid to the N strain was significantly less than that of the S strain. When high concentrations of the DNV, which result in 100% infection in S strain, are administered to the $F_2$ hybrid and to the backcrossed hybrid to the N strain, the susceptible and nonsusceptible larvae segregate at a 3 : 1 and a 1 : 1 ratio, respectively. These results indicate that nonsusceptibility to BmDNV-1 infection is controlled genetically by a recessive gene that is not sex linked (63).

Similar results were obtained on BmDNV-2; i.e., the nonsusceptibility to DNV-2 was controlled genetically by an another recessive gene that is not sex linked (69). It was already stated that BmDNV-2 may be a mixture of two similar DNVs. From this result, at least one gene susceptible to BmDNV-2 which may consist of two similar DNVs, seems to locate on the same locus. This is an important result since most practical Japanese strains are susceptible to BmDNV-2.

Recently, a new dominant gene considered nonsusceptible to BmDNV-1 was found in a silkworm strain in Japan (71). This gene was not located on the sex, second, twelfth, or seventeenth chromosome. However, the mechanism of susceptibility and nonsusceptibility is not yet clear. Therefore, rearing of the silkworm strain which contains homozygously the nonsusceptible recessive gene is one of the best ways to avoid the damage by DNV infection.

## Epizootiology

The epizootiology of densonucleosis is important in finding ways to avoid economic damage that results from the disease in sericulture. Furthermore, when some DNVs are used as viral insecticides, attention should be directed to the relation between DNV and the environment.

On the occurrence of BmDNV-1 in sericulture farms in two separate districts of Japan, epizootiological investigations were made immunologically (23). In Saitama and Nagano prefectures, an enzootic of densonucleosis was noted at a few farms in the Nagano Prefecture. The enzootic was mainly due to the rearing of silkworm strains that were nonsusceptible or highly resistant to BmDNV-1 infection. However, the DNV antigen was detected generally in the dusts of every farm in the Nagano Prefecture, even in farms where nonsusceptible strains were reared and had no occurrence of densonucleosis.

It has also been found that the BmDNV-1 detected in the dust originated from mulberry leaves contaminated with the virus, which had multiplied chronically in the mulberry pyralid, *Glyphodes pyloalis*, infesting the mulberry plantations (23). The virus obtained from the midgut of the mulberry pyralid was indistinguishable serologically from BmDNV-1 and was also pathogenic to susceptible strains of BmDNV-1. In Saitama Prefecture, on the other hand, the DNV antigen was not detected in the dust of any farm, nor did larvae of the mulberry pyralid contain BmDNV-1 antigen in the midgut.

These results suggest that the epizootic of densonucleosis in sericultural farms is caused by two major factors: (1) the rearing of silkworm strains susceptible to BmDNV-1, and (2) the infestation of mulberry pyralid infected with DNV in the mulberry plantations. The results also suggest that the virus isolated as BmDNV-1 is originally derived from a virus of the mulberry pyralid.

Subsequently, it was reported that BmDNV-2 (Saku isolate) may be also derived from a virus of other insects (64). On the other hand, Seki reported that on BmDNV-2 (Yamanashi isolate) he could not find any DNV-2 antigen in many *G. pyloalis* found in Yamanashi Prefecture (72). Therefore, he concluded that BmDNV-2 (Yamanashi isolate) which persisted in the rearing environment gave rise to the year-to-year infection cycle of the silkworm.

## DIAGNOSTIC PROCEDURES

### Symptoms

Most DNVs known so far are polytrophic in tissue tropism. In Gm, *Aedes*, *Diatraea*, *Casphalia*, and *Pseudoplusia* DNVs, almost all of the tissues—i.e., fat body, nervous system, silk gland, muscular membrane, Malpighian tubule, gonad, and molting gland—are susceptible (4,10). On the other hand, Bm, *Sibine*, and *Periplaneta* DNVs infect only midgut epithelia (10).

All BmDNVs multiply only in the nuclei of columnar cells of the midgut epithelium, and no pathological changes are observed in other tissues. When silkworm larvae are infected per os with the BmDNV-1, they usually die after seven days showing body flaccidity as a major sign. The alimental canal of the diseased larvae is pale yellow in color with little internal content. This symptom is similar to that of IFV infection, but the histopathological features differ between the two viruses (12).

In the BmDNV-3, the infected nucleus hypertrophies, and at the last stage of infection, the nucleus becomes more than 2.5 times larger than normal (16). The symptoms of BmDNV-2 are similar to those of BmDNV-3 (18).

On the mosquito larva *A. aegypti* infected with *Aedes* DNV, the cells of the hypodermis, imaginal disc, and Malpighian tubule are shown to be affected, but most obvious pathological changes occur in the fat body cells (28). Infected nuclei are much larger than healthy nuclei.

In the case of the lepidopteran insect *S. fusca*, a notable modification caused by the *Sibine* DNV infection develops in the digestive tract and often forms a tumorlike mass (31). In the diseased larvae, the transparent digestive wall becomes opaque, whitish, and thickened. This thickening of the wall is the most characteristic sign that the disease is at an advanced stage.

The cockroach *P. fuliginosa* infected with the *Periplaneta* DNV displays the following symptoms (32). Prior to death, the hind legs become paralyzed and their movements are uncoordinated. The abdomen is swollen with enlarged fat body colored milky white in contrast to the brownish white tissues observed in an uninfected cockroach. Over half of the infected insects develop ulcers in the hind gut.

In the dragonfly *Leucorrhnia dubia* infected with a densolike virus, the nymphs become sluggish and flaccid, but there is no external sign of the disease (73).

In the case of the butterfly *A. vanillae* infected with the *Agraulis* DNV, the larvae become flaccid and are discolored after death (33). They are killed mainly at the prepupal or pupal stage, and the dead chrysalis turns light brown. If the insect survives, the prepupa frequently fails to shed its larval skin or the adult fails to eclose. Adults that eclose have deformed wings.

On the lepidopteran palm pest *C. extranea* infected with the *Casphalia* DNV, the larval color changes from green to yellowish brown, and the transparent gut becomes opaque (25).

## Diagnosis

Diagnosis from the symptoms of densonucleosis is difficult, since insects infected with DNV become flaccid, but they do not form occlusion body. The infected insects show no specific outer or inner symptoms, as noted. Therefore, on diagnosis for densonucleosis, especially in the silkworm *Bombyx mori*, several methods are applied.

### Histopathological Method

This is the most reliable method for diagnosis of densonucleosis. Generally, samples of infected tissue are fixed with Carnoy's fluid. After they are embedded in paraffin, the sections are stained with Feulgen reaction or methyl green. When an insect is infected with DNV, characteristic features on hypertrophy of the nucleus are observed under a light microscope (12). However, this method is fairly complicated compared to the other methods.

### Immunodiffusion Method

According to Watanabe and Maeda (62), the microgel diffusion method is useful for simple diagnosis for densonucleosis. A microscope slide is covered with a layer of 1% agarose buffered with barbital sodium acetate. The wells of 4 mm in diameter are made with a cork borer. The antiserum against DNV, diluted 16-fold to 36-fold with phosphate-buffered saline (PBS), was placed in the central well, and a suspension of macerated tissue for each test larva was placed in the peripheral wells. The slides were held at room temperature for two days in petri dishes, the bottoms of which were layered with a wet filter paper to prevent dessication. When antigen–antibody reaction occurred, a characteristic line was formed between the antibody and macerated tissue.

### Latex Agglutination Test

Agglutination tests using antibody-sensitized latex particles were used for the specific detection of BmDNV-2 (74). In this case, 0.5 μg/ml of DNV protein could be detected. The sensitivity of the latex agglutination test was 5 times higher than that of immunodiffusion or ring test but was 5–20 times lower than that of enzyme-linked immunosorbent assay (ELISA). The test is completed within 5 min. Extracts from DNV-infected silkworm larvae agglutinated latex particles specifically, while there was no agglutination by extract from healthy larvae. This method is useful for conventional and rapid diagnosis for densonucleosis.

### Immunofluorescence Test

For the preparation of fluorescent antibody, gamma globulin was separated from the DNV antibody by precipitation with 33% saturated ammonium sulfate. Conjugation of fluorescein isothiocyanate (FITC) with antibody is performed by following method (75). Five milliliters of 3% gamma globulin solution is added with 1.75 ml of PBS and 0.75 ml of 0.5 *M* carbonate buffer (pH 9.5) containing 1.5 mg of FITC. The reaction mixture is incubated at 4°C for 4 h with continuous stirring. Free FITC remaining in the mixture is removed by Sephadex G25 column chromatography, and a further purification of FITC–antibody conjugate is performed by DEAE cellulose column chromatography.

Tissue smear is prepared on glass slides, fixed with acetone for 10 min at room temperature, and stained with fluorescent antibody for 1 h at 37°C in a moistened chamber (75). After staining, the smears are washed three times with PBS for 15 min and mounted with Elvanol for observation under a fluorescent microscope. Based on some data on the diagnosis for IFV of the silkworm, this method may have almost the same sensitivity as the latex agglutintation test.

### ELISA and Dot Immunobinding Assay (DIBA)

To detect BmDNV-2, ELISA and DIBA were applied by Arakawa and Shimizu (76). Both methods showed the highest sensitivity for detection of DNV. When using purified DNV, the sensitivities were 10 ng/ml by DIBA and 6 ng/ml by ELISA.

## REFERENCES

1. Siegl, G., Bates, R. C., Berns, K. I., Carter, B. J., Kelly, D. C., Kurstak, E., and Tattersall, P. Characteristics and taxonomy of Parvoviridae. *Intervirology* 23, 61, 1985.
2. Kurstak, E. and Tijssen, P. Animal parvoviruses: Comparative aspects and diagnosis, in *Comparative Diagnosis of Viral Diseases*, Kurstak, E. and Kurstak, C., Eds., Academic Press, New York, 1981, 4.

3. Kurstak, E., Vernoux, J. P., and Brakier-Gingras, L. Etude biophysique de l'acide désoxyribonucléique du virus de la densonucléose. *Arch. Gesamte Virusforsch.* 40, 274, 1973.
4. Kurstak, E. Small densonucleosis virus (DNV). *Adv. Virus Res.* 17, 207, 1972.
5. Kurstak, E. and Tijssen, P. Densonucleosis viruses (Parvoviridae), in *The Atlas of Insects and Plant Viruses*, Maramorosch, K., Ed., Academic Press, New York, 1977, 67.
6. Kurstak, E. and Vago, C. Transmission du virus de la densonucléose par le parasitime d'un Hyménoptère. *Rev. Can. Biol.* 26, 311, 1967.
7. Garzon, S. and Kurstak, E. Ultrastructural studies on the morphogenesis of the densonucleosis virus (Parvovirus). *Virology* 70, 517, 1976.
8. Tijssen, P. and Kurstak, E. Studies on the structure of the two infectious types of densonucleosis virus. *Intervirology* 11, 261, 1979.
9. Tijssen, P. and Kurstak, E. Biochemical, biophysical, and biological properties of densonucleosis virus (Parvovirus). *J. Virol.* 37(1), 17, 1981.
10. Kawase, S. Pathology associated with densoviruses, in *Viral Insecticides for Biological Control*, Maramorosch, K. and Sherman, K. E., Eds., Academic Press, London, 1985, 197.
11. Meynadier, G., Vago, C., Plantevin, G., and Atger, P. Virose d'un type habituel chez le Lépidoptère, *Galleria mellonella* L. *Rev. Zool. Agric. Appl.* 63, 20, 1964.
12. Shimizu, T. Pathogenicity of an infectious flacherie virus of the silkworm, *Bombyx mori*, obtained from sericultural farms in the suburbs of Ina city. *J. Sericult. Sci. Jpn.* 44, 45, 1975.
13. Matsui, M. Smaller virus particles than the flacherie virus found in the diseased silkworm larvae, *Bombyx mori* L. *Jpn. J. Appl. Entomol. Zool.* 17, 113, 1973.
14. Furuta, Y. Studies on the previously unreported virus infecting the silkworm, *Bombyx mori*. *J. Sericult. Sci. Jpn.* 42, 443, 1973.
15. Kawase, S., Nakagaki, M., Bando, H., and Furuta, Y. Studies on the chemical properties of "flacherie virus isolated by Furuta" in the silkworm, *Bombyx mori*. *J. Sericult. Sci. Jpn.* 51, 58, 1982.
16. Iwashita, Y. and Chun, C. Y. The development of a densonucleosis virus isolated from silkworm larvae, *Bombyx mori*, of China, in *Ultrastructure and Functioning of Insect Cells*, Akai, H., King, R. C., and Morohoshi, S., Eds., Soc. Insect Cells, Tokyo, Japan, 1982, 161.
17. Kurihara, Y., Watanabe, H., Maeda, S., and Simizu, T. Chemical characteristics of a previously undescribed densonucleosis virus isolated from the silkworm, *Bombyx mori*. *J. Sericult. Sci. Jpn.* 53, 33, 1984.
18. Seki, H. and Iwashita, Y. Histopathological features and pathogenicity of a densonucleosis virus of the silkworm, *Bombyx mori*, isolated from sericultural farms in Yamanashi prefecture. *J. Sericult. Sci. Jpn.* 52, 400, 1983.
19. Watanabe, H., Kawase, S., Shimizu, T., and Seki, H. Differences in serological characteristics of densonucleosis viruses in the silkworm, *Bombyx mori*. *J. Sericult. Sci. Jpn.* 55, 75, 1986.
20. Iwashita, Y. Personal communication, 1986.
21. Watanabe, H. Unpublished result, 1986.

22. Watanabe, H. Characteristics of densonucleosis in the silkworm, *Bombyx mori*. *Jpn. Agric. Res. Q.* 15, 133, 1981.
23. Watanabe, H. and Shimizu, T. Epizootiological studies on the occurrence of densonucleosis in the silkworm, *Bombyx mori*, reared at sericultural farms. *J. Sericult. Sci. Jpn.* 49, 485, 1980.
24. Chao, Y. C., Young, S. Y., Kim, J. S., and Scott, H. A. A newly isolated densonucleosis virus from *Pseudoplusia includens* (Lepidoptera: Noctuidae). *J. Invertebr. Pathol.* 46, 70, 1985.
25. Fediere, G. Recherche sur des viroses épizootiques du Lépidoptère *Limacoidae ravageus* des palmacées, thesis, University of Science and Technology, Languedoc, 1983.
26. Nakagaki, M. and Kawase, S. DNA of a new parvo-like virus isolated from silkworm, *Bombyx mori*. *J. Invertebr. Pathol.* 35, 124, 1980.
27. Rivers, C. F. and Longworth, J. F. A non-occluded virus of *Junonia coenia* (Nymphalidae: Lepidoptera). *J. Invertebr. Pathol.* 20, 369, 1972.
28. Lebedeva, O. P., Kuznetsova, M. A., Zelenko, A. P., and Gudz-Gorban, A. P. Investigation of a virus disease of the densonucleosis type in a laboratory culture of *Aedes aegypti*. *Acta Virol.* 17, 253, 1973.
29. Meynadier, G., Matz, G., Veyrunes, J. C., and Bres, N. Virose de type densonucléose chez les Orthoptères. *Ann. Soc. Entomol. Fr. (N.S.)* 13, 487, 1977.
30. Meynadier, G., Galichet, P. F., Veyrunes, J. C., and Amargier, A. Mise en évidence d'une densonucléose chez *Diatraea saccharalis* (Lep.: Pyralidae). *Entomophaga* 22, 115, 1977.
31. Meynadier, G., Amargier, A., and Genty, P. Une virose de type densonucléose chez le lépidoptère *Sibine fusca* Stoll. *Oleagineux* 32, 357, 1977.
32. Suto, C. Characterization of a virus newly isolated from the smoky-brown cockroach, *Periplaneta fuliginosa* (Serville). *Nagoya J. Med. Sci.* 42, 13, 1979.
33. Kelly, D. C., Ayres, M. D., Spencer, L. K., and Rivers, C. F. Densonucleosis virus 3: A recent insect parvovirus isolated from *Agraulis vanillae* (Lepidoptera: Nymphalidae). *Microbiologica* 3, 455, 1980.
34. Sun, F. L., Ma, G. H., and Chen, M. S. A new insect virus of *Pieris rapae* L. I. Isolation and characterization of the virus. *Acta Microbiol. Sin.* 21, 41, 1981.
35. Liang, D. G., Cai, Y. E., Liu, D. Y., Zhang, Q. L., Hu, Y. Y., He, H. J., and Zhao, K. B. *The Atlas of Insect Viruses in China*. Hunan Science and Technology Press, Wuhan, China, 1986.
36. Suto, C., Kawamoto, F., and Kumada, N. A new virus isolated from the cockroach, *Periplaneta fuliginosa* (Serville). *Microbiol. Immunol.* 23, 207, 1979.
37. Johnson, F. B. Parvovirus proteins, in *The Parvoviruses*, Berns, K. I., Ed., Plenum Press, New York, 1984, 259.
38. Nakagaki, M. and Kawase, S. Capsid structure of the densonucleosis virus of the silkworm, *Bombyx mori*. *J. Sericult. Sci. Jpn.* 51, 420, 1982.
39. Nakagaki, M. and Kawase, S. Structural proteins of densonucleosis virus isolated from the silkworm, *Bombyx mori*, infected with the flacherie virus. *J. Invertebr. Pathol.* 36, 166, 1980.
40. Kawase, S., Cai, Y. M., Bando, H., and Seki, H. Chemical properties of the

Yamanashi isolate of the *Bombyx* densonucleosis virus. *J. Sericult. Sci. Jpn.* 53, 341, 1984.

41. Hu, Y. Y., Seki, H., and Kawase, S. Electron microscopic studies on DNA from a densonucleosis virus (Yamanashi isolate) of the silkworm, *Bombyx mori. Appl. Entomol. Zool.* 21, 613, 1986.
42. Kelly, D. C. and Elliott, R. M. Polyamines contained by two densonucleosis viruses. *J. Virol.* 21, 408, 1977.
43. Bando, H., Nakagaki, M., and Kawase, S. Polyamines in densonucleosis virus from the silkworm, *Bombyx mori. J. Invertebr. Pathol.* 42, 264, 1983.
44. Furuta, Y. Studies on the small flacherie virus of the silkworm, *Bombyx mori* L. I. Difference between SFV and IFV in infectivity and the sensitivity to formaldehyde and heat. *J. Sericult. Sci. Jpn.* 46, 297, 1977.
45. Seki, H. Effects of physicochemical treatments on a silkworm densonucleosis virus (Yamanashi isolate) of the silkworm, *Bombyx mori. Appl. Entomol. Zool.* 21, 515, 1986.
46. Kawase, S. and Kang, S. K. On the nucleic acid of a newly isolated virus from the flacherie-diseased silkworm larvae. *J. Sericult. Sci. Jpn.* 45, 87, 1976.
47. Tijssen, P., van den Hurk, J., and Kurstak, E. Biochemical, biophysical, and biological properties of densonucleosis virus. I. Structural proteins. *J. Virol.* 17, 686, 1976.
48. Bando, H., Kondo, N., and Kawase, S. Molecular homology among the structural proteins of densonucleosis virus from the silkworm, *Bombyx mori. Arch. Virol.* 80, 209, 1984.
49. Iwashita, Y. Personal communication, 1986.
50. Bando, H. and Kawase, S. Changes in polyamine and DNA concentrations in the midgut epithelium of the silkworm, *Bombyx mori* (Lepidoptera: Bombycidae) infected with a densonucleosis virus. *Appl. Entomol. Zool.* 21, 47, 1986.
51. Bachrach, U., Don, S., and Wiener, H. Occurrence of polyamines in myxovirus. *J. Gen. Virol.* 22, 451, 1974.
52. Gosule, L. C. and Schellman, J. A. Compact form of DNA induced by spermidine. *Nature* 259, 333, 1976.
53. Sheppard, S. L., Burnes, A. T. H., and Boyle, S. M. Polyamines in encephalomyocarditis virus. *J. Virol.* 34, 266, 1980.
54. Kobayashi, M., Hashimoto, Y. Seki, H. and Watanabe, H. In vitro translation of RNA from the midgut of the silkworm, *Bombyx mori,* infected with a densonucleosis virus *J. Invertebr. Pathol.* 52, 259, 1988.
55. Bando, H., Kusuda, J., Gojobori, T., Maruyama, T., and Kawase, S. Organization and nucleotide sequence of a densonucleosis virus genome imply a host-dependent evolution of the parvoviruses. *J. Virol.* 61, 553, 1987.
56. Bando, H., Kusuda, J., and Kawase, S. Molecular cloning and characterization of *Bombyx* densovirus genomic DNA. *Arch. Virol.* 93, 139, 1987.
57. Goldberg, M. L. Ph.D. diss., Stanford University, Stanford, CA, 1979.
58. Corden, J., Wasylyk, B., and Buchwalder, A. Promotor sequences of eukaryotic protein-coding genes. *Science* 209, 1406, 1980.
59. Kozak, M. Point mutations define a sequence flanking the AUG codon that modulates translation by eukaryotic ribosomes. *Cell* 44, 283, 1986.

60. Watanabe, H., Maeda, S., Matsui, M., and Shimizu, T. Histopathology of the midgut epithelium of the silkworm, *Bombyx mori*, infected with a newly-isolated virus from the flacherie diseased larvae. *J. Sericult. Sci. Jpn.* 45, 29, 1976.
61. Buchatsky, L. P. and Raikova, A. P. Electron microscope study of mosquito densonucleosis virus maturation. *Acta Virol.* 23, 170, 1979.
62. Watanabe, H. and Maeda, S. Multiplication of a densonucleosis virus in the silkworm, *Bombyx mori* L., reared at high temperature. *Jpn. J. Appl. Entomol. Zool.* 23, 151, 1979.
63. Sato, F. and Inoue, H. Double infection of viruses in the midgut of the silkworm, *Bombyx mori*. *Bull. Sericult. Exp. Sta.* 27, 427, 1978.
64. Shimizu, T. and Watanabe, H. Failure to harvest cocoons attributed to an epizootic densonucleosis caused by a new strain of the virus in sericultural farms. *J. Sericult. Sci. Jpn.* 53, 436, 1984.
65. Kawase, S., Bando, H., and Cai, Y. M. Purification of a densonucleosis virus from silkworm feces. *J. Sericult. Sci. Jpn.* 52, 547, 1983.
66. Kobayashi, M., Hashimoto, Y., Mori, H., and Nagamine, T. Changes in DNA, RNA, protein and glycogen in the midgut of the silkworm, *Bombyx mori* (Lepidoptera: Bombycidae), during the infection with a densonucleosis virus. *Appl. Entomol. Zool.* 21, 486, 1986.
67. Seki, H. The serological properties and infectivity of Yamanashi isolate of the silkworm densonucleosis virus. *J. Sericult. Sci. Jpn.* 53, 69, 1984.
68. Watanabe, H. and Maeda, S. Genetic resistance to peroral infection with a densonucleosis virus in the silkworm, *Bombyx mori*. *J. Sericult. Sci. Jpn.* 47, 209, 1978.
69. Seki, H. Mode of inheritance of the resistance to the infection with the densonucleosis virus (Yamanashi isolate) in the silkworm, *Bombyx mori*. *J. Sericult. Sci. Jpn.* 53, 472, 1984.
70. Seki, H. Interstrain difference in the resistance to the infection with densonucleosis virus (Yamanashi isolate) in the silkworm, *Bombyx mori*. *J. Sericult. Sci. Jpn.* 54, 445, 1985.
71. Eguchi, R., Furuta, Y., and Ninaki, O. Dominant nonsusceptibility to densonucleosis virus in the silkworm, *Bombyx mori*. *J. Sericult. Sci. Jpn.* 55, 177, 1986.
72. Seki, H. Epizootiology of the silkworm densonucleosis caused by the Yamanashi isolate of the virus. *J. Sericult. Sci. Jpn.* 55, 421, 1986.
73. Charpentier, R. A nonoccluded virus in nymphs of the dragonfly, *Leucorrhinia dubia* (Odonata, Anisotera). *J. Invertebr. Pathol.* 34, 95, 1979.
74. Shimizu, S. and Arakawa, A. Latex agglutination test for the detection of the cytoplasmic polyhedrosis virus and the densonucleosis virus of the silkworm, *Bombyx mori*. *J. Sericult. Sci. Jpn.* 55, 153, 1986.
75. Maeda, S. and Watanabe, H. Immunofluorescence observation of the infection of densonucleosis virus in the silkworm, *Bombyx mori*. *Jpn. J. Appl. Entomol. Zool.* 22, 98, 1970.
76. Arakawa, A. and Shimizu, S. A dot immunoblotting assay for the detection of densonucleosis virus of the silkworm, *Bombyx mori*. *J. Sericult. Sci. Jpn.* 54, 500, 1985.

# Index